MONOGRAPHS AND TEXTBOOKS IN PURE AND APPLIED MATHEMATICS

1. *K. Yano,* Integral Formulas in Riemannian Geometry (1970)
2. *S. Kobayashi,* Hyperbolic Manifolds and Holomorphic Mappings (1970)
3. *V. S. Vladimirov,* Equations of Mathematical Physics (A. Jeffrey, editor; A. Littlewood, translator) (1970)
4. *B. N. Pshenichnyi,* Necessary Conditions for an Extremum (L. Neustadt, translation editor; K. Makowski, translator) (1971)
5. *L. Narici, E. Beckenstein, and G. Bachman,* Functional Analysis and Valuation Theory (1971)
6. *S. S. Passman,* Infinite Group Rings (1971)
7. *L. Dornhoff,* Group Representation Theory (in two parts). Part A: Ordinary Representation Theory. Part B: Modular Representation Theory (1971, 1972)
8. *W. Boothby and G. L. Weiss (eds.),* Symmetric Spaces: Short Courses Presented at Washington University (1972)
9. *Y. Matsushima,* Differentiable Manifolds (E. T. Kobayashi, translator) (1972)
10. *L. E. Ward, Jr.,* Topology: An Outline for a First Course (1972)
11. *A. Babakhanian,* Cohomological Methods in Group Theory (1972)
12. *R. Gilmer,* Multiplicative Ideal Theory (1972)
13. *J. Yeh,* Stochastic Processes and the Wiener Integral (1973)
14. *J. Barros-Neto,* Introduction to the Theory of Distributions (1973)
15. *R. Larsen,* Functional Analysis: An Introduction (1973)
16. *K. Yano and S. Ishihara,* Tangent and Cotangent Bundles: Differential Geometry (1973)
17. *C. Procesi,* Rings with Polynomial Identities (1973)
18. *R. Hermann,* Geometry, Physics, and Systems (1973)
19. *N. R. Wallach,* Harmonic Analysis on Homogeneous Spaces (1973)
20. *J. Dieudonné,* Introduction to the Theory of Formal Groups (1973)
21. *I. Vaisman,* Cohomology and Differential Forms (1973)
22. *B.-Y. Chen,* Geometry of Submanifolds (1973)
23. *M. Marcus,* Finite Dimensional Multilinear Algebra (in two parts) (1973, 1975)
24. *R. Larsen,* Banach Algebras: An Introduction (1973)
25. *R. O. Kujala and A. L. Vitter (eds.),* Value Distribution Theory: Part A; Part B: Deficit and Bezout Estimates by Wilhelm Stoll (1973)
26. *K. B. Stolarsky,* Algebraic Numbers and Diophantine Approximation (1974)
27. *A. R. Magid,* The Separable Galois Theory of Commutative Rings (1974)
28. *B. R. McDonald,* Finite Rings with Identity (1974)
29. *J. Satake,* Linear Algebra (S. Koh, T. A. Akiba, and S. Ihara, translators) (1975)
30. *J. S. Golan,* Localization of Noncommutative Rings (1975)
31. *G. Klambauer,* Mathematical Analysis (1975)
32. *M. K. Agoston,* Algebraic Topology: A First Course (1976)
33. *K. R. Goodearl,* Ring Theory: Nonsingular Rings and Modules (1976)
34. *L. E. Mansfield,* Linear Algebra with Geometric Applications: Selected Topics (1976)
35. *N. J. Pullman,* Matrix Theory and Its Applications (1976)
36. *B. R. McDonald,* Geometric Algebra Over Local Rings (1976)
37. *C. W. Groetsch,* Generalized Inverses of Linear Operators: Representation and Approximation (1977)
38. *J. E. Kuczkowski and J. L. Gersting,* Abstract Algebra: A First Look (1977)
39. *C. O. Christenson and W. L. Voxman,* Aspects of Topology (1977)
40. *M. Nagata,* Field Theory (1977)
41. *R. L. Long,* Algebraic Number Theory (1977)
42. *W. F. Pfeffer,* Integrals and Measures (1977)
43. *R. L. Wheeden and A. Zygmund,* Measure and Integral: An Introduction to Real Analysis (1977)
44. *J. H. Curtiss,* Introduction to Functions of a Complex Variable (1978)
45. *K. Hrbacek and T. Jech,* Introduction to Set Theory (1978)

Additional Volumes in Preparation

PROBLEMS AND EXAMPLES IN DIFFERENTIAL EQUATIONS

PROBLEMS AND EXAMPLES IN DIFFERENTIAL EQUATIONS

Piotr Biler
Tadeusz Nadzieja

University of Wrocław
Wrocław, Poland

Marcel Dekker, Inc.   **New York • Basel • Hong Kong**

Library of Congress Cataloging-in-Publication Data

Biler, Piotr.
   Problems and examples in differential equations / Piotr Biler,
Tadeusz Nadzieja.
            p.   cm. -- (Monographs and textbooks in pure and applied
mathematics)
   Includes bibliographical references and index.
   ISBN 0-8247-8637-8
   1. Differential equations--Problems, exercises, etc.
1. Nadzieja, Tadeusz.   II. Title.   III. Series.
QA371.B55   1992
515'.35--dc20   92-17583
                                                                  CIP

This book is printed on acid-free paper.

MARCEL DEKKER, INC.
270 Madison Avenue, New York, New York 10016

Current printing (last digit):
10 9 8 7 6 5 4 3 2 1

PRINTED IN THE UNITED STATES OF AMERICA

To Andrzej Krzywicki
our master and friend

Preface

The exercises in this volume are of rather advanced level. You will find here original problems from graduate courses in pure and applied mathematics and even small research topics, significant theorems and information on recent results. The subjects considered are classic as well as modern ones.

The core of this collection is problems, but there are also many examples and counterexamples that are scattered in monographs and not easily accessible. We lay stress on examples, not on the general theory, since we think that an example is more attractive than an overwhelming and sometimes dry theorem (see the motto to the Appendix to Chapter X in [R-S]!).

We have prepared this volume retaining much of the spirit and style of *Problems in Mathematical Analysis* by Alfred Witkowski and Piotr Biler. Repeating some phrases from the preface to [B-W] we may say that ''The answers and the hints (or rather suggestions) are in general extremely laconic because there is no unique way to the result. Active readers will find their own methods and would not have this pleasure if the full reasoning were given (moreover, this would make the volume enormous).''*

We hope that the numerous references given here will be useful as suggestions to further reading and sources of other interesting exercises.

---

*''*Le secret d'ennuyer est celui de tout dire.*'' —Voltaire

We tried to give, whenever possible, English language references. We transformed slightly the original formulation of problems in some cases in order to obtain the version most frequently referred to in the literature.

Warning: This book sometimes forces potential readers to consult other books and journals to compare results and see modifications and generalizations. In our opinion, it is the essential way to approach mathematics: solve problems knowing the roots of advanced topics in elementary background.

Hearty thanks are due to Alain Bachelot, Jean-Michel Ghidaglia, Jan Goncerzewicz, Danielle Hilhorst, Andrzej Krzywicki, Andrzej Lasota, Wojciech Okrasiński, Michel Pierre, and Jean-Claude Saut for helpful discussions on various aspects of differential equations, for comments and for supplying us with some problems.

We are grateful to Jean-Marie Strelcyn for sharing with us interesting references and his enthusiasm, Alfred Witkowski for encouragement and Kryspin Porembski for technical assistance. Last but not least we acknowledge the excellent cooperation of the staff of Marcel Dekker, Inc., during the production of this volume.

Piotr Biler
Tadeusz Nadzieja

Contents

PROBLEMS AND
EXAMPLES IN
DIFFERENTIAL EQUATIONS

PROBLEMS

1

ORDINARY DIFFERENTIAL EQUATIONS

Existence, Uniqueness, Differential Equations on $\mathbb{R}$, Inequalities

1.1 An absolutely continuous function $x(t)$ on $\mathbb{R}$ is a solution, in the sense of Carathéodory, of the problem $x' = f(x)$, $x(0) = x_0$, if for almost every $t \in \mathbb{R}$, $x'(t) = f(x(t))$ and $x(0) = x_0$. Show that the problem $x' = \bullet_Q(x)$, $x(0) = a$, where 1_Q is the characteristic function of the set of rational numbers, has a unique solution if a is irrational, and has no solutions if a is rational.

1.2 Does there exist an absolutely continuous function $x : [0,1] \to \mathbb{R}$, $x(0) = 0$, satisfying almost everywhere (a.e.) the equation $x'(t) = f(x(t))$, where $f(x) = -1$ if $x \geq 0$, $f(x) = 1$ if $x < 0$?

1.3 Show that the differential equation $|x'| + |x| + 1 = 0$ has no solutions.

1.4 Find an example of a C^∞ function $g : \mathbb{R} \to \mathbb{R}$ such that the problem $g(x') = x$, $x(0) = 0$, has no solution.

1.5 Solve the equation $t^2 x' = x - t$ with the condition $x(0) = 0$. Does it have an analytic solution?

1.6 Give an example of a homogeneous first-order ordinary differential equation with analytic coefficients (on the whole real line) all of whose solutions vanish of infinite order at the origin.

3

1.7 If $x(t)$ is analytic at $t = 0$ and satisfies $x' \exp(t(x')^2) = 1$, $x(0) = 0$, determine its power series expansion. The problem arises as an approximation to a nonlinear heat flow equation.

1.8 Using the Schauder fixed-point theorem, prove the Peano theorem on the local existence of solutions of the initial value problem $x' = f(t,x)$, $x(0) = x_0$, where f is a continuous function defined on (a subset of) $\mathbb{R}^{n+1}$ with values in $\mathbb{R}^n$.

1.9 Consider the problem $x' = f(x)$, $x(0) = x_0$, $x \in \mathbb{R}^n$. Show that uniqueness of solutions of this problem implies their continuous dependence on the initial conditions.

1.10 Prove the fiber contraction theorem: Let (X,d) and (Y,ρ) be complete metric spaces and let $F : X \times Y \to X \times Y$ be a map of the form $F(x,y) = (G(x),H(x,y))$. Assume that:
(i) $G : X \to X$ has an attracting fixed point x_0; that is, $G(x_0) = x_0$, and $\lim_{n \to \infty} G^n(x) = x_0$ for every $x \in X$.
(ii)   The map $x \mapsto H(x,y)$ is continuous in x, for every $y \in Y$.
(iii)   For every $x \in X$ the map $H_x : Y \to Y$ defined by $H_x(y) = H(x,y)$ is a λ-contraction, with $\lambda < 1$; that is, $\rho(H_x(y_1),H_x(y_2)) \leq \lambda\rho(y_1,y_2)$ for all $x \in X$ and $y_1,y_2 \in Y$.
Then, if y_x denotes the unique attracting fixed point of H_x, the point $p = (x_0,y_{x_0})$ is an attracting fixed point of F.

1.11 Using the fiber contraction theorem (see the preceding problem) prove the theorem on existence and smooth dependence on initial conditions of the solution of the problem $x' = f(x)$, $x(0) = x_0$, $x \in \mathbb{R}^n$, where f is a (sufficiently) smooth function.

1.12 Let $x_0 : [-1,1] \to \mathbb{R}$ be a continuous function. Show that the sequence of successive approximations $x_n(t) = 2 \int_0^t | x_{n-1}(s) |^{1/2} \, ds$ converges to a solution of the problem $x' = 2 | x |^{1/2}$.

1.13 Find a continuous function $f : \mathbb{R}^2 \to \mathbb{R}$ such that the sequence of successive approximations for the problem $x' = f(t,x)$, $x(0) = x_0$, is not convergent and no convergent subsequences converge to a solution of this problem.

1.14 Assume that f is a continuous function on the set $R = \{(t,x) : 0 \leq t \leq a, \ | x | \leq b\}$ and $| f(t,x_1) - f(t,x_2) | \leq | x_1 - x_2 | /t$ for each $(t,x_1),(t,x_2) \in R$, $t \neq 0$. Prove that the sequence of successive approximations converges to the unique solution of the initial value problem $x' = f(t,x)$, $x(0) = 0$.

1.15 Prove the following generalization of the result in the preceding problem. Assume that the function f satisfies the condition $| f(t,x_1) - f(t,x_2) | \leq (t^{-1} + g(t)) | x_1 - x_2 |$, where $g(t)$ is continuous and integrable on the interval

(0,a]. Show that the solution of the problem $x' = f(t,x)$, $x(0) = 0$, is unique and the sequence of successive approximations, when applied to this problem, converges to this unique solution.

1.16 Let f be a continuous function on the rectangle $R = \{(t,x) : 0 \leq t \leq a, \ |x| \leq b\}$. Assume that $f(t,0) \geq 0$ and $f(t,x_1) \leq f(t,x_2)$ if $x_1 \leq x_2$. Prove that the successive approximations for the problem $x' = f(t,x)$, $x(0) = 0$, converge to a solution on $[0,\alpha)$, where $\alpha = \min(a,b/M)$, $M = \max |f|$.

1.17 Let $U \subset \mathbb{R}^n$ be an open set, let $f : [a,b] \times U \to \mathbb{R}^n$ be continuous, and let $x_0 \in U$. Prove that there exists ε, $0 < \varepsilon < b - a$, such that the successive approximations for the problem $x' = f(t,x)$, $x(a) = x_0$, converge uniformly in $[a, a + \varepsilon]$ if there exists a continuous function $\omega : [a,b] \times [0,c] \to \mathbb{R}^+$ such that $|f(t,x_1) - f(t,x_2)| \leq \omega(t, |x_1 - x_2|)$, where $\omega(t,\cdot)$ is increasing, and $x \equiv 0$ is the unique solution of the initial value problem $x' = \omega(t,x)$, $x(a) = 0$.

1.18 Define the sequence $(x_n(t))$, $|t| < 1$, in the following way: $x_0(t) = 0$, $x_n(t) = 1 + \int_0^t (x_{n-1}(s))^2 \, ds$. Show that x_n is a polynomial of degree $2^{n-1} - 1$ and (x_n) converges to a function x which is the solution of the problem $x' = x^2$, $x(0) = 1$.

1.19 Find an example of a continuous function $f(t,x)$ such that the sequence of the Euler polygonal approximations constructed for the problem $x' = f(t,x)$, $x(0) = 0$, does not converge.

1.20 Prove the existence of the maximal solution of the problem $x' = f(t,x)$, $x(0) = x_0$ with a continuous function f, according to the following scheme: Prove that the set H of solutions has a countable dense subset (x_n), $n \in \mathbb{N}$. Define the sequence (v_n), $n \in \mathbb{N}$, by induction: $v_1 = x_1$, $v_{n+1} = \max(x_n, v_n)$. Then prove that each $v_n \in H$ and that (v_n) converges to the maximal solution.

1.21 Prove that every solution of the problem $x' = f(t,x)$, $x(0) = x_0$, $(t,x) \in \mathbb{R}^2$ with a continuous function f, is less than or equal to every solution of $x' = f(t,x) + 1/n$, $x(0) = x_0$, $n \in \mathbb{N}$. Using this fact prove the existence of the maximal solution of this problem.

1.22 Let $x(t)$ be the solution of the initial value problem $x' = f(t,x)$, $x(0) = x_0$, $(t,x) \in \mathbb{R}^2$. The Euler method for finding approximate values of $x(t)$ is $x_{k+1} = x_k + hf(t_k,x_k)$. However, the quantity x_{k+1} is not computed exactly; one always introduces an error ε_k with $|\varepsilon_k| < \varepsilon$; that is, the computer computes number y_k such that $y_{k+1} = y_k + hf(t_k,y_k) + \varepsilon_k$, $y_0 = x_0$. Suppose that $|f_x| < L$ and $|f_t + ff_x| < D$ for all $(t,x) \in \mathbb{R}^2$. Show that

$$E_{k+1} = |x(t_{k+1}) - y_{k+1}| \leq (1 + hL)E_k + \frac{Dh^2}{2} + \varepsilon$$

and

$$E_k \le \frac{(Dh/2 + \varepsilon/h)(\exp(\alpha L) - 1)}{L}$$

for $kh \le \alpha$. Choose h so that the error E_k is minimized.

1.23 Find the solution, for $t \ge 0$, of the problem $x' = \max(t,x)$, $x(0) = 0$.

1.24 Assume that $f : \mathbb{R} \to \mathbb{R}$ is a continuous function and $f(x_0) \ne 0$. Show that the problem $x' = f(x)$, $x(t_0) = x_0$, has a locally unique solution.

1.25 Show that the differential equation $x' = x^{2/3}$ has infinitely many solutions satisfying $x(0) = 0$ on every interval $[0,\beta]$. For what values of α are there infinitely many solutions on $[0,\alpha]$ satisfying $x(0) = -1$?

1.26 Let $f(x)$ be a continuous function on $\mathbb{R}$, $f(0) = 0$. Show that the initial value problem $x' = f(x)$, $x(0) = 0$, has the unique solution $x(t) \equiv 0$ unless there exists an $\varepsilon > 0$ such that either $f(x) \ge 0$ for $0 \le x \le \varepsilon$ and $1/f(x)$ is integrable over $[0,\varepsilon]$, or $f(x) \le 0$ for $-\varepsilon \le x \le 0$ and $1/f(x)$ is integrable over $[-\varepsilon,0]$.

1.27 Find a C^1 function $f : \mathbb{R}^{n+1} \to \mathbb{R}^n$, $n > 1$, and a solution $x(t)$ of $x' = f(t,x)$ such that $x(t_1) = x(t_2)$ for some $t_1 \ne t_2$ and $x'(t_1)$, $x'(t_2)$ are linearly independent.

1.28 Define $f : \mathbb{R}^2 \to \mathbb{R}$ by $f(t,x) = -2t$ for $x \ge t^2$, $f(t,x) = -2x/t$ for $|x| < t^2$, $f(t,x) = 2t$ for $x \le -t^2$. Show that the problem $x' = f(t,x)$, $x(0) = 0$, has a unique solution. Can one apply the Picard theorem on successive approximations to this problem?

1.29 Let $f,g : \mathbb{R} \to \mathbb{R}$ be continuous functions and f satisfy the Lipschitz condition. Prove that the problem $x' = f(x)$, $y' = g(x)y$, $x(t_0) = x_0$, $y(t_0) = y_0$, has a unique solution.

1.30 Let $f : \mathbb{R}^2 \to \mathbb{R}$ be a continuous function. Assume that $f(t,x) < 0$ if $tx > 0$, and $f(t,x) > 0$ if $tx < 0$. Show that the problem $x' = f(t,x)$, $x(0) = 0$, has a unique solution.

1.31 Consider a homogeneous differential equation $x' = f(x/t)$, where $f(k) = k$. Show that
(i) If $f'(k) < 1$, then there is no solution tangent to $x = kt$ at the origin except the obvious one $x = kt$.
(ii) If $f'(k) > 1$, then there is an infinite number of such solutions.

1.32 Let $f : \mathbb{R}^{n+1} \to \mathbb{R}^n$ be a continuous function. Suppose that $(f(t,x_1) - f(t,x_2), x_1 - x_2) \le 0$ for $t \in \mathbb{R}$, $x_1, x_2 \in \mathbb{R}^n$. Show that the initial value problem

$x' = f(t,x)$, $x(t_0) = x_0$, has a unique solution on every interval $[t_0, t_0 + \varepsilon]$, $\varepsilon > 0$.

1.33 Let f be a C^2 function on $\mathbb{R}^3$. Suppose that $f(t_0,x_0,y_0) = 0$ and $f_y(t_0,x_0,y_0) \neq 0$. Prove that there exists a unique function $x(t)$ on some interval containing t_0 which satisfies $f(t,x(t),x'(t)) = 0$, $x(t_0) = x_0$, $x'(t_0) = y_0$.

1.34 Find an example of a C^2 function f on $\mathbb{R}^3$ such that the problem $f(t,x(t),x'(t)) = 0$, $x(t_0) = x_0$, $x'(t_0) = y_0$, has nonunique solutions for some $(t_0,x_0,y_0) \in \mathbb{R}^3$.

1.35 Consider the equation $x' = x(\log x)^a$, $x(0) = 0$. For what values of a is there a unique solution?

1.36 Prove that the differential equation $x' = t - 1/x$ has a unique solution on $[0,\infty)$ which is positive everywhere and tends to 0 at ∞.

1.37 Show that for the equation $tx' + ax = f(t)$, where $a > 0$, $\lim_{t \to 0} f(t) = b$, there exists a unique solution which is bounded as $t \to 0$. What happens if $a < 0$?

1.38 Consider the initial value problem $x' = f(t,x)$, $x(0) = 0$, $0 \leq t \leq a$, $|x| \leq b$. Assume that f is continuous and $|f(t,x) - f(t,y)| \leq h(t) |x - y|$, where h is an integrable function on $(0,a]$. Show that this problem has a unique solution.

1.39 The Kamke uniqueness criterion: Let $f(t,x)$ be a continuous function on $\mathbb{R}^{n+1}$ into $\mathbb{R}^n$ and $g(t,x)$ be a continuous scalar function on $\mathbb{R}^2$. Assume that $g(t,0) = 0$ and that the only solution $x(t)$ of $x' = g(t,x)$ on any interval $(t_0,t_0 + \varepsilon]$ satisfying the condition $x(t) \to 0$ and $x(t)/(t - t_0) \to 0$ as $t \to t_0+0$ is $x(t) \equiv 0$. Show that under the assumption $|f(t,x_1) - f(t,x_2)| \leq g(t, |x_1 - x_2|)$ the initial value problem $x' = f(t,x)$, $x(t_0) = x_0$, has at most one solution on any interval $[t_0, t_0 + \varepsilon]$, $\varepsilon > 0$. Show that the above condition cannot be replaced by $x(t) \to 0$ and $x'(t) \to 0$ as $t \to t_0+0$.

1.40 Show the following uniqueness theorem: Let $\omega(t,x)$ be a nonnegative function defined on $0 < t < a$, $x \geq 0$, which is Lebesgue measurable in t for fixed x and continuous nondecreasing in x for fixed t. Let there exist a function $M(t)$ defined on $0 < t < a$, and Lebesgue integrable on $\gamma < t < a$ for every $\gamma > 0$, such that $\omega(t,x) \leq M(t)$. Suppose that for each α, $0 < \alpha < a$, $x(t) \equiv 0$ is the only absolutely continuous function on $0 \leq t < \alpha$ which satisfies $x'(t) = \omega(t,x(t))$ a.e., and such that the right one-sided derivative $x'_+(0)$ exists, $x(0) = x'_+(0) = 0$. Then, if $f(t,x)$ is continuous on $|t - t_0| \leq a$, $|x - x_0| \leq b$, and satisfies for $t \neq t_0$ $|f(t,x) - f(t,y)| \leq$

$\omega(\mid t - t_0 \mid, \mid x - y \mid)$, there exists at most one solution of $x' = f(t,x)$ on $\mid t - t_0 \mid \le a$ for which $x(t_0) = x_0$.

1.41 The Nagumo uniqueness criterion: Show that if f is a continuous function on $\mathbb{R}^{n+1}$ into $\mathbb{R}^n$ and $\mid f(t,x_1) - f(t,x_2) \mid \le \mid x_1 - x_2 \mid /(t - t_0)$, $t > t_0$, $x_1, x_2 \in \mathbb{R}^2$, then the initial value problem $x' = f(t,x)$, $x(t_0) = x_0$, has at most one solution.

1.42 Show that there exists a continuous real-valued function f on $\mathbb{R}^2$ with the following properties:
(i) $\mid f(t,x) - f(t,y) \mid \le C \mid x - y \mid /t$ for $t > 0$ and some $C > 1$;
(ii)   the problem $x' = f(t,x)$, $x(0) = 0$, has more than one solution.

1.43 Consider two initial value problems:
(i) $x' = f(t,x)$, $x(0) = 0$;
(ii) $x' = g(t,x)$, $x(0) = 0$,
where t, $x \in \mathbb{R}$, and f, g are continuous functions. Assume that $f(t,x) - f(t,y) = g(t,x) - g(t,y)$ for all t, x, $y \in \mathbb{R}$ and that problem (i) has a unique solution. Does problem (ii) have a unique solution also?

1.44 Let $f : \mathbb{R}^2 \to \mathbb{R}$ be a continuous function. Show that the problem $x' = f(t,x) + \lambda$, $x(0) = 0$, $\lambda \in \mathbb{R}$, has nonunique solutions for countably many values of the parameter λ only.

1.45 Let A be any closed subset of $\mathbb{R}$. Show that there exists a continuous function $f(t,x)$ on $\mathbb{R}^2$, $f(t,0) = 0$, such that the initial value problem $x' = f(t,x)$, $x(t_0) = 0$, has a locally unique solution if and only if $t_0 \notin A$.

1.46 Show that if C is any nowhere dense closed subset of $\mathbb{R}$, then there exists a function $f(t,x)$ such that the set of points (t_0,x_0) for which the problem $x' = f(t,x)$, $x(t_0) = x_0$, has a nonunique solution contains all points in the tx plane whose t coordinate is in C.

1.47 The initial value problem $x' = f(t,x)$, $x(0) = 0$ with a continuous function f, has two solutions. Show that it has infinitely many solutions.

1.48 Does there exist a continuous function $f(t,x)$, t, $x \in \mathbb{R}$, such that for each t_0, x_0 the problem $x' = f(t,x)$, $x(t_0) = x_0$, has a unique solution on $(-\infty,t_0)$ and nonunique solutions on (t_0,∞)?

1.49 Let $f : \mathbb{R}^2 \to \mathbb{R}$ be a continuous function. Suppose that there exist two solutions φ_1, $\varphi_2 : [0,1] \to \mathbb{R}$ of the differential equation $x' = f(t,x)$ such that Graph $\varphi_1 \cap$ Graph $\varphi_2 = \{(0,p), (1,q)\}$ and Graph $\varphi_1 \cup$ Graph φ_2 is the boundary of a domain D (homeomorphic to the disk). Prove that for each $z \in D$ there exists a solution φ of this equation such that Graph φ contains $(0,p)$, $(1,q)$, and z.

1.50 Let $f(t,x)$ be a continuous function for $t_0 \le t \le t_0 + a$ and $x \in \mathbb{R}^n$ and let S_c, $t_0 < c < a$, be the set of points x_c for which there is a solution $x(t)$ of the problem $x' = f(t,x)$, $x(t_0) = x_0$, such that $x(c) = x_c$. Show by an example that S_c need not be connected if $n = 2$ and not all solutions exist on $[t_0, t_0 + c]$.

1.51 The Kneser theorem. Let $f(t,x)$ be continuous function on $P = \{(t,x) \in \mathbb{R}^{n+1}, t_0 \le t \le t_0 + a, |x - x_0| \le b\}$. Let $|f(t,x)| \le M$, $\alpha = \min(a, b/M)$, and $t_0 < c \le t_0 + \alpha$. Finally, let S_c be the set of points x_c for which there is a solution $x(t)$ of the problem $x' = f(t,x)$, $x(t_0) = x_0$, such that $x(c) = x_c$. Prove that S_c is a closed connected set. Give an example that S_c need not be convex if $n > 1$.

1.52 Prove that the intersection of the set of all solutions of the problem $x' = f(t,x)$, $x(t_0) = x_0$, $x \in \mathbb{R}^2$, with the hyperplane $t = $ const may not be simply connected.

1.53 Consider the problem $x' = f(t,x)$, $x(t_0) = x_0$, where f is a continuous function on $\mathbb{R}^{n+1}$ into $\mathbb{R}^n$. Assume that all solutions of this problem are defined on $[t_0, t_0 + \alpha]$. Show that the set of all solutions is a connected compact subset of the space $C[t_0, t_0 + \alpha]$ with the supremum norm.

1.54 Let $f(t,x)$ be a continuous function, $t \in \mathbb{R}$, $x \in \mathbb{R}^2$. Define the set $K_{p,s} = \{x \in \mathbb{R}^2 :$ there exists a solution $x(t)$ of the problem $x' = f(t,x)$, $x(0) = p$, and $x(s) = x\}$. Show that there exist f and $p \in \mathbb{R}^2$ such that $K_{p,s} = \{(x_1,x_2) \in \mathbb{R}^2 : x_1, x_2 \in [-1,1], x_1 \ne 0, x_2 = \sin(1/x_1)\}$ for some s. Prove that there does not exist f such that $K_{p,s} = \{z \in \mathbb{C} : z = (1 - 1/\varphi) \exp(i\varphi)$ for $2\pi \le \varphi < \infty\} \cup \{z \in \mathbb{C} : |z| = 1\}$.

1.55 Show that the initial value problem $x' = x^2 + t$, $x(0) = 0$, has no solution defined on the whole interval $(0,3)$.

1.56 Prove that there is a solution $x(t)$ of the problem $x' = \sin(tx)$, $x(0) = 0$, defined on the whole $\mathbb{R}$ such that $x(t) > 0$ for all t and $\lim_{t \to \pm\infty} x(t) = 0$.

1.57 Given the initial value problem $x' = 1/(t^2 + x^2)$, $x(0) = x_0 \ne 0$, prove that:
(i)   the solutions of this problem are unique and defined on the whole $\mathbb{R}$;
(ii) $\lim_{t \to \pm\infty} x(t)$ exists and is finite.
Give bounds depending on x_0 for $\lim_{t \to \infty} x(t) - \lim_{t \to -\infty} x(t)$.

1.58 Study the existence of global solutions, that is, defined on the whole $\mathbb{R}$, of the equations
(i) $x' = |x|^a$;
(ii) $x' = (x^2 + e^t)^a$;
(iii) $x' = |x|^{a-1} + tx^{2a/3}$;
(iv) $x' = (x^2 + y^2 + 2)^{-a}$, $y' = x(1 + y^2)^a$.

1.59 Study the existence of solutions on $t_0 \le t < \infty$ of the equations
(i) $x' = t^3 - x^3$;
(ii) $x' = tx + e^{-x}$.

1.60 Show that if f and f_x are continuous functions on $\mathbb{R}^2$ and $f_x(t,x) \leq k(t)$ for a continuous function k, then the equation $x' = f(t,x)$ has solutions defined on $t_0 \leq t < \infty$.

1.61 Suppose that the problem $x' = f(t,x)$, $x(t_0) = x_0$, $x_0 \in \mathbb{R}^n$, has local solutions for each (t_0, x_0) and the inequality $(x, f(t,x)) \leq k(t)| x |^2$ is satisfied for all $| x | \geq R$ and a continuous function k. Show that this problem has a solution defined on (t_0,∞).

1.62 Let $f : \mathbb{R}^2 \to \mathbb{R}$ be a continuous function decreasing in the second variable. Prove that the initial value problem $x' = f(t,x)$, $x(t_0) = x_0$, has a solution defined on $[t_0,\infty)$.

1.63 Let $f, F : \mathbb{R}^2 \to \mathbb{R}$ be continuous functions with continuous partial derivatives with respect to the second variable. Prove that $x' = f(t,x) + F(t,x)$ has a bounded solution if $f(t,0) \equiv 0$, $f_x \leq m < 0$, and $| F(t,x) | \leq h(t)$ with $\int_0^\infty h(s)\, ds < \infty$.

1.64 Let f be a continuous function on $R = \{(t,x) : 0 \leq t \leq a, x \in \mathbb{R}\}$. Assume that there exists a continuous function $\varphi : [0,\infty) \to \mathbb{R}$ such that $| f(t,x) | \leq \varphi(| x |)$, and for some $b > 0$, $\int_b^\infty (\varphi(r))^{-1}\, dr = \infty$. Prove that every solution of the equation $x' = f(t,x)$ is defined on the whole interval $[0,a]$.

1.65 Let $f(t,x)$ be a continuous function for $t \geq 0$, $x \in \mathbb{R}^n$ and satisfy $| f(t,x) | \leq \varphi(t)| x |$ where $\varphi(t)$ is a continuous function for $t \geq 0$ such that $\int_0^\infty \varphi(t)\, dt < \infty$. Show that every solution of the problem $x' = f(t,x)$, $x(0) = x_0$, exists for $t \geq 0$ and $\lim_{t \to \infty} x(t) = p$, $p \neq 0$, unless $x(t) \equiv 0$.

1.66 Let f,g be continuous functions on $[t_0, t_0 + a] \times \mathbb{R}$ such that $f(t,x) \leq g(t,x)$ and let $x(t)$ be a solution of the problem $x' = f(t,x)$, $x(t_0) = x_0$, on $[t_0,\delta)$, $\delta \leq t_0 + a$, satisfying $x(t) \to \infty$ as $t \to \delta$. Prove that the maximal solution $m(t)$ of the problem $x' = g(t,x)$, $x(t_0) = x_0$, has the maximal interval of existence $[t_0,\omega)$, where $\omega \leq \delta$ and $m(t) \to \infty$ as $t \to \omega$.

1.67 Let $f : \mathbb{R}^n \to \mathbb{R}^n$ and $V : \mathbb{R}^n \to \mathbb{R}$ be C^1 functions. Assuming that $(\nabla V(x), f(x)) \leq 0$ and $V(x) \geq | x |^2$ for each $x \in \mathbb{R}^n$, prove that each solution of the equation $x' = f(x)$ is defined for all $t > 0$.

1.68 Consider the differential equation $x' = f(t,x)$, $t \in \mathbb{R}$, $x \in \mathbb{R}^n$, where f is a C^1 function. Suppose that $| f(t,x) | \leq \gamma(t)| x | + \beta(t)$, where β and γ are C^1 positive functions. Show that the solutions of this equation exist on the whole $\mathbb{R}$.

1.69 Given the differential equation $x' = f(t,x)$, $t \in \mathbb{R}$, $x \in \mathbb{R}^n$, with a C^1 function f, define for $T > 0$ $\varphi_T(s) = \max\{ | f(t,x) | : t \in [0,T], | x | \leq s\}$. Assume that the problem $s' = \varphi_T(s)$, $s(0) = | x_0 |$, admits a solution defined on $[0,\infty)$

for all $T > 0$. Show that the problem $x' = f(t,x)$, $x(0) = x_0$, has a solution defined on $[0,\infty)$.

1.70 Let $V : \mathbb{R}^n \to \mathbb{R}$ be a C^2 function, and suppose that $V^{-1}((-\infty,c])$ is compact for every $c \in \mathbb{R}$ and $\nabla V(x) \neq 0$, except for a finite number of points $p_1, p_2, \ldots, p_m$. Prove that:
(i)   every solution $x(t)$ of $x' = -\nabla V(x)$ is defined for all $t \geq 0$;
(ii) $x(t)$ tends to one of the equilibrium points p_k, $k = 1, \ldots, m$, as $t \to \infty$.

1.71 Consider the nonlinear system $x_j' + x_j \sum_{k=1}^{n} a_{jk} x_k = \sum_{k,m=1}^{n} b_{jkm} x_k x_m$, $j = 1, \ldots, n$, $b_{jkm} > 0$, such that the above equations imply a relation of the form $\sum_{j=1}^{n} a_j x_j = c$. Show that under these assumptions every solution for which $x_j(0) \geq 0$ may be continued throughout $0 \leq t < \infty$ and the solutions remain nonnegative and uniformly bounded.

1.72 Prove the R. E. Vinograd theorem: For any differential equation $x' = f(x)$, $x \in \mathbb{R}^2$, with Lipschitz function f defined in an open domain $G \subset \mathbb{R}^n$, there exists a differential equation $x' = g(x)$ such that the solutions of this equation are defined on the whole $\mathbb{R}$ and its trajectories in G are the same as those of the first equation.

1.73 Show that the vector fields $X = y^2 \, \partial/\partial x$ and $Y = x^2 \, \partial/\partial y$ are complete on $\mathbb{R}^2$ but $X + Y$ is not.

1.74 Let X and Y be vector fields on a Hilbert space H which satisfy the following conditions:
(i) X, Y are bounded and Lipschitz on bounded sets;
(ii)   there is a constant $\beta \geq 0$ such that $(Y(x), x) \leq \beta \mid x \mid^2$ for all $x \in H$;
(iii) there is a locally Lipschitz monotone increasing function $c(s) > 0$, $s \geq 0$, such that $\int^\infty (c(s))^{-1} \, ds = \infty$ and if $x(t)$ is an integral curve of X, $\mid x(t) \mid' \leq c(\mid x(t) \mid)$.
Prove that X, Y, and $X + Y$ have trajectories defined for all $t \geq 0$.

1.75 The Alekseev formula. Let f, $F : \mathbb{R}^2 \to \mathbb{R}$ be continuous functions with continuous partial derivatives with respect to the second variable. Denote by $x(t,t_0,x_0)$ the unique solution of the problem $x' = f(t,x)$, $x(t_0) = x_0$, and by $y(t,t_0,x_0)$ the unique solution of the problem $x' = f(t,x) + F(t,x)$, $x(t_0) = x_0$. Prove that the following relation holds:

$$y(t,t_0,x_0) = x(t,t_0,x_0) + \int_{t_0}^{t} \frac{\partial(x(t,s,y(s,t_0,x_0)))}{\partial x_0} F(s,y(s,t_0,x_0)) \, ds$$

Remark: A similar formula holds in $\mathbb{R}^n$.

1.76 Show that if $f \in C^1(\mathbb{R})$ and $\mid f(x) - \cos x \mid \leq 1$ for all $x \in \mathbb{R}$, then every solution of the differential equation $x' = f(x)$ is bounded.

1.77 Consider the equation $x' = ax + b$ with real parameters a and b and its solution $x = -a/b$. Explain why this solution is not continuous as a and/or b tends to zero while the right-hand side of the equation depends on a and b in a very regular way.

1.78 Determine all twice differentiable functions such that for every $x > 0$ the length of the graph of f on $[0,x]$ is equal to the area of the set between this graph and the x-axis.

1.79 Compare the solutions of the equations
(i) $x' + ax = 0$;
(ii) $x'(1 - a\varepsilon) + ax = 0$;
(iii) $a\varepsilon^2 x''/2 + (1 - a\varepsilon) x' + ax = 0$;
with appropriate initial conditions.

1.80 Show that each integral curve of the equation

$$x' = \left(\frac{x^2 + 1}{t + 1}\right)^{1/3}$$

has two horizontal asymptotes.

1.81 Determine constants a,m such that the differential equation $(x')^3 + m(x')^2 = a(x + mt)$ has a singular interval.

1.82 Solve the differential equation $x = t(x')^2 + (x')^3$ and determine the number of integral curves passing through any given point in the plane.

1.83 Determine all functions $P(t,x)$, $G(t,x)$ such that the differential equation $P(t,x) \, dt + G(t,x) \, dx = 0$ admits two integration factors of the form $f(t + x)$, $g(t - x)$. Solve this equation.

1.84 Determine n so that the equation

$$(at^2 + 2btx + cx^2)(t^2 + x^2)^{-n} (x \, dt - t \, dx) = 0$$

is exact.

1.85 Show that any solution of the differential equation $x' = P(t,x)/Q(t,x)$ with polynomials P and Q is ultimately monotone, together with all its derivatives, and satisfies one or the other of the relations $x(t) \simeq at^b \exp(R(t))$, $x(t) \simeq at^b(\log t)^{1/c}$, where R is a polynomial in t and c is an integer.

1.86 Prove that the differential equation

$$tx' - x = \frac{x^2 - t^2}{at^2 + bt + c}$$

with real constants a, b, c has solutions independent of a, b, c. Find the general solution. Show that this general solution is rational if and only if $4/(b^2 - 4ac)$ is the square of an integer.

1.87 Show that if f and g are smooth functions and a distribution x satisfies the equation $x' + fx = g$, then x is a smooth function which verifies the above equation in the usual sense.

1.88 Assume that $x'(t) + x^2(t) = \mathcal{O}(1/t)$ as $t \to 0^+$. Show that either $x(t) = \mathcal{O}(\log(1/t))$ or $x(t) = -1/t$ as $t \to 0^+$.

1.89 Solve the initial value problem

$$r' + \frac{(r^2 + a^2)^{1/2}}{t} = b, \qquad t > 1$$

where $r(1) = r_0$ and a, b are constants. This equation arises in a problem of charge transfer by dissociated water within a layer of a strong 1,1 valent electrolyte solution.

1.90 Show that for arbitrary four solutions of the Riccati equation $x' + p(t)x^2 + q(t)x + r(t) = 0$ the cross ratio $(x_4 - x_1)(x_3 - x_2)/(x_3 - x_1)(x_4 - x_2)$ is a constant.

1.91 Show that knowing three solutions of a Riccati equation x_1, x_2, x_3, one knows all the solutions.

1.92 Show that the Riccati equation $x' = x^2 + q(t)x + r(t)$ with continuous coefficients can have at most two periodic solutions.

1.93 Find the general solution of the Riccati equation $x' + a(t)x^2 + b(t)x + c(t) = 0$ where the coefficients are proportional, that is, $a(t):b(t):c(t) = m:n:p$ for some constants m, n, p.

1.94 Solve the Riccati equation $x' + ax^2 = bt^n$ in terms of continued fractions.

1.95 Show that three solutions $x_1 < x_2 < x_3$ determine the coefficients in the Riccati equation $x' = p(t)x^2 + q(t)x + r(t)$. Show that the two solutions determine the Riccati equation of the type $x' = x^2 + q(t)x + r(t)$.

1.96 Consider the Riccati equation $x' = a(t)x^2 + 2b(t)x + a(t)u^2(t)$ and suppose that this equation admits two solutions $x_1(t)$, $x_2(t) = c^2x_1(t)$ with some constant $c \neq 1$. Give a necessary and sufficient condition on a, b, u guaranteeing existence of two such solutions and find the general solution of the equation in this case.

1.97 Consider the Riccati equation $x' + x^2 + p(t)x + q(t) = 0$ with continuous coefficients p and q. Show that (locally in t, because in general the solutions of the Riccati equation are only local in t)

$$x(t) = \min_u (\exp(-\int_{t_0}^{t} (2u(s) + p(s))\, ds)$$

$$\times (x_0 + \int_{t_0}^{t} (u^2(\tau) - q(\tau)) \exp(\int_{t_0}^{\tau} (2u(s) + p(s))\, ds)\, d\tau));$$

the minimum is taken over all continuous functions u.

1.98 Find the Carleman linearization of the Riccati equation $x' = -x + x^2$, that is, the equations for the functions $x_k(t) = x(t)^k$.
Remark: Another application of this linearization procedure is illustrated by the following example: Writing the equation $x'' + x + \varepsilon x^3 = 0$ in the form of the system $x' = v$, $v' = -x - \varepsilon x^3$, obtain an infinite system of linear equations for the quantities $x^j v^k$.

1.99 It may be proved that for the Riccati equation $x' = a(t) + b(t)x + c(t)x^2$ with differentiable 2π-periodic functions, a, b, c, the following alternative holds: Either
(i)   the equation does not have periodic solutions, or
(ii)   it has exactly one 2π-periodic solution, or
(iii) it has exactly two 2π-periodic solutions, or
(iv)   all the solutions defined on $[-\pi,\pi]$ are 2π-periodic (they have 2π-periodic continuations).
Verify that the following equations illustrate these properties:
(i) $x' = (x^2 + 1)(2 + \sin t)$,
(ii) $x' = x^2(2 + \sin t)$,
(iii) $x' = (x^2 - 1)(2 + \sin t)$,
(iv) $x' = x^2 \sin t$.

1.100 Let functions g and h be continuous for $t \geq 0$. Show that every function v satisfying the differential inequality $v' + g(t) v \geq h(t)$, $v(0) = c$, is not less than u, where $u' + g(t)u = h(t)$, $u(0) = c$. Draw as a corollary that for small $t > 0$ the solution of the equation $v' + g(t)v = v^2$, $v(0) = c$, can be written in the form $v = \max_w (c \exp(-\int_0^t(g(s) - 2w(s)) \, ds) - \int_0^t \exp(-\int_0^s(g(\tau) - 2w(\tau) \, d\tau)w^2(s) \, ds))$; the maximum is taken over all continuous functions w.

1.101 Consider the problem $t^{n+1}x'(t) + g(t)x(t) + f(t) = 0$, $x(1) = b$, $b \geq 0$, $f, g \in C[0,1]$, and $f > 0$. Denote by x_b the solution of this problem. Show that $x_0 > 0$, and $x_b(t) \geq x_0(t)$ on $[0,1]$.

1.102 Let $f : [a,b] \times \mathbb{R} \to \mathbb{R}$ be a continuous function. Prove that if $v : [a,b] \to \mathbb{R}$ is a C^1 function such that $v' \leq f(t,v)$, $v(a) \leq x_0$, and if x is the maximal solution of $x' = f(t,x)$, $x(0) = x_0$, then $v \leq x$ on $[a,b]$.

1.103 Let $f: [a,b] \times \mathbb{R} \to \mathbb{R}$ be a Lipschitz function and $v' \leq f(t,v)$, $x' = f(t,x)$ in $[a,b]$, $v(a) \leq x(a)$. Prove that for any $t_0 \in (a,b]$ either $v(t_0) < x(t_0)$ or $v \equiv x$ in $[a,t_0]$.

1.104 Prove a generalized singular Gronwall lemma: If $x \in L^\infty(0,T)$ satisfies $x(t) \leq C_1 + C_2 \int_0^t (t - s)^{-1/2}x(s) \, ds$, then $x(t) \leq M(T,C_1,C_2)$ on $[0,T]$.

1.105 Assume that $x(t) \leq a(t) + b \int_0^t (t - s)^{c-1}x(s) \, ds$ with a $c > 0$. Prove that $x(t) \leq a(t) + \theta \int_0^t E_c'(\theta(t - s)) a(s) \, ds$, where $\theta = b\Gamma(c)^{1/c}$, $E_c(z) = \sum_{n=0}^\infty z^{nc}/\Gamma(nc + 1)$. This generalizes the result of the preceding problem.

1.106 Prove the uniform Gronwall lemma: Let g, h, y be three positive locally integrable functions on (t_0,∞) such that y' is locally integrable on (t_0,∞), and which satisfy $y' \le gy + h$ for $t \ge t_0$, $\int_t^{t+r} g(s)\,ds \le a_1$, $\int_t^{t+r} h(s)\,ds \le a_2$, $\int_t^{t+r} y(s)\,ds \le a_3$ for $t \ge t_0$, where r, a_1, a_2, a_3 are positive constants. Then $y(t + r) \le (a_3/r + a_2)\exp(a_1)$ for all $t \ge t_0$.

1.107 Prove the Bihari lemma: Suppose that u and f are nonnegative continuous functions on $\mathbb{R}^+$ and $u(t) \le C + \int_0^t f(s)\Phi(u(s))\,ds$ for some $C > 0$ and a continuous positive nondecreasing function Φ. Then $u(t) \le \Psi^{-1}(\int_0^t f(s)\,ds)$, where $\Psi(u) = \int_c^u dv/\Phi(v)$. Of course, if Φ is defined for $0 < u < u_0$, then one should assume that $\int_0^t f(s)\,ds < \Psi(u_0 - 0)$ for all $0 \le t < \infty$.

1.108 Let nonnegative functions v, u, vu, and wu^a be locally integrable for $x \ge 0$. If $u_0 > 0$ and $0 \le a < 1$, then the inequality $u(x) \le u_0 + \int_0^x v(s)u(s)\,ds + \int_0^x w(s)u^a(s)\,ds$ implies

$$u(x)\exp\left(-\int_0^x v(s)\,ds\right)$$
$$\le \left(u_0^{1-a} + (1-a)\int_0^x w(s)\exp(-(1-a)\int_0^s v)\,ds\right)^{1/(1-a)}$$

1.109 Let $1 \le p < \infty$ and the inequality $u(x) \le v(x) + w(x)(\int_0^x u^p(s)\,ds)^{1/p}$ holds a.e. in the interval $[a,b]$ for nonnegative functions u, v, $w \in L^p$. Show that

$$\left(\int_a^x u^p(s)\,ds\right)^{1/p} \le \frac{(\int_a^x v^p(s)\,\varepsilon(s)\,ds)^{1/p}}{(1 - (1 - \varepsilon(x))^{1/p}}$$

where $\varepsilon(x) = \exp(-\int_a^x w^p(s)\,ds)$.

1.110 The Nagumo theorem. If f is a continuous function such that $f(0) = 0$, $\lim_{t \to 0} f(t)/t = 0$, then the inequality $f(t) \le \int_0^t f(s)s^{-1}\,ds$ implies $f(t) \equiv 0$. Suppose in addition that $f(t)/t^2 \in L^1(0,a)$ for some $a > 0$. If g satisfies the same set of conditions, then $f(t) \le g(t) + \int_0^t f(s)s^{-1}\,ds$ implies $f(t) \le g(t) + t\int_0^t g(s)s^{-2}\,ds$.

Linear Differential Equations

1.111 Let $\lambda_1, \ldots, \lambda_n$ be the eigenvalues of an $n \times n$ matrix A. Show that $\exp(tA) = a_1(t)B_1 + a_2(t)B_2 + \cdots + a_n(t)B_n$, where the functions $a_k(t)$ are defined recurrently: $a_1(t) = \exp(\lambda_1 t)$, $a_k(t) = \int_0^t \exp(\lambda_k(t - u))a_{k-1}(u)\,du$, $k = 2, \ldots, n$, and $B_1 = I$, $B_k = (A - \lambda_1 I) \cdots (A - \lambda_{k-1}I)$, $k = 2, \ldots, n$.

1.112 Characterize the canonical form of hyperbolic linear automorphisms A of $\mathbb{R}^n$ such that $A = \varphi_1$ for some flow $\{\varphi_t\}$ on $\mathbb{R}^n$.

1.113 Let $x(t) \ne 0$ be a solution of a linear equation $x' = Ax$ in $\mathbb{R}^n$. Show that exactly one of the following alternatives holds:
(i) $\lim_{t \to \infty} x(t) = 0$ and $\lim_{t \to -\infty} |x(t)| = \infty$;
(ii) $\lim_{t \to \infty} |x(t)| = \infty$ and $\lim_{t \to -\infty} x(t) = 0$;

(iii)   there exist constants $M, N > 0$ such that $M \leq \mid x(t) \mid \leq N$ for all $t \in \mathbb{R}$;

(iv) $\lim_{t \to \infty} \mid x(t) \mid = \infty$ and $\lim_{t \to -\infty} \mid x(t) \mid = \infty$.

1.114 Suppose that all solutions of a linear equation $x' = Ax$ are periodic, with the same period. Show that A is semisimple and its characteristic polynomial is a power of $t^2 + a^2$ with $a \in \mathbb{R}$.

1.115 Prove that the flow associated with a linear equation with constant coefficients cannot have an asymptotically stable closed orbit.

1.116 Consider linear equations $x' = Ax$ and $x' = Bx$. Assume that 0 is an asymptotically stable solution of these equations and $AB = BA$. Prove that 0 is an asymptotically stable solution of the equation $x' = (A + B)x$.

1.117 Assume that a linear operator A on $\mathbb{R}^4$ has eigenvalues $\pm ai, \pm bi, a > 0, b > 0$. Show that:

(i)   If a/b is a rational number then every solution of $x' = Ax$ is periodic.

(ii)   If a/b is irrational then there is a nonperiodic solution $x(t)$ such that $M \leq \mid x(t) \mid \leq N$ for suitable constants $M, N > 0$.

1.118 Let A be a linear operator on $\mathbb{R}^n$ all of whose eigenvalues have real part 0. Show that $0 \in \mathbb{R}^n$ is a stable equilibrium of $x' = Ax$ if and only if A is semisimple and that 0 is never asymptotically stable.

1.119 Suppose that all eigenvalues of a matrix A have nonpositive real parts. Show that

(i)   If A is semisimple then every solution $x(t)$ of $x' = Ax$ is bounded.

(ii)   Find an example of a not semisimple matrix A such that the equation of $x' = Ax$ has a solution of $x(t)$ such that $\lim_{t \to \infty} \mid x(t) \mid = \infty$.

1.120 Suppose that at least one eigenvalue of the linear operator A on $\mathbb{R}^n$ has positive real part. Prove that for any $a \in \mathbb{R}^n$, $\varepsilon > 0$, there is a solution $x(t)$ of the equation $x' = Ax$ such that $\mid x(0) - a \mid < \varepsilon$ and $\lim_{t \to \infty} \mid x(t) \mid = \infty$.

1.121 Let A be a linear operator on $\mathbb{R}^n$. Prove that the equation $x' = Ax$ has a T-periodic solution if and only if A has an eigenvalue of the form $2\pi i k/T, k \in \mathbb{Z}$.

1.122 Prove that a linear vector field on $\mathbb{R}^n$ is hyperbolic if and only if for each $p \in \mathbb{R}^n$, the ω-limit set of the point p is either the singleton $\{0\}$ or it is empty.

1.123 Let $\Phi(t)$ be an $n \times n$ matrix whose elements are C^1 functions and $\Phi(t)$ is nonsingular for all $t \in \mathbb{R}$. Prove that there exists a unique matrix $A(t)$ such that $\Phi(t)$ is the fundamental matrix of the equation $x' = A(t)x$.

1.124 $A(t)$ is an antisymmetric real matrix whose elements are continuous functions. Prove that fundamental matrices $\Phi(t)$ of $x' = A(t)x$ satisfy $\Phi^*(t)\Phi(t) = $ const, and vice versa.

1.125 $A(t)$ is a matrix whose elements are continuous functions and $(\int_0^t A(s)\ ds)A(t) = A(t)(\int_0^t A(s)\ ds)$. Prove that $\Phi(t) = \exp(\int_0^t A(s)\ ds)$ is a fundamental matrix for the equation $x' = A(t)x$.

1.126 Consider the system $x' = a(t)x + b(t)y$, $y' = c(t)x + d(t)y$, and assume that $b(t)d(t) < 0$ for all t. Show that between each pair of zeros of a component of $(x(t),y(t))$ there exists a zero of another component.

1.127 Prove that if the linear systems $x' = A(t)x$, $x' = -A^*(t)x$ are stable, and a continuous function f satisfies $|f(t,x)| \le c(t)|x|$ with $\int_0^\infty c(t)dt < \infty$, then there is a constant $C > 0$ such that for every solution of the system $x' = A(t)x + f(t,x)$, $|x(t)| \le C|Y(t)||x(0)|$, where $Y(t)$ is the fundamental matrix for the equation $x' = A(t)x$.

1.128 Let $A(t) = \sum_{m=1}^\infty A_m t^m$ with constant matrices A_m and let $x(t) = \sum_{m=1}^\infty a_m t^m$ be a solution of the equation $x' = A(t)x$. Prove that $(m+1)a_{m+1} = \sum_{j=0}^m A_{m-j}a_j$.

1.129 Assume that for every fundamental matrix $Y(t)$ of the linear system $x' = G(t)x$, with a continuous matrix $G(t)$, there exists $\lim_{t\to\infty} Y(t) = Y_\infty$ and Y_∞ is nonsingular. Show that for every solution $x(t)$ of this system $\lim_{t\to\infty} x(t) = x_0$ exists and for every vector $x_0 \in \mathbb{R}^n$ there is a solution $x(t)$ such that $\lim_{t\to\infty} x(t) = x_0$.

1.130 Consider a linear system $x' = P(t)x$ with a continuous, bounded matrix P. The system is said to satisfy the Perron condition if for every continuous and bounded function f the inhomogeneous system $x' = P(t)x + f(t)$ has a solution $x = x(t)$, $x(t_0) = 0$, bounded for $t \ge t_0$. Prove that the Perron condition is equivalent to the following inequality satisfied by the fundamental matrix $X(t)$, $X(0) = I$: $|X(t)X(t)^{-1}| \le C \exp(-\alpha(t-s))$ for some $C,\alpha > 0$ and all $t_0 \le s \le t$.

1.131 Prove the Liapunov inequality: For the linear differential equation $x' = A(t)x$ with a bounded $n \times n$ matrix $A(t)$ the sum of the characteristic exponents is not less than $\lim \sup_{t\to\infty} t^{-1} \int_0^t \text{Re Tr } A(s)\ ds$. Give an example where this inequality is strict.

1.132 Prove the Ważewski inequality: For every solution of the linear differential equation $x' = A(t)x$ the inequality $|x(0)| \exp(\int_0^t \lambda(s)\ ds) \le |x(t)| \le |x(0)| \exp(\int_0^t \Lambda(s)\ ds)$ holds, where $\lambda(t)$, $\Lambda(t)$ are the minimal and maximal eigenvalues of the matrix $(A(t) + A^*(t))/2$.

1.133 Prove the Lappo-Danilevskiĭ theorem: If for every $0 \le a \le t < \infty$, $A(t) (\int_a^t A(s)\ ds) = (\int_a^t A(s)\ ds)A(t)$ and the limit matrix $A = \lim_{t\to\infty} t^{-1} \int_a^t A(s)\ ds$ has all its eigenvalues strictly negative, then 0 is a globally asymptotically stable solution of $x' = A(t)x$.

1.134 Prove the Levinson theorem: Suppose that all solutions of the linear system $x' = Ax$ are bounded on $[0,\infty)$. Then all the solutions of the perturbed system $y' = (A + B(t))y$, where B is a continuous matrix function and $\int_0^\infty |B(t)| \, dt < \infty$, are also bounded. Moreover, $\lim_{t\to\infty} |x(t) - y(t)| = 0$ with a bijective correspondence between the initial data $x(0)$ and $y(0)$.

1.135 Give an example that the boundedness of all solutions of $x' = A(t)x$, together with the conditions $|B(t)| \to 0$, $\int_0^\infty |B(t)| \, dt < \infty$, is not sufficient to ensure the boundedness of all solutions of $x' = (A(t) + B(t))x$ as $t \to \infty$.

1.136 Consider the system $x' = (A + B(t) + C(t))x$ where A is a constant matrix all of whose characteristic roots have nonpositive real parts, while those with zero real parts are simple (this may be replaced by the condition that they correspond to simple factors), $B(t) \to 0$ as $t \to \infty$, $\int_0^\infty |dB/dt| \, dt < \infty$, $\int_0^\infty |C(t)| \, dt < \infty$, and the characteristic roots of $A + B(t)$ have nonpositive real parts for $t \geq 0$. Prove that under these conditions, all solutions of this system remain bounded as $t \to \infty$.

1.137 Consider the systems $y' = A(t)y + \varphi(t)$, $x' = A(t)x + \Phi(t,x)$ with continuous functions φ and Φ. Prove the Bellman theorem: A necessary and sufficient condition in order that all solutions of the nonlinear system are bounded for every function Φ with either
(i) $|\Phi(t,x)| \leq M$, $t \geq 0$, x arbitrary, or
(ii) $|\Phi(t,x)| \leq f_1(t)$, $\int_0^\infty f_1(t) \, dt < \infty$, or
(iii) $|\Phi(t,x)| \leq f_2(t)$, $\int_0^\infty f_2^2(t) \, dt < \infty$,
is that the linear system has only bounded solutions for all vector functions φ satisfying respectively the condition
(i') $|\varphi(t)| \leq N$ for $t \geq 0$, or
(ii') $\int_0^\infty |\varphi(t)| \, dt < \infty$, or
(iii') $\int_0^\infty |\varphi(t)|^2 \, dt < \infty$.
Conversely, if condition (i'), (ii'), or (iii') is satisfied, and the function Φ satisfies respectively the further condition
(i") $|\Phi(t,x)| \leq C_1 |x|$, or
(ii") $|\Phi(t,x)| \leq C_1 f_3(t) |x|$, $\int_0^\infty f_3(t) \, dt < \infty$, or
(iii") $|\Phi(t,x)| \leq C_1 f_4(t) |x|$, $\int_0^\infty f_4^2(t) \, dt < \infty$,
for all $t \geq 0$ and C_1 sufficiently small, whenever $|x| \leq C_2$, C_2 sufficiently small, then every solution of the nonlinear system with sufficiently small initial condition $x(0)$ is bounded on $[0,\infty)$. In particular, under the conditions (i'), (i") $\lim_{t\to\infty} |x(t)| = 0$.

1.138 Let $f : \mathbb{R}^{n+1} \to \mathbb{R}^n$ be a continuous function and let there exist a constant a and a function $g(t)$ such that $|f(t,x)| \leq g(t) |x|$ for $|x| \leq a$ and $t > 0$

and let $\int_0^\infty g(s) \, ds < \infty$. Assume that the matrix A has eigenvalues with negative real part only. Prove that there exists a constant M such that any solution $x(t)$ of the equation $x' = Ax + f(t,x)$ satisfies $|x(t)| \leq M \mid x(0) \mid$ if $|x(0)| \leq a/M$.

1.139 Let $B(t)$ be a continuous matrix such that $\int_0^\infty |B(t)| \, dt < \infty$. Prove that any solution $x(t) \neq 0$ of the equation $x' = B(t)x$ tends to a limit different from zero as $t \to \infty$. Moreover, for a given constant vector v, there is a unique solution which tends to v as $t \to \infty$.

1.140 Find a solution of the equation $tx'(t) = Ax(t)$ with a given $x(1) \in \mathbb{R}^n$, where A is a constant matrix with eigenvalues $1, 2, \ldots, n$.

1.141 Prove that $x(t) = \exp(Af(t))v$ solves the equation $x'(t) - f'(t)Ax = 0$, where A is a constant matrix with distinct nonzero eigenvalues.

1.142 Prove the Dollard-Friedman theorem: Suppose that $A(t)$ is a continuous matrix on $[R,\infty)$ such that $A = A_1 + A_2$, where
(i) $|A_1(\cdot)| \in L^1(R,\infty)$;
(ii)   there exists $B(t) = -\lim_{r \to \infty} \int_t^r A_2(s) \, ds$;
(iii) $|B(\cdot)A(\cdot)| \in L^1(R,\infty)$.
Then there exists a unique continuously differentiable function $x(t,x_0)$ such that $x'(t) = A(t)x(t)$ and $\lim_{t \to \infty} x(t,x_0)) = x_0$. Each nonzero solution of the above differential equation has a nonzero limit as t tends to ∞ and $|x(t,x_0) - x_0| \leq 2(\sup_{s \geq t} |B(s)| + \int_t^\infty |A_1(s)| \, ds + \int_t^\infty |B(s)A(s)| \, ds) |x_0|$ for all t such that the expression in parentheses is less than $1/2$.

Second-Order Linear Equations

1.143 Do there exist continuous functions p,q on $(-a,a)$, $a > 0$, such that the function
(i) $y_1(x) = x^2 \sin x$,
(ii) $y_2(x) = 1 - \cos x$
satisfies the equation $y'' + p(x)y' + q(x)y = 0$ on this interval?

1.144 Analyze completely the possibility of solving the problem $tx'' + ax' + btx = 0$, $x(0) = x_0$. Show that, in general (this should be made precise), the Cauchy problem, obtained by adjoining the condition $x'(0) = x_1$, cannot be solved.

1.145 Using the maximum principle, prove the uniqueness of solutions of the initial value problem $u'' + g(x)u' + h(x)u = f(x)$, $u(a) = b$, $u'(a) = c$, with the functions g and h bounded. Show by constructing an example that the boundedness of g is essential.

1.146 The differential equation $t^2x'' + \alpha tx' + \beta x = 0$, where α, β are constants, $t > 0$, is known as the Euler equation. Letting $D = (\alpha - 1)^2 - 4\beta$, show that its general solution is
(i) $x(t) = c_1 t^a + c_2 t^b$ if $D > 0$;
(ii) $(c_1 + c_2 \log t)t^a$ if $D = 0$;
(iii) $t^{\mathrm{Re}\, a} (c_1 \cos(\mathrm{Im}\, a \log t) + c_2 \sin(\mathrm{Im}\, a \log t))$ if $D < 0$,
where a, b are the solutions of the algebraic equation $r^2 + (\alpha - 1)r + \beta = 0$.

1.147 Find the general solution to $x'' + 2cx' - (1 - 2 \operatorname{sech}^2 t)x = 0$.

1.148 Given a linear differential equation $y'' + p(x)y' + q(x)y = 0$, find a necessary and sufficient condition on p, q for the existence of two solutions y_1, y_2 such that:
(i) $y_2(x) = xy_1(x)$,
(ii) $y_2(x) = 1/y_1(x)$.

1.149 Determine the constant a such that the differential equation $x^2(x + 1)y'' + x(ax + 4)y' + (6x + 2)y = 0$ has rational solutions only.

1.150 Study oscillatory properties of solutions of the equation $x'' + mx/t^2 = 0$.

1.151 Show that for arbitrary function $x = x(t)$, $x'' + a_1(t)x' + a_2(t)x = W(x_1^2(x/x_1)'/W)'/x_1$, where x_1 and x_2 are two linearly independent solutions of the homogeneous equation and W is the Wronskian of x_1 and x_2. Generalize to obtain the formula

$$\frac{d^n x}{dt^n} + a_1(t) \frac{d^{n-1}x}{dt^{n-1}} + \cdots + a_n(t)x$$

$$= \frac{W_n}{W_{n-1}} \frac{d}{dt} \left(\left(\frac{W_{n-1}^2}{W_{n-2}W_n} \right) \cdots \frac{d}{dt} \left(\left(\frac{W_2^2}{W_1 W_3} \right) \frac{d}{dt} \left(\left(\frac{W_1^2}{W_0 W_2} \right) \frac{d}{dt} \left(\frac{x}{W_1} \right) \right) \right) \right).$$

1.152 Prove the Darboux lemma: If g is a nontrivial solution of $-y'' + qy = \lambda y$ for $\lambda = \mu$ and if f is a nontrivial solution for $\lambda \neq \mu$, then $f' - fg'/g =: [g,f]/g$ is a nontrivial solution of $-y'' + (q - 2(\log g)'')y = \lambda y$ for the same λ. Also, for $\lambda = \mu$, the general solution of the latter equation is given by $g^{-1}(a + b\int_0^x g^2(s)\, ds)$, where a and b are arbitrary constants.

1.153 Let x_1 and x_2 be solutions of the equation $x'' + a_1(t)x' + a_2(t)x = 0$. *Show that $x_1(t)x_2'(t) - x_2(t)x_1'(t) = C \exp(-\int^t a_1(s)\, ds)$ where C is a constant. Prove that $x_1(t)$ and $x_1(t) \int^t x_1^{-2}(s)\, ds$ are independent solutions on any interval where $x_1(t) \neq 0$.*

1.154 By considering the differential equation $ay = -xy' + y''$, show that when $a > 0$, $x > 0$,

$$\frac{\int_0^\infty \exp(-xt - t^2/2)\, t^a\, dt}{\int_0^\infty \exp(-xt - t^2/2)t^{a-1}\, dt} = \frac{a}{x + (a + 1)/(x + (a + 2)/x + \cdots)}$$

(this is the continued fraction).

1.155 Consider the second-order linear differential equation $y = Q_0 y' + P_1 y''$, where Q_0, P_1 are functions of x. After differentiation this equation becomes $y' = Q_1 y'' + P_2 y^{(3)}$, where $Q_1 = (Q_0 + P_1')/(1 - Q_0')$, $P_2 = P_1/(1 - Q_0')$. This process is repeated infinitely and a set of relations $y^{(n)} = Q_n y^{(n+1)} + P_{n+1} y^{(n+2)}$ is obtained, where $Q_n = (Q_{n-1} + P_n')/(1 - Q_{n-1}')$, $P_{n+1} = P_n/(1 - Q_{n-1}')$. Observe that

$$\frac{y}{y'} = Q_0 + \frac{P_1}{y'/y''} = Q_0 + \frac{P_1}{Q_1 + P_2(y''/y^3)} = \cdots = Q_0$$

$$+ \frac{P_1}{Q_1 + (P_2/Q_2 + (\cdots + P_n/(Q_n + R_n)))}$$

where $R_n = P_{n+1}/(y^{(n+1)}/y^{(n+2)})$. The continued fraction

$$\frac{1}{Q_0 + (P_1/Q_1 + (P_2/Q_2 + \cdots))}$$

should represent the logarithmic derivative of a solution of the equation. Prove that this continued fraction converges and has the value y'/y if $y \neq 0$ and
(i) $P_n \to P$, $Q_n \to Q$ as $n \to \infty$,
(ii) the roots r_1, r_2 of the equation $r^2 = Qr + P$ are of unequal modulus,
(iii) if $|r_2| < |r_1|$ then $\lim_{n\to\infty} |y^{(n)}|^{1/n} < |r_2|^{-1}$ provided $r_2 \neq 0$ (when $r_2 = 0$ the last condition is replaced by the condition that the limit is finite).

1.156 Show that the system $y'' + py' + qy = r$, $a_1 y(a) + a_2 y(b) + a_3 y'(a) + a_4 y'(b) = A$, $b_1 y(a) + b_2 y(b) + b_3 y'(a) + b_4 y'(b) = B$, on the interval $[a,b]$ is selfadjoint if $a_2 b_4 - a_4 b_2 = (a_1 b_3 - a_3 b_1) \exp(\int_a^b p\, dx)$.

1.157 Prove that if all the solutions of $x'' + a(t)x = 0$ belong to $L^2(0,\infty)$ and $\sup |b(t)| < \infty$, then all the solutions of $x'' + (a(t) + b(t))x = 0$ belong to $L^2(0,\infty)$. Moreover, the space $L^2(0,\infty)$ in the assumption and in the conclusion may be replaced by $(L^p \cap L^q)(0,\infty)$ with $1 < p < \infty$ and $p^{-1} + q^{-1} = 1$.

1.158 Show that there are unbounded solutions of the equation $x'' + (1 + \varepsilon \cos t)x = 0$ for every $|\varepsilon|$ sufficiently small. If $k > 0$, then all solutions of the equation $x'' + kx' + (1 + \varepsilon \cos t)x = 0$ tend to zero as $t \to \infty$ provided $|\varepsilon|$ is sufficiently small.

1.159 Estimate the solution (and its derivative) of the equation $t^2x'' + 2tx' - 2x = f(t)$ which is bounded if $t \to 0$ and $t \to \infty$, assuming that $0 \le f(t) \le m$.

1.160 What can be said about the function p if all the solutions of the equation $x'' + p(t)x' + q(t)x = 0$ tend to zero, together with their first derivatives, as $t \to \infty$?

1.161 Prove the Liapunov integral stability criterion: If a continuous nonnegative T-periodic function p satisfies the inequality $0 < T \int_0^T p(t)\, dt \le 4$, then all the solutions of the equation $x'' + p(t)x = 0$ are bounded (together with their first derivatives) for $t \in \mathbb{R}$. Give an example showing that the constant 4 in the above condition cannot be replaced by a larger one.

1.162 Prove that if $a(t)$ increases to ∞ as $t \to \infty$, then all solutions of the equation $x'' + a(t)x = 0$ are bounded on $[0, \infty)$.

1.163 Prove the Osgood theorem: If $0 < c_1 < f(t) < c_2$, then there is a unique solution of $x'' = f(t)x$ which is bounded as $t \to \infty$, and this solution approaches zero as $t \to \infty$.

1.164 Prove the Cacciopoli-Hukuhara-Nagumo theorem: If $f(t)$ is a real function such that $\int_0^\infty |f(t) - c|\, dt < \infty$ for some $c > 0$, then all solutions of the equation $x'' + f(t)x = 0$ are bounded in $[0, \infty)$. Give an example showing that the condition $\lim_{t \to \infty} f(t) = c$ is not sufficient to guarantee the boundedness (as was erroneously stated by Fatou).

1.165 Prove a generalization of the preceding theorem: All solutions of $x'' + (1 + \varphi(t) + \psi(t))x = 0$ are bounded, provided that $\int_0^\infty |\varphi(t)|\, dt < \infty$, $\int_0^\infty |\psi'(t)|\, dt < \infty$, $\psi(t) \to 0$ as $t \to \infty$.

1.166 Show that there are no bounded solutions of the equation $x'' + (1 + (1 + t^4)^{-1})x = \cos t$.

1.167 What conditions on $f(t)$ will ensure that all solutions of $x'' + x = f(t)$ are bounded as $t \to \infty$? Is $f(t) \to f_0$ as $t \to \infty$ sufficient?

1.168 For what values of a and b are all solutions of the equation $x'' + (1 + t^{-b} \sin t^a)x = 0$ bounded as $t \to \infty$?

1.169 Show that all solutions of the equation

$$x'' + \left(1 + \frac{b(\sin 2t)}{t}\right)x = 0$$

cannot be bounded as $t \to \infty$.

1.170 Prove the Gusarov theorem: If $x'' + (1 + f(t))x = 0$ with $1 + f(t) \geq a^2 > 0$, $\int_0^\infty |f''(t)| \, dt < \infty$, then $x(t)$ is bounded as $t \to \infty$.

1.171 Prove the Wintner theorem: If f is a real continuous function, $\lim_{t \to \infty} f(t) = c$ for some $c > 0$, and f is of bounded variation (in particular, monotone), then all solutions of the equation $x'' + f(t)x = 0$ are bounded in $[0,\infty)$ and asymptotic to solutions of the limit equation $x'' + cx = 0$.

1.172 Consider two equations $x'' - 2t^{-1}x' + x = 0$, $x'' + 2t^{-1}x' + x = 0$, having monotone coefficients. Show on these examples that an immediate extension of the result in the preceding exercise for general second-order equations is not true.

1.173 Study the asymptotic behavior of solutions to the equations
(i) $t^2 x'' + x \log^2 t = 0$,
(ii) $x'' - 4t^2 x = 0$,
(iii) $tx'' + x = 0$.

1.174 Consider the differential equation $x'' + ax' + bx = 0$, $a, b > 0$. Show that constants c, d can be found such that $x^2 + cxx' + d(x')^2 > 0$ for $x^2 + (x')^2 > 0$ and $(x^2 + cxx' + d(x')^2)' \leq -\varepsilon(x^2 + cxx' + d(x')^2)$ for some $\varepsilon > 0$. Conclude that $x^2 + (x')^2 = \mathcal{O}(e^{-ht})$ for some $h > 0$ as $t \to \infty$.

1.175 Let f be integrable and $\int_1^\infty t|f(t)| \, dt < \infty$. Show that $x'' + f(t)x = 0$ has the solutions $x_1(t)$ and $x_2(t)$ such that $x_1(t) \to 1$, $x_1'(t) \to 0$, $x_2(t)/t \to 1$, $x_2'(t) \to 1$ as $t \to \infty$.

1.176 Calculate the asymptotic behavior as $t \to \infty$ for the solutions of the initial value problem $y'' + y = 1/t$, $\pi \leq t < \infty$, $y(\pi) = y'(\pi) = 0$.

1.177 Prove the Ascoli theorem: If $\varphi(t)$ is monotone and tends to a^2 as $t \to \infty$, then for $x'' + \varphi(t)x = 0$, $\lim_{t \to \infty} \max_{0 \leq s \leq t} |x(s)| = c$, $\lim_{t \to \infty} \max_{0 \leq s \leq t} |x'(s)| = ac$.

1.178 Assume that $b(t) = \mathcal{O}(t^{1-a})$, $a > 0$. Show that the equation $x'' + (1 + b(t))x = 0$ has two solutions of the form $x_1(t) = \cos t + \mathcal{O}(t^{-a})$, $x_2(t) = \sin t + \mathcal{O}(t^{-a})$.

1.179 Show that for the linear equation $x'' + p(t)x = 0$, where $p(t) > 0$ is continuously differentiable, $p' \in L^1(a,\infty)$, $p(\infty) \neq 0$, the general solution has the asymptotics $x(t) = c_1 \cos (\int_a^t p^{1/2}(s) \, ds) + c_2 \sin(\int_a^t p^{1/2}(s) \, ds) + \eta(t)$, $\eta(t), \eta'(t) = o(1)$ as $t \to \infty$. Derive analogous asymptotic formulas for the equation $x'' + p(t)x = 0$ where $-p$ satisfies the conditions above.

1.180 Prove the Nagumo-Hartman lemma: Let $\Phi : [0,\infty) \to \mathbb{R}$ be a continuous function, strictly positive and such that $\int_0^\infty s \, ds/\Phi(s) = \infty$, and let α, k, M be positive constants. Let $y : [0,T] \to \mathbb{R}^n$ be a mapping of class C^2 such that

(i) $\sup_{t\in[0,T]} | y(t) | \leq M$,

(ii) $| y''(t) | \leq \Phi(| y'(t) |)$,

(iii) $| y''(t) | \leq \alpha(| y(t) |^2)'' + K$.

Then there exists a constant N (depending only on Φ, α, M, and K) such that $\sup_{t\in[0,T]} | y'(t) | \leq N$. Note that condition (iii) is superfluous if $n = 1$.

1.181 Suppose that $x_1'' + q_1(t)x_1 = 0$ and $x_2'' + q_2(t)x_2 = 0$, $x_1(t) \neq 0$ for all $t \in [a,b]$ and that the inequality

$$-\frac{x_1'(a)}{x_1(a)} + \int_a^t q_1(s)\, ds \geq \left| -\frac{x_2'(a)}{x_2(a)} + \int_a^t q_2(s)\, ds \right|$$

holds for $t \in [a,b]$. Show that x_2 does not vanish on $[a,b]$ and

$$-\frac{x_1'(t)}{x_1(t)} \geq \left| \frac{x_2'(t)}{x_2(t)} \right|$$

for $t \in [a,b]$.

1.182 Show that the equation $(p(t)x')' + q(t)x = 0$, where p, q are continuous functions on $[a,b]$, $p > 0$, can be transformed by the Prüfer transformation $r = (x^2 + p^2(x')^2)^{1/2}$, $\varphi = \arctan(x/(px'))$, into

$$\varphi' = \frac{\cos^2 \varphi}{p(t)} + q(t) \sin^2 \varphi, \qquad r' = -\left(q(t) - \frac{1}{p(t)} \right) r \sin \varphi \cos \varphi$$

1.183 Prove the Sturm comparison theorem: Consider two equations $(p_i(t)x')' + q_i(t)x = 0$, $i = 1,2$, with continuous functions p_i, q_i on an interval $[a,b]$ such that $p_1(t) \geq p_2(t) > 0$, $q_1(t) \leq q_2(t)$. Let $x_1(t) \neq 0$ be a solution of the first equation and let $x_1(t)$ vanish at exactly n points $a < t_1 < t_2 < \cdots < t_n \leq b$ in $[a,b]$. Let $x_2(t)$ be a solution of the second equation satisfying $p_1(a)x_1'(a)/x_1(a) \geq p_2(a)x_2'(a)/x_2(a)$. Show that $x_2(t)$ has at least n zeros in $(a,t_n]$.

1.184 Let $(p(t)x')' + q_1(t)x = 0$ and $(p(t)y')' + q_2(t)y = 0$. Assume that $q_2(t) > q_1(t)$ on (a,b) and for some $\delta > 0$ $x(t) > 0$, $y(t) > 0$ on $(a, a + \delta)$, and $\lim_{t\to a+} p(t)(x'(t)y(t) - x(t)y'(t)) \geq 0$. Prove that if $x(t_1) = 0$ for some $t_1 \in (a,b)$ then there is a $t_2 \in (a,t_1)$ such that $y(t_2) = 0$.

1.185 Let $x_1(t)$, $x_2(t)$ be linearly independent solutions of the equation $(p(t)x')' + q(t)x = 0$, where p, q are continuous functions on $[a,b]$, $p > 0$. Show that the zeros of x_1 separate and are separated by those of x_2.

1.186 Let $y(t)$ be a nontrivial solution of the equation $y'' + q(t)y = 0$. Assume that $q(t) > 0$ for $t > 0$ and $\int_0^\infty q(t)\, dt = \infty$. Prove that there exists a sequence $t_n \to \infty$ such that $y(t_n) = 0$.

1.187 Show that the distance between two consecutive zeros of any solution of the Airy equation $x'' + tx = 0$ tends to zero as t goes to infinity.

1.188 Consider the equation $(p(t)x')' + (ar(t) + bq(t))x = 0$, where a, b are constants, $p(t)$ is C^1 on $\mathbb{R}$, and $r(t)$, $q(t)$ are continuous on $\mathbb{R}$. Assume that $p > 0$, $r > 0$, $a > 0$, $\int_0^1 q(t)\, dt = 0$, and the functions p, r, q are periodic with period 1. Show that the solution $x(t)$ of this equation satisfying $x(t + 1) = \lambda x(t)$, where λ is constant, must vanish at least at one point in the interval $[0,1]$.

1.189 Prove that if u is a solution of the equation $u'' + g(x)u' + h(x)u = 0$ with

$$h(x) - \frac{g'(x)}{2} - \frac{(g(x))^2}{4} \leq M$$

then the distance between zeros of u is at least $\pi M^{-1/2}$. If

$$h(x) - \frac{g'(x)}{2} - \frac{(g(x))^2}{4} \geq m > 0$$

then the distance between consecutive zeros of u is at most $\pi m^{-1/2}$.

1.190 Consider the Bessel equation

$$x'' + \frac{x'}{t} + \left(1 - \frac{\mu^2}{t^2}\right) x = 0$$

where μ is a real parameter. Show that the zeros of the solution $x(t)$ on $t > 0$ form a sequence (t_n) such that $(t_n - t_{n-1}) \to \pi$ as n goes to infinity.

1.191 Show that the positive zeros of the Bessel functions $J_n(x)$, $J_{n+1}(x)$, $J_{n+2}(x)$ follow one another cyclically in that order if $n > -1$ and in the reverse order if $n < -1$.

1.192 Let y be any solution of the equation $(K(x)y')' - G(x)y = 0$. If throughout the interval (a,b), $K > 0$, $K' \neq 0$, $G < 0$, $(K'/G)' < 1$, then the zeros of y, y', y'' follow one another cyclically in that order if $K' > 0$ and in the reverse order if $K' < 0$.

1.193 Prove that if all solutions of the equation $x'' + \varphi(t)x = 0$ are oscillatory (that is, they have an infinite number of zeros in the interval $[0,\infty)$), and if $\psi(t) \geq \varphi(t)$, then all solutions of $y'' + \psi(t)y = 0$ are oscillatory.

1.194 Prove the nonoscillation theorem of de la Vallée–Poussin: The second-order operator $Ly = y'' + p(x)y + q(x)$ has the nonoscillation property on the interval $[a,b]$ (that is, every solution of $Ly = 0$ has at most one zero in (a,b)) if and only if there exists a function z with absolutely continuous derivative z' such that $z > 0$, $Lz \le 0$ for $x \in (a,b]$.

Remark: The well-known Kneser nonoscillation criterion $(y'' + p(x)y = 0$ has the nonoscillation property if $0 < p(x) < 1/(4x^2))$ is a corollary of the above theorem.

Second-Order Nonlinear Equations

1.195 Find the differential equation of central conics $x^2 + 2hxy + by^2 = k$, where b and h are variable parameters, $k \in \mathbb{R}$.

1.196 Find the differential equation which has as its general solution the function

$$y = \frac{1 + kx}{x + 2kx^2} \qquad k \in \mathbb{R}$$

1.197 Let $x(t)$ be a solution of the equation $x'' + x - 2x^3 = 0$ and let $D = x'^2(0) + x^2(0) - x^4(0)$. Show that, if $D < 1/4$ then this solution is periodic, and unbounded if $D > 1/4$.

1.198 Consider the equation $x' + x = (x'')^2$. For $x(0)$ sufficiently small and $x'(0)$ suitably chosen, is there a solution which exists for all $t \ge 0$ and approaches 0 like e^{-t} as $t \to \infty$?

1.199 Let $x(t)$ be the solution of the problem $x'' = -\lambda x' + V'(x)$, $x(0) = 0$, $x'(0) = 10$, with $V(x) = \frac{1}{2}x^2\exp(-x^2)$. Show that there exists λ_0 such that $\lim_{t \to \infty} x(t) = \infty$ for $\lambda < \lambda_0$ and $\lim_{t \to \infty} x(t) = 0$ for $\lambda > \lambda_0$.

1.200 Given the equation $x'' = a^2 - x^2$, $a > 0$, study the behavior of the solutions, in particular, periodic and nonperiodic solutions.

1.201 Determine the general solution of the equation $(f'' + 2)f = (f' + 2x) \times (f' + x)$.

Remark: This problem arises in modeling the Helmholtz equation in two dimensions.

1.202 Show that for every $x_0 \in (0,1)$ there exist $\alpha > 0$ and $\beta > 0$ such that the problem $yy'' + (y')^2 + yy'/x + \alpha xy' = 0$, $y(x_0) = 0$, $y'(x_0) = \beta$, has a solution $y(x)$ which satisfies $\lim_{x \to \infty} y(x) = 1$.

Remark: This problem arises in the study of infiltration of water.

1.203 Prove that each equation of the form $(y - x)y'' + F(y') = 0$ has a first integral of the form $(y - x)f(y') = $ const.

1.204 Show that the solutions of the equations $y'' + \sigma y^n = 0$ and $yy'' + C(y')^2 + \sigma = 0$ lead to the solutions of the equations $y'' + \sigma y^m = 0$ and $yy'' + \lambda(y')^2 + \sigma = 0$, where $m = (-5 - 3n)/(3 + n)$ and $\lambda = -C/(1 + 3C)$.

1.205 It is well known that each solution of the polynomial differential equation $P(t,x,x') = 0$ satisfies an upper bound of the form $x(t) = \mathcal{O}(\exp(t^k))$ for some k as $t \to \infty$. Show that no such general bound can exist for second-order polynomial equations.

1.206 Given the equation $x'' + f(x)x' + g(x) = 0$, where $f(x) = (n + 2)bx^n - 2a$, $g(x) = x(c + (bx^n - n)^2)$, show that the solution is $x = \cos(p + \omega t)(qe^{-nat} + nbe^{-nat} \int_0^t e^{nas} \cos^n(p + ws) \, ds)^{-1/n}$, $c = w^2$.

1.207 Study properties of the solutions of the equation of the isothermic equilibrium of a star $(x^2 y')' = x^2 e^{-y}$.

1.208 Study integrability cases of the Meshcherskiĭ system

$$x'' + \frac{\mu(t)x}{(x^2 + y^2)^{3/2}} = 0 \qquad y'' + \frac{\mu(t)y}{(x^2 + y^2)^{3/2}} = 0$$

1.209 Find a function which is a first integral simultaneously for the systems:
(i) $x'' + ax + by = 0$, $y'' + dy + bx = 0$;
(ii) $x'' + ax + by + cy' = 0$, $y'' + dy + bx - cx' = 0$.

1.210 Solve the system of differential equations $x'' = x^3 + xy^2$, $y'' = 2yx^2$, where $x(0) = y'(0) = 1$, $x'(0) = y(0) = 0$.

1.211 Find a necessary and sufficient condition on the functions p, q for existence of solutions y, z of the equations $y'' = p(x)y$, $z'' = q(x)z$, such that $y(x)z(x) \equiv 1$.

1.212 Assume that a smooth function $V : \mathbb{R} \to \mathbb{R}$ is bounded from below. Show that each solution of the equation $x'' = -V'(x)$ is defined on the whole $\mathbb{R}$. Moreover, if $\lim_{x \to \pm\infty} V(x) = \infty$, then the solutions are bounded.

1.213 Suppose that $f(t,x)$ is defined for $0 \le t < \infty$, $0 < x < \infty$, in such a way that f is continuous, $f(t,x) \ge 0$, $f(t,0) \equiv 0$, $\partial f/\partial x$ is positive and bounded. Prove that for every $x_0 > 0$ the problem $x'' = f(t,x)$, $x(0) = x_0$, has a solution $x(t)$ which exists on $\mathbb{R}^+$, $x(t) \ge 0$ and $x'(t) \le 0$ for $t \ge 0$.

1.214 Prove the F. John theorem: Consider the equation $x'' = F(x, \cos at)$, where F satisfies the following hypotheses: $F(x,z)$ is defined for $x \in \mathbb{R}$, $-1 \le z \le 1$, F is continuous, $F_x < 0$, $F_z > 0$, $\lim_{x \to \pm\infty} F(x,z) = \mp\infty$ uniformly with respect to z; $|F(x,z)/x| \le C$ for $|x| \ge |x_0|$ or $|F_z(x,z)/F(x,z')| \to 0$

as $|x| \to \infty$ uniformly with respect to z and z'. Under these hypotheses every solution is defined on the entire real axis.

1.215 Prove the one-dimensional maximum principle: Suppose that u satisfies the differential inequality $Lu = u'' + g(x,u)u' \geq 0$ for $a < x < b$, with a bounded function g. If $u(x) \leq M$ in (a,b) and if the maximum M of u is attained at an interior point c of (a,b), then $u \equiv M$. Show that the boundedness assumption on g is essential.

1.216 Assume that $f(t,x,z)$ is a C^1 function in $t \geq 0$, $x \geq 0$, $z \leq 0$, $f(t,0,0) = 0$, and $f(t,x,z) \geq 0$. Prove that there exists a constant c_0 such that if $0 < c < c_0$, then the problem $x'' = f(t,x,x')$, $x(0) = c$, has at least one solution $x(t)$ such that $x(t) \geq 0$ and $x'(t) \leq 0$ for $t \geq 0$. Show that it is not always possible to take $c_0 = \infty$.

1.217 Consider the equation $x'' + p(t)x' + q(t)f(x) = 0$, where p, f are continuous functions and q is continuously differentiable. Suppose that $0 < q(t) \leq M$, $p(t) \geq -q'(t)/2q(t)$, and $\int_0^{\pm\infty} f(x)\, dx = \infty$. Prove the boundedness of x and x' on $\mathbb{R}^+$.

1.218 Show that for every continuous odd function g such that $g(x)/x \to \infty |x| \to \infty$, there exists a function p, $|p(t)| \leq 1$, such that the equation $x'' + g(x) = p(t)$ has at least one unbounded solution as $t \to \infty$.

Differential Equations in the Plane

1.219 Integrate the system $x' = (ax + by)y$, $y' = -(ax + by)x$.

1.220 Let $F(x,y)$ and $G(x,y)$ be C^1 functions on $\mathbb{R}^2$. Assume that F is even, G is odd in x. Show that each solution of the problem $x' = F(x,y)$, $y' = G(x,y)$, $x(0) = 0$, $x(T) = 0$, is $2T$-periodic.

1.221 Show that there is a unique nontrivial periodic solution of the problem $x' = y - f(x)$, $y' = -g(x)$, when all of the following conditions are assumed:
(i) f and g are C^1;
(ii) $g(-x) = -g(x)$ and $xg(x) > 0$ for all $x \neq 0$;
(iii) $f(-x) = -f(x)$ and $f(x) < 0$ for $0 < x < a$;
(iv)   for $x > a$, $f(x)$ is positive and increases to ∞.

1.222 Let f and g be C^1 vector fields on $\mathbb{R}^2$ such that $(f(x),g(x)) = 0$ for all $x \in \mathbb{R}^2$. Assuming that f has a (nontrivial) closed orbit, prove that g has a singular point.

1.223 Consider the vector field on $\mathbb{R}^2$ $(P(x,y),Q(x,y))$, where P and Q are relatively prime polynomials of degree at most 2, which are not both linear. Show that if A and B are two singular points, then the paths intersecting the half line $[A,\infty)$ cross it in the same sense as the paths intersecting $[B,\infty)$ and in the opposite sense to the paths intersecting (A,B).

1.224 Consider two-dimensional autonomous system $x' = P(x,y)$, $y' = Q(x,y)$, where P and Q are relatively prime polynomials of degree at most 2, which are not both linear. Show that:
(i) Three critical points can never be collinear.
(ii) The interior of a closed orbit is a convex region.
(iii) There exists a unique critical point in the interior of each closed path.
(iv) Two closed orbits are oppositely oriented if their interiors have no common point.
(v) Two closed orbits are similarly oriented if their interiors have a common point.
(vi) A critical point in the interior of a closed orbit must be either a focus or a center.

1.225 Find all functions $f: \mathbb{R}^2 \setminus \{0\} \to (0,\infty)$ of class C^∞ such that the flow generated by the vector field $(-xf(x,y), -yf(x,y))$ preserves the Lebesgue measure.

1.226 Assume that X is a vector field on $\mathbb{R}^2$ and div $X = 0$. Show that there does not exist any isolated periodic orbit of X.

1.227 Find two vector fields X and Y on $\mathbb{R}^2$ such that X has a first integral, X and Y are topologically conjugate, and Y does not have any first integral.

1.228 Prove the Dulac criterion of nonexistence of periodic orbits: Let P, Q, B : $G \to \mathbb{R}$ be C^2 functions defined on a simply connected domain $G \subset \mathbb{R}^2$. Assume that $\operatorname{div}(BP, BQ) > 0$. Show that there is no periodic orbit of the vector field (P,Q) inside G.

1.229 Let γ be a periodic orbit of an analytic vector field on $\mathbb{R}^2$. Show that there exists a neighborhood U of γ such that either U is a union of periodic trajectories or U does not contain any periodic orbit other than γ.

1.230 Let C be any closed subset of $[0,1]$. Show that there exist functions $u(r)$, $v(r)$ which are Lipschitz continuous on $[0,1]$, $u^2(r) + v^2(r) > 0$ for $r \neq 0$, and the solution of the problem $x' = xu(r) - yv(r)$, $y' = xv(r) + yu(r)$, $x(0) = x_0$, $y(0) = y_0$, $r = (x^2 + y^2)^{1/2}$, is a closed orbit if $(x_0^2 + y_0^2)^{1/2} \in C$ and is a spiral if $(x_0^2 + y_0^2)^{1/2} \notin C$.

1.231 Let γ be a closed orbit of a C^1 vector field on the plane. Assume that all the fixed points within γ are hyperbolic. Show that there must be an odd number $2n + 1$ of them, n are saddles, and $n + 1$ are either sinks or sources.

1.232 Prove the Mikołajska theorem: Consider a system of differential equations such that $x' = f(t,x,y)$, $y' = g(t,x,f(t,x,y))$, where f, g are continuous functions and there exist unique global solutions. If $|f(t,x,y)| \leq |y| \varphi(t)$, $|g(t,x,f(t,x,y))| \leq |x| \psi(t)$, and $\lim_{t\to\infty} \int_0^t \varphi(s) \int_0^s \psi(\tau) \, d\tau \, ds < \infty$, then there exists a family of solutions $(x(t),y(t))$ such that $\lim_{t\to\infty} x(t) = c$, c a real parameter.

1.233 Consider a C^1 vector field on the plane $X(x,y) = (f(x,y),g(x,y))$. Assume that the origin is the unique critical point, $DX(x,y)$ has eigenvalues which have negative real parts everywhere in $\mathbb{R}^2$, and one of the four functions f_x, f_y, g_x, g_y vanishes identically in $\mathbb{R}^2$. Show that every trajectory of this vector field approaches the origin as $t \to \infty$.

1.234 Let X be a C^1 vector field on the plane and x a regular point such that $x \in \omega(x)$. Show that x is a periodic orbit. Is it true in $\mathbb{R}^n$ if $n > 2$?

1.235 Let γ be a nonperiodic orbit of the vector field on the plane. Prove that $\omega(\gamma) \cap \alpha(\gamma)$ is empty or is a singular point.

1.236 Let γ be a closed orbit enclosing an open set U contained in a domain W of a planar dynamical system. Show that U contains an equilibrium.

1.237 Let $F(x)$ be a C^1 vector field on the plane such that $F(0) = 0$, $DF(0)$ has an eigenvalue $a + bi$, $b \neq 0$. Show that there exists $\delta > 0$ such that the trajectory $x(t)$ of F passing through x_0, $|x_0| < \delta$, satisfies exactly one of these properties:
(i) $x(t)$ is periodic and 0 belongs to the domain enclosed by this orbit;
(ii) $\omega(x_0) = \{0\}$, $0 \notin \alpha(x_0)$;
(iii) $\alpha(x_0) = \{0\}$, $0 \notin \omega(x_0)$;
(iv) $\omega(x_0)$ is a periodic orbit and 0 belongs to the domain enclosed by this orbit;
(v) $\alpha(x_0)$ is a periodic orbit and 0 belongs to the domain enclosed by this orbit.
Show examples that each of these possibilities may hold.

1.238 Find an example of a differential equation in the plane such that for some point p its ω-limit set contains infinitely many trajectories.

1.239 Sketch the phase portrait of a flow on the plane such that:
(i)   the nonwandering set is the whole plane, but
(ii) for some $x \in \mathbb{R}^2$, x is neither an α-limit point nor an ω-limit point of any point.

1.240 Give an example of a vector field on $\mathbb{R}^2$ with finitely many singular points and such that the set of orbitally unstable orbits is a dense subset of the plane.

1.241 Sketch the phase portrait of the system $x' = x^2(y - 1)(4 - x^2)$, $y' = y^2(x - 1)(y^2 - x)$.

1.242 Sketch the phase portrait of the system $x' = (1 - x^2)(x - \alpha y)$, $y' = (1 - y^2)(y + \alpha x)$, with $\alpha > 1$.

1.243 Study the trajectories of the dynamical system $x' = (1 - x^2)(y + x(1 - x^2))$, $y' = -x + (1 - x^2)y$.

1.244 Describe the behavior near the origin of trajectories of the homogeneous system $x' = a_0 y^m + a_1 y^{m-1} x + \cdots + a_m x^m$, $y' = b_0 y^m + b_1 y^{m-1} x + \cdots + b_m x^m$.

1.245 Show that the system of differential equations

$$x' = f\left(\frac{x}{t}, \frac{y}{t}\right), \qquad y' = g\left(\frac{x}{t}, \frac{y}{t}\right)$$

has a first integral of the form $F(x/t, y/t) = $ const. Determine a necessary and sufficient condition for the trajectories of this system to lie on spheres centered at the origin.

1.246 Find the general solution of the system $x' = (y(x^2 + y^2) + x)(x^2 + y^2)$, $y' = (-x(x^2 + y^2) + y)(x^2 + y^2)$.

1.247 Find the general solution of the following system of differential equations: $x' = r(x \sin t - y \cos t)(r \cos t - x)$, $y' = r(x \sin t - y \cos t)(r \sin t - y)$. The problem arises in the study of trailer-truck jackknifing.

1.248 Prove that the quadratic system in the plane $x' = \lambda x - y - 10x^2 + (5 + \delta)xy + y^2$, $y' = x + x^2 + (-25 + 8\varepsilon - 9\delta)xy$, where $\delta = -10^{-13}$, $\varepsilon = -10^{-52}$, $\lambda = -10^{-200}$, has at least four limit circles.

1.249 Study a generalized Volterra model for predator-prey coexistence $x' = ax - x^2 - by\varphi(x)$, $y' = -cy + dy\varphi(x)$, where φ is a positive increasing smooth function.

1.250 The Emden-Fowler equation from astrophysics $(\xi^2 \eta')' + \xi^a \eta^b = 0$ is transformed by the change of variables $x = \xi\eta'/\eta$, $y = \xi^{a-1}\eta^b/\eta'$, $t = \log|\xi|$, into the system $x' = -x(1 + x + y)$, $y' = y(a + 1 + bx + y)$. Sketch the phase portrait of this system.

1.251 The Blasius equation of fluid mechanics $\eta^{(3)} + \eta\eta'' = 0$ is transformed by the change of variables $x = \eta\eta'/\eta''$, $y = (\eta')^2/\eta\eta''$, $t = \log|\eta'|$, into the system $x' = x(1 + x + y)$, $y' = y(2 + x - y)$. Sketch the phase portrait of this system.

1.252 Show that the system $x' = y$, $y' = -ax - by + \alpha x^2 + \beta y^2$, does not have nontrivial periodic orbits.

1.253 Show that the system $x' = y$, $y' = -x + (1 - x^2 - y^2)y$, has only one closed orbit γ and for any $0 \neq p \in \mathbb{R}^2$, $\omega(p) = \gamma$.

1.254 Sketch the phase portrait of the system $x' = -xy$, $y' = x/2 - y^2$.

1.255 Show that the differential equation $x' = y + x(1 + \beta y)(x^2 + y^2 + 1)$, $y' = -x + (y - \beta x^2)(x^2 + y^2 + 1)$, does not have a periodic solution.

1.256 Show that there exists a homeomorphism of some neighborhood of $0 \in \mathbb{R}^2$ which maps the trajectories of the system $x' = -y$, $y' = x - x^5$, onto the trajectories of the system $x' = -y$, $y' = x$.

1.257 Consider the system $x' = -\omega y + x(1 - x^2 - y^2) - y(x^2 + y^2)$, $y' = \omega x + y(1 - x^2 - y^2) + x(x^2 + y^2) - K$, where ω and K are constants. Assume that γ is a periodic orbit and for each $p \in \gamma$, $|p| > 2^{-1/2}$. Show that 0 belongs to the domain enclosed by γ.

1.258 Let $\{\varphi_t\}$ be the flow generated by the following system: $x' = \mu_1 x - x(x^2 + y^2) - xy^2$, $y' = \mu_2 y - y(x^2 + y^2) - yx^2$. Prove that for all values of μ_1, μ_2 there is a closed simply connected set $D \neq \emptyset$, $\neq \mathbb{R}^2$ such that $\varphi_t(D) \subset D$ for all $t > 0$. Show that for $\mu_1 = \mu_2 > 0$, the line $x = y$ separates two distinct domains of attraction.

1.259 Show that the system $x' = x(ax + by + c)$, $y' = y(dx + ey + f)$, does not have periodic solutions.

1.260 Show that there is one nontrivial periodic solution of the Van der Pol equation $x' = y - x^3 + x$, $y' = -x$, and every nonequilibrium solution tends to this periodic solution.

1.261 Let $\gamma = (\varphi(t), \psi(t))$ be a T-periodic orbit of the vector field (P,Q) on the plane, and let S be a Poincaré section at $(\varphi(0), \psi(0))$ and f the Poincaré map. Prove that if S is orthogonal to γ, then $f'(\varphi(0), \psi(0)) = \exp(\int_0^T (P_x(\varphi(s), \psi(s)) + Q_y(\varphi(s), \psi(s)))\,ds)$.

1.262 Find the closed orbits of the following system: $r' = r(\mu + vr^2 - r^4)$, $\varphi' = 1 - r^2$, for different values of μ and v. Discuss their stability in terms of the Poincaré map.

1.263 Determine values of the parameter a such that the system $\varphi' = 1$, $r' = (r - 1)(a + \sin^2 \varphi)$ has a stable limit cycle.

1.264 Study periodic solutions of the system $x' = -y + (x^2 + y^2 - 1)x \sin(x^2 + y^2 - 1)^{-1}$, $y' = x + (x^2 + y^2 - 1)y \sin(x^2 + y^2 - 1)^{-1}$ for $x^2 + y^2 \neq 1$ and $x' = -y$, $y' = x$ for $x^2 + y^2 = 1$.

1.265 Show that the system $x' = (1 - x^2 - y^2)x - y$, $y' = x + (1 - x^2 - y^2)x$ has a unique closed orbit and compute its Poincaré map.

1.266 Sketch the phase portrait of the system $x' = y + x^7 + 4x^4y^2 - y$, $y' = ax^ny - x^7 + y^2$, where $n \geq 0$, $a \neq 0$.

1.267 Sketch the phase portrait of the system $x' = xy$, $y' = y^2 - 6x^2y + x^4$.

1.268 Sketch the phase portrait of the system $x' = xy$, $y' = x^2 + y^2$.

1.269 Sketch the phase portrait of the system $x' = xy$, $y' = y^2 - x^4$.

1.270 Study the system describing dynamics of a chemical reactor

$$x' = -x \exp\left(-\frac{1}{y}\right) + \lambda(a - x), \qquad y' = x \exp\left(-\frac{1}{y}\right) + \beta(b - y).$$

1.271 Sketch trajectories of the Perron system $x' = x - 4y \mid xy \mid^{1/2}$, $y' = -y + 4x \mid xy \mid^{1/2}$.

1.272 Sketch trajectories of the Digel system

$$x' = -xr^4 - yr^2 \cos\left(\frac{1}{r}\right) - yr^3 \sin\left(\frac{1}{r}\right), \qquad y' = -yr^4 + xr^2 \cos\left(\frac{1}{r}\right)$$

$$+ xr^3 \sin\left(\frac{1}{r}\right).$$

1.273 Study trajectories of the homogeneous system $x' = (-3x^4 + 6x^2y^2 + y^4)(x^2 + y^2)^{-3/2}$, $y' = 8xy^3(x^2 + y^2)^{-3/2}$.

1.274 Sketch the phase portraits of the systems

$$x' = -ax - by + xy, \qquad y' = bx - ay + \frac{x^2 - y^2}{2}$$

for $a = 0$ and $a > 0$, with real b. When are these systems structurally stable?

1.275 Study the system

$$x' = y, \qquad y' = y \sin\left(\frac{\pi}{x^2 + y^2}\right)$$

(stationary points, periodic orbits, Liapunov function, phase portrait).

1.276 Sketch the phase portrait of the system $x' = -y^3(x^2 - 1)(2 + xy)$, $y' = x^3(y^2 - 1)(2 - xy)$.

1.277 Consider the equation $x'' - \beta(x')^2 x'' + \lambda x' + x = 0$ which describes the current x' in an electrical circuit with a capacitance, a resistance, and an induction coil. Study the conservative case $\lambda = 0$ (describe the phase portrait) and the stability of the origin for $\lambda > 0$.

1.278 Sketch the phase portraits of the following two nonlinear oscillators and their linearizations:

(i) $x'' + x + \varepsilon | x' | x' = 0$,
(ii) $x'' + x + \varepsilon x^2 = 0$.

1.279 Study the phase portrait of the quadratically damped oscillator $x'' + h(x')^2 + x = 0$. Determine separatrices for this equation.

1.280 Sketch the phase portrait for the damped pendulum $x'' + 2\alpha x' + \sin x = 0$, $0 < \alpha \ll 1$ and the pendulum $x'' + \sin x = \beta$, subject to applied torque $\beta > 0$. Consider also the damped pendulum with torque $x'' + 2\alpha x' + \sin x = \beta$. Both undamped systems possess homoclinic and periodic orbits. Can the damped systems possess any such orbit?

1.281 Discuss the phase plane trajectories of the equation $x'' + x = a/(b - x)$, where a and b are positive constants. This equation depicts, approximately, the motion of a magnet suspended by a spring above a large fixed iron plate.

Flows, Singular Points, Bifurcations

1.282 Let $f : \mathbb{R} \to \mathbb{R}$ be a C^1 function and $f'(x) > 0$. Show that there exists a C^1 function $F : \mathbb{R} \to \mathbb{R}$ such that $f(x) = \varphi_1(x)$, where $\{\varphi_t\}$ is the flow generated by $x' = F(x)$.

1.283 Let $\{\varphi_t\}$ be a continuous flow on $\mathbb{R}$. Show that there exist a C^1 flow $\{\psi_t\}$ and a homeomorphism $h : \mathbb{R} \to \mathbb{R}$ such that $h \circ \varphi_t = \varphi_t \circ h$.

1.284 Prove the Poincaré-Bendixson theorem: Let X be a C^1 vector field on two-dimensional sphere S^2. Assume that X has finitely many singular points. Prove that for each $p \in S^2$, $\omega(p)$ is a periodic orbit or is a union of singular points and regular trajectories whose ω- and α-limit sets are singular points.

1.285 Prove the Poincaré-Bendixson theorem for vector fields defined on the projective plane.

1.286 Assume that $\{\varphi_t\}$ is a continuous flow on $\mathbb{R}^n$. Show that if the ω-limit set of a point $p \in \mathbb{R}^n$ is bounded then it is connected. Give an example that the ω-limit set need not be connected.

1.287 Let $\{\varphi_t\}$ be a continuous flow on a metric space M. Assume that A is a compact minimal set in M. Show that if A has an interior point, then A is an open set.

1.288 Assume that a differential equation $x' = f(x)$, $x \in \mathbb{R}^n$, defines the flow $\{\varphi_t\}$. Show that:
(i)   If the trajectory $\gamma(p)$ of a point p is recurrent—that is, for every $\varepsilon > 0$ there exists $t(\varepsilon) > 0$ such that for each t_0 $\gamma(p)$ is contained in the ε-

neighborhood of the set $\{\varphi_t(p) : t \in [t_0, t_0 + t(\varepsilon)]\}$—then p is Poisson stable, that is, $p \in \omega(p)$.

(ii)   If a set A is minimal and bounded and $p \in A$, then $\gamma(p)$ is recurrent.

(iii) If the trajectory of some p is bounded, then there exists a recurrent trajectory.

(iv)   If $\gamma(p)$ is recurrent, then cl $\gamma(p)$ is a minimal bounded set.

1.289 Let A be a closed, bounded, invariant set of the flow $\{\varphi_t\}$ generated by a differential equation $x' = f(x)$, $x \in \mathbb{R}^n$. Assume that for each $\varepsilon > 0$ and $p \in A$ there exist $T > 0$ and $q \in A$ such that $|p - q| \leq \varepsilon$ and A is contained in the ε-neighborhood of the set $\{\varphi_t(q) : t \in [0,T]\}$. Show that there exists $x \in A$ whose trajectory is dense in A.

1.290 Give an example of a dynamical system with a locally disconnected minimal set of almost periodic solutions.

1.291 The Denjoy-Siegel theorem reads: If X is a C^r, $r \geq 2$, nonsingular vector field on the two-dimensional torus, then one of the following statements about trajectories of X is true:

(i)   there exists a periodic trajectory;

(ii)   all trajectories are ergodic (that is, dense). Construct an example which shows that the hypothesis $r \geq 2$ is essential.

1.292 Show that there exists a C^1 vector field on the three-dimensional sphere without periodic trajectories.

1.293 Consider a C^1 flow $\{\varphi_t\}$ on $\mathbb{R}^n$ and suppose that there exists a C^1 function U such that for each $p \in \mathbb{R}^n$, $(dU(\varphi_t(p))/dt) |_{t=0} > 0$. Show that for each $p \in \mathbb{R}^n$ there exists a neighborhood V of p and $t_0 > 0$ such that $\varphi_t(V) \cap V = \varnothing$ for $t > t_0$.

1.294 Prove that a gradient vector field on a Riemannian manifold has no non-constant recurrent trajectories.

1.295 Assume that a trajectory γ of the vector field on a manifold is a compact set. Show that γ is either a singular point or a periodic orbit.

1.296 Let $\{\varphi(t,p)\}$ be a C^1 flow on a manifold. Assume that $\varphi(t_n,q) \to q$ for some sequence $t_n \to \infty$. Show that there exists a sequence $\{k_n\}$, $k_n \in \mathbb{N}$, $k_n \to \infty$, such that $\varphi(k_n,q) \to q$.

1.297 Let X be a C^1 vector field on a manifold M. Assume that the set $\cup\{\omega(x) : x \in M\}$ is the union of finitely many orbits of X. Show that the Birkhoff center (cl $\cup\{\gamma(p) : \alpha(p) = \omega(p)\}$, $\gamma(p)$ is the trajectory of p) is the union of periodic orbits and singular points.

1.298 Let $\{\varphi_t\}$ be a continuous flow on a manifold and let A be an invariant, asymptotically stable set with a dense trajectory and containing Liapunov stable trajectories. Prove that A is a torus and the flow on A is almost periodic.

1.299 Let X and Y be C^2 vector fields defined in an open subset U of $\mathbb{R}^n$ and φ_t^X, φ_t^Y be the local flows generated by X and Y, respectively. Prove that for every $x_0 \in U$ there is a neighborhood V of x_0 and $\varepsilon > 0$ such that $\varphi_t^{X+Y}(x) = \lim_{n \to \infty} (\varphi_{t/n}^X \varphi_{t/n}^Y)^n(x)$ uniformly for x in V and $|t| \leq \varepsilon$. Prove also that $\varphi_t^{[X,Y]}(x) = \lim_{n\to\infty}(\varphi^Y_{-\sqrt{t/n}} \varphi^X_{-\sqrt{t/n}} \varphi^Y_{\sqrt{t/n}} \varphi^X_{\sqrt{t/n}})^n(x)$, where $[X,Y]$ denotes the commutator of X and Y.

1.300 Let X and Y be vector fields on a manifold and let $\{\varphi_t^s\}$ be the flow generated by the vector field $X + sY$, $s \in \mathbb{R}$. Show that

$$\left. \frac{d\varphi_t^s(x)}{ds} \right|_{s=0} = \int_0^t D\varphi_u^0 \left(Y(\varphi^0_{-u+t}(x)) \right) du$$

1.301 Let X, Y be C^1 vector fields on two-dimensional sphere and $\{\varphi_t\}$, $\{\psi_t\}$ the flows generated by X and Y, respectively. Assume that for each $t, s \in \mathbb{R}$, $\varphi_t \circ \psi_s = \psi_s \circ \varphi_t$. Prove that X and Y have a common singular point.

1.302 The point x is chain recurrent for the flow $\{\varphi(t,x)\}$ if for every $\varepsilon > 0$, there are points $x = x_0, x_1, \ldots, x_n = x$ and times $t_1, t_2, \ldots, t_n \geq 1$ such that the distance from $\varphi(t_j, x_{j-1})$ to x_j is less than ε. Give an example of a flow with the set of chain recurrent points different from the set of nonwandering points.

1.303 Let X be a vector field on a manifold M, ω a volume form on M, and $\{\varphi_t\}$ the flow generated by X. Prove the Liouville formula

$$\det_\omega(D\varphi_t) = \exp \left(\int_0^t (\mathrm{div}_\omega X)(\varphi_s(x)) \, ds \right).$$

1.304 Let X be a vector field on a manifold M and ω be a volume form on M. Prove that X preserves some volume form if and only if there exists a function $k : M \to \mathbb{R}$ of class C^1 such that $Xk = -\mathrm{div}_\omega X$.

1.305 Show that the solutions of the Lorenz system $x' = \sigma(y - x)$, $y' = \rho x - y - xz$, $z' = -\beta z + xy$, $\sigma, \rho, \beta > 0$, exist on the whole $\mathbb{R}$. Prove also that there exists a bounded set of volume zero toward which all trajectories tend.

1.306 Consider the system of differential equations $x' = y$, $y' = -xz$, $z' = xy$ and its solution $x(t), y(t), z(t)$. Show that $z(t)$ can be expressed as an elliptic integral.

1.307 Study the trajectories of the following systems in $\mathbb{R}^3$:
(i) $x' = -y(x^2 + y^2) - 3xz^2 + 2x(x^2 + y^2)$,
       $y' = x(x^2 + y^2) - 3yz^2 + 2y(x^2 + y^2)$,
       $z' = -2x^3 + z(x^2 + y^2)$;

(ii) $x' = -y - x(x^2 + y^2)^{1/2},$
       $y' = x - y(x^2 + y^2)^{1/2},$
       $z' = z - (x^2 + y^2)^{1/2};$

(iii) $x' = -y(x^2 + y^2)^3 + x(x^2 + y^2 - 4z^2)(z^2 - x^2 - y^2)^2,$
       $y' = x(x^2 + y^2)^3 + y(x^2 + y^2 - 4z^2)(z^2 - x^2 - y^2)^2,$
       $z' = z(z^2 - 4(x^2 + y^2))(z^2 - x^2 - y^2)^2.$

1.308 Find two integrals of the system $ax' + (b - c)yz = 0,$ $by' + (c - a)zx = 0,$ $cz' + (a - b)xy = 0.$

1.309 Study trajectories of the system $x' = yz - x^2,$ $y' = zx - y^2,$ $z' = xy - z^2,$ in $\mathbb{R}^3$.

1.310 Let f be a C^1 vector field on $\mathbb{R}^n$. Assume that for each $\varepsilon > 0$ and each neighborhood U of a point p there exists a point $q \in U$ such that the trajectory of f starting at q is periodic and its period is less than ε. Show that p is a singular point.

1.311 Assume that $x(t)$ is a solution of the differential equation $x' = f(x),$ $x \in \mathbb{R}^n$, and $\lim_{t \to \infty} x(t) = p$. Show that p is a singular point.

1.312 Let f be a C^1 vector field on some open subset Ω of $\mathbb{R}^n$ and let $\{\varphi_t\}$ be the flow generated by f. Assume that for some $T > 0$ and for every neighborhood U of a point $p \in \Omega$ there exists $q \in U$ such that $\varphi_t(q) \in U$ for all $t \in [0,T]$. Prove that p is a singular point.

1.313 Assume that f is a C^1 vector field on $\mathbb{R}^n$ and $f(0) = 0$. Suppose that an eigenvalue of $Df(0)$ has positive real part. Show that in every neighborhood of 0 there is a point q such that if $x(t)$ is the solution of initial value problem $x' = f(x),$ $x(0) = q$, then $| x(t) |$ is increasing on some interval $[0,t_0],$ $t_0 > 0$.

1.314 Find a function $F : \mathbb{R}^2 \to \mathbb{R}^2$ such that 0 is a sink for the differential equation $x' = F(x)$ and a center for the linear equation $x' = DF(0)x$.

1.315 Consider the vector field on $\mathbb{R}^2$ $x' = -x \sin \alpha - y \cos \alpha + (x^2 + y^2 - 1)^2$ $(x \cos \alpha - y \sin \alpha),$ $y' = x \cos \alpha - y \sin \alpha + (x^2 + y^2 - 1)^2(x \sin \alpha + y \cos \alpha)$ which is obtained by rotation by angle α of the vector field $(-y + x(x^2 + y^2 - 1)^2,$ $x + y(x^2 + y^2 - 1)^2)$. Study the character of the singular point $(0,0)$ of this vector field.

1.316 Prove that the origin is a center for the Delaunay system of the theory of lunar motion $r' = M(1 + M_1r^2 + M_2r^4) \sin \theta,$ $\theta' = N(1 + N_1r^2 + N_2r^4 + N_3r^6) + r^{-1}M(1 + P_1r^2 + P_2r^4) \cos \theta,$ where $x = r \cos \theta,$ $y = r \sin \theta,$ $N \neq 0,$ M is an even function of x and y, $M = \mathbb{O}(x^2 + y^2)$.

1.317 Prove that the origin is a center for the system $x' = y + ax^3 + bx^2y - 3cxy^2 + dy^3,$ $y' = -x^3 - 3ax^2y - bxy^2 + cy^3$.

1.318 Prove that the quadratic system $x' = -y + xy$, $y' = x + 2y^2 - cx^2$ has a center at the origin surrounded by a one-parameter family of periodic trajectories. The period is not a monotone function of the amplitude for certain $c > 1.4$.

1.319 Consider the system $y' = -x + P(x,y)$, $x' = y + Q(x,y)$, where P, Q are analytic functions whose expansions begin with terms of degree higher than one, P contains only terms of even degree in x, and Q contains only terms of odd degree in x. Show that the origin is a center.

1.320 Show that if the singular point $(0,0)$ of the equation $(ax + by)\,dx + (mx + ny)\,dy = 0$ is a center, then this equation is of total differential type. If there exists a continuous nonconstant integration factor and $an \neq bm$, then this singular point is a saddle.

1.321 Classify the singular point $x = x' = 0$ of the equation $x'' + \varepsilon x^2 x' + x = 0$, where ε is a real parameter.

1.322 Find stable and unstable manifolds of the trivial solution for the system $x' = x$, $y' = -y + x^2$.

1.323 Verify that any center manifold of the system $x' = -x + y^2$, $y' = -yz - y^3$, $z' = 0$ cannot be C^∞.

1.324 Verify that in the Lanford example $x' = -x + h(y)$, $y' = -z$, $z' = 0$, where h is an analytic function, $h(y) = o(y)$ near the origin, any center manifold of the form $\{x = \varphi(y,z)\}$ cannot be analytic.

1.325 Study trajectories of the system $x' = xy$, $y' = -y + x^2$ in a neighborhood of the origin.

1.326 Show that the (analytic) system $x' = -x^2$, $y' = -y + x^2$ has no analytic center manifold.

1.327 Show that for the Kelley system $x' = x^2$, $y' = -y$ there exist a unique analytic center manifold and many C^∞ center manifolds.

1.328 Prove that the vector field $(ax + y^2, by)$ on $\mathbb{R}^2$ can be linearized at $(0,0)$ by a C^2 change of coordinates if and only if $a \neq 2b$.

1.329 Give an example that there exist a linear operator A on $\mathbb{R}^2$ and a C^∞ function $g : \mathbb{R}^2 \to \mathbb{R}^2$, $g(q) = o(\,|\,q\,|^n)$ for each $n \geq 0$ such that the vector field $X(q) = Aq + g(q)$ cannot be linearized at $(0,0)$ by a C^2 change of coordinates.

1.330 Let $\{\varphi_t\}$ be the flow generated by a differential equation $x' = f(x)$, $x \in \mathbb{R}^2$. Assume that $f(0) = 0$ and the origin is asymptotically stable. Show that there exist a neighborhood U of 0 and C^1 diffeomorphism $h : U \to U$, $h(0) = 0$, such that $h \circ \varphi_t(x) = \exp(tDf(0))h(x)$ for each $x \in U$. Is it true in $\mathbb{R}^n$, $n > 2$?

1.331 Let 0 be a hyperbolic singular point of C^1 vector fields X and Y on $\mathbb{R}^n$. Assume that there exists a diffeomorphism $h : \mathbb{R}^n \to \mathbb{R}^n$ which maps trajectories of X onto trajectories of Y and $h(0) = 0$. Show that the eigenvalues of $DX(0)$ are proportional to those of $DY(0)$.

1.332 Construct an example of a differential equation $x' = Ax + f(t,x)$ such that any homeomorphic change of variables transforming it into the linear system $y' = Ay$ is not differentiable (it is assumed that the matrix A does not have eigenvalues with zero imaginary part).
Remark: This example shows that, in general, the change of variables in the conclusion of the Hartman-Grobman theorem is not smooth.

1.333 Study the system $x' = y$, $y' = xy$ in the plane. Is it possible to eliminate the nilpotent singular point (at the origin) by a small perturbation, say $x^2 + y^2 + xy^2 + \cdots$, of the second equation?

1.334 Study the system $x' = x^2 + y^2 - 1$, $y' = 5(xy - 1)$, in particular the behavior of trajectories at infinity.

1.335 Let $F(x,y)$ be a C^1 vector field on $\mathbb{R}^2$ and (x_0,y_0) be a Liapunov stable singular point. Show that $\text{index}_F (x_0,y_0) = 1$.

1.336 Show that the vector fields in the complex plane defined by $z' = z^k$ and $z' = z^{-k}$ have unique singular points $z = 0$ with indexes k and $-k$, respectively.

1.337 Let $f(x,y) = (f_1(x,y),f_2(x,y))$ be a C^1 vector field on the plane and (x_0,y_0) an isolated singular point. Assume that for some $i, j, \partial f_i/\partial x_j \neq 0$. Show that the index of (x_0,y_0) is -1, 0, or 1.

1.338 Let $f(x,y)$ be a C^1 vector field on $\mathbb{R}^2$. Assume that (x_0,y_0) is a singular point and $\det Df(x_0,y_0) \neq 0$. Show that the index of the point (x_0,y_0) is $\text{sgn} \det(Df(x_0,y_0))$.

1.339 Let $X = (f(x,y),g(x,y))$ be a C^1 vector field on an open subset Q of $\mathbb{R}^2$ with the Jacobian determinant different from 0 whenever $X = 0$. Let C be a positively oriented Jordan curve in Ω with its interior G in Ω and $X \neq 0$ on C. Show that there is at most a finite number of stationary points $p_1, \ldots ,p_k$ in G and the index of X with respect to C is $n_+ - n_-$, where n_+ (n_-) is the number of these points at which the Jacobian determinant is positive (negative, respectively).

1.340 Let $F(x,y) = (\varphi(x,y),\psi(x,y))$ be a C^1 vector field on $\mathbb{R}^2$ and p_0 be an isolated singular point. Define the vector field $G(x,y) = (\varphi(x,y),\psi(x,y) -$

$a(x,y)\varphi(x,y))$, where $a(x,y)$ is a continuous function. Show that p_0 is an isolated singular point of the vector field G and $\text{index}_F \, p_0 = \text{index}_G \, p_0$.

1.341 Let $(\varphi(x,y),\psi(x,y))$ be a C^∞ vector field on $\mathbb{R}^2$. Assume that $\varphi(x_0,y_0) = \psi(x_0,y_0) = 0$ and $\partial\varphi(x_0,y_0)/\partial y \neq 0$. The last condition implies the existence, in some neighborhood of y_0, of a function $\alpha(y)$ such that $\varphi(\alpha(y), y) = 0$ and $\alpha(y_0) = x_0$. Let $\alpha'(y_0) = \alpha''(y_0) = \cdots = \alpha^{(k-1)}(y_0) = 0$ and $\alpha^{(k)}(y_0) \neq 0$. Show that (x_0,y_0) is an isolated singular point and $\text{index}_{(\varphi,\psi)} \, (x_0,y_0) = \text{sgn}(\partial\varphi(x_0,y_0)/\partial x) \, \text{sgn}(\alpha^{(k)}(y_0))(1 - (-1)^k)/2$.

1.342 Let $\Phi(x,y) = (a_m x^m + a_{m-1}x^{m-1}y + \cdots + a_0 y^m, b_n x^n + b_{n-1}x^{n-1}y + \cdots + b_0 y^n)$ be a vector field on $\mathbb{R}^2$ and $(0,0)$ be a unique singular point of Φ. Assume that m is odd and n is even. Show that $\text{index}_\Phi \, (0,0) = 0$.

1.343 Let $\Phi(x,y) = (\varphi(x,y),\psi(x,y))$ be a C^1 vector field on $\mathbb{R}^2$ and $(0,0)$ an isolated singular point. Assume that $\varphi(-x,-y) = \varphi(x,y)$ and $\psi(-x,-y) = -\psi(x,y)$. Show that $\text{index}_\Phi \, (0,0) = 0$.

1.344 Let $F(x,y) = (a_1 x^2 + 2a_2 xy + a_3 y^2, b_1 x^2 + 2b_2 xy + b_3 y^2)$ be a vector field on $\mathbb{R}^2$. Define

$$D_1 = \det\begin{bmatrix} a_1 & a_2 \\ b_1 & b_2 \end{bmatrix}, \quad D_2 = \det\begin{bmatrix} a_2 & a_3 \\ b_2 & b_3 \end{bmatrix}, \quad D_3 = \det\begin{bmatrix} a_1 & a_3 \\ b_1 & b_3 \end{bmatrix}$$

and $D = 4D_1 D_2 - D_3^2$. Show that $(0,0)$ is an isolated singular point if and only if $D \neq 0$, $\text{index}_F \, (0,0) = 0$ if $D < 0$, and if $D > 0$ then $\text{index}_F \, (0,0) = 2 \, \text{sgn} \, D_1$.

1.345 Show that index of an isolated singular point of a vector field $(y,f(x,y))$ is $1, -1$, or 0.

1.346 Sketch phase portraits for the family of systems $x' = \mu + x^2 - xy$, $y' = y^2 - x^2/2 - 1$, and show that a saddle connection exists for $\mu = 0$. What happens for $\mu > 0$ and for $\mu < 0$?

1.347 Study phase portraits of the following systems: $x' = P_\alpha(x,y) = x \cos \alpha + y \sin \alpha - (x \cos \alpha - y \sin \alpha)(x^2 + y^2)$, $y' = Q_\alpha(x,y) = x \sin \alpha - y \cos \alpha -(x \sin \alpha + y \cos \alpha)(x^2 + y^2)$, where α is a parameter from $[0,\pi)$. Show that $\alpha = \pi/2$ is a bifurcation point and that there exists $\varepsilon > 0$ such that for each $\alpha < \pi/2 - \varepsilon$ and $\beta > \pi/2 + \varepsilon$ there exists a homeomorphism h which transforms trajectories of the vector fields (P_α,Q_α) onto trajectories of the vector field (P_β,Q_β). Of course, h cannot preserve the orientation of trajectories.

1.348 Describe simplest bifurcations of equilibria in the following systems:
(i) $x' = \mu - x^2$ (saddle-node);
(ii) $x' = \mu x - x^2$ (transcritical);

(iii) $x' = \mu x - x^3$ (pitchfork);

(iv) $x' = -y + x(\mu - (x^2 + y^2))$, $y' = x + y(\mu - (x^2 + y^2))$ (Hopf).

1.349 Show that the equation $x'' + (x^2 - \mu)x' + 2x + x^3 = 0$ has a point of bifurcation at $\mu = 0$ and has periodic solutions for $\mu > 0$.

1.350 Verify that the system $x' = y$, $y' = \mu(1 - x^2)y - x$, satisfies the hypotheses of the (Poincaré-Andronov-)Hopf bifurcation theorem. Explain why there is no Hopf bifurcation in the following systems:

(i) $r' = -r(r + \mu)^2$, $\theta' = 1$;

(ii) $r' = r(\mu - r^2)(2\mu - r^2)^2$, $\theta' = 1$;

(iii) $r' = r(r + \mu)(r - \mu)$, $\theta' = 1$;

(iv) $r' = \mu r(r + \mu)^2$, $\theta' = 1$;

(v) $r' = -\mu^2 r(r + \mu)^2(r - \mu)$, $\theta' = 1$.

1.351 Prove the existence of a stable limit cycle for the Van der Pol equation $x'' + \mu(x^2 - 1)x' + x = 0$ using the Hopf bifurcation theorem.

1.352 Show that $\mu = 0$ is a point of bifurcation of the system $x' = -\mu x - y$, $y' = x + y^3$, and for $\mu > 0$ there is an unstable periodic solution.

1.353 Consider the system on $\mathbb{R}^2$: $x' = -y + x - \lambda(x^2 + y^2)x$, $y' = x + y - \lambda(x^2 + y^2)y$. Show that there exists a periodic orbit for $\lambda > 0$ and this orbit goes to infinity as $\lambda \to 0$ (this orbit bifurcates from infinity).

1.354 Sketch the phase portrait of the system $x' = y - x(\lambda x^2 + y^2 - 1)$, $y' = -\lambda x - y(\lambda x^2 + y^2 - 1)$. Show that for $\lambda = 0$ there is no periodic orbit and for $\lambda > 0$ there is one periodic orbit which bifurcates from the lines $y = 1$ and $y = -1$.

1.355 Study the asymptotic behavior as $\mu \to \infty$ of the following equations:

$$x' = x\left(\frac{1}{\mu} - x - x^2\right) + \delta \quad \text{and} \quad x' = x\left(\frac{1}{\mu} - x^2\right) - \delta$$

In particular, describe stationary solutions and their stability.

1.356 Study the phenomenon of secondary bifurcation from a double eigenvalue on the example of the system $x' = \mu x - x^2 + y^2$, $y' = \mu y - cxy$, $c \neq 0$, $c \neq 1$.

1.357 Study the system $x' = \mu x - x^2 - 2xy$, $y' = (\mu - \sigma)y + xy + y^2$, where σ is a fixed parameter and μ is a bifurcation parameter.

1.358 Study periodic solutions of the system $x' = \mu x - y - xr^{2k-1} \sin(1/r)$, $y' = x + \mu y - yr^{2k-1} \sin(1/r)$, where $r^2 = x^2 + y^2$, μ is a real parameter, and k is a positive integer.

1.359 Study bifurcations of the system of differential equations $u'' + \lambda(u + v(u^2 + v^2)) = 0$, $v'' + \lambda(v - u(u^2 + v^2)) = 0$ with the boundary conditions $u(0) = u(1) = v(0) = v(1) = 0$ and a real parameter λ.

1.360 Study the asymptotic behavior of the system $x' = y$, $y' = c(x^2 - ax + \lambda)/(a - x)$, where λ is a real parameter.

1.361 Prove that the limit cycle of the system $x' = -\omega y + x(a - (x^2 + y^2)^{1/2})$, $y' = \omega x + y(a - (x^2 + y^2)^{1/2})$ is structurally stable. Show that the limit cycle of the system $x' = -\omega y + x(a - (x^2 + y^2)^{1/2})^3$, $y' = \omega x + y(a - (x^2 + y^2)^{1/2})^3$ is not structurally stable.

1.362 Study bifurcations of the solutions of the Fitz-Hugh–Nagumo system $x' = \eta + x + y - x^3/3$, $y' = \rho(a - x - by)$, where a, η are real numbers, $b, \rho \in (0,1)$.

1.363 Show that for the Chafee system $r' = r(\varepsilon - r^2)^{n_1}(2\varepsilon - r^2)^{n_2}\cdots(m\varepsilon - r^2)^{n_m}$, $\theta' = 1$, $z' = z$, where $n_1 + n_2 + \cdots + n_m$ is an even number, there exist m families of periodic orbits bifurcating from the origin as ε crosses 0.

1.364 Study bifurcations of periodic orbits for the Lorenz system $x' = -\sigma x + \sigma y$, $y' = -xz + rx - y$, $z' = xy - bz$, where σ, b are fixed numbers and r is a real parameter.

1.365 The Liapunov bifurcation theorem can be stated as follows: Suppose that a Hamiltonian system $p' = H_q$, $q' = -H_p$ can be transformed into $x' = Ax + f(x)$, where A is a constant matrix, f is a power series with no linear terms, $x = (p,q) \in \mathbb{R}^{2n}$. Let the matrix A have k pairs of imaginary eigenvalues $\pm i\lambda_j$ such that $\lambda_j/\lambda_i, j, i = 1, \ldots ,k, i \neq j$, is not an integer. Then the above system has k real one-parameter periodic solutions of periods $\tau_j(R)$, R being the amplitude, which satisfy the relations $\lim_{k \to 0} \tau_j(R) = 2\pi/\lambda_j$, that is, the periods of the linearized equation $x' = Ax$. Construct examples showing the importance of the assumptions on the structure of the system and that $\lambda_j/\lambda_i, i \neq j$, is not an integer.

Stability Theory and Related Topics

1.366 Show that 0 is a Liapunov stable solution of the equation $x' = a(t)x$, with a continuous function of $a(t)$, if and only if $\lim \sup_{t \to \infty} \int_0^t a(s)\, ds < \infty$.

1.367 Assume that $a(t)$ is a continuous T-periodic function. Show that 0 is a Liapunov stable solution of the equation $x' = a(t)x$ if and only if $\int_0^T a(s)\, ds \leq 0$ and is asymptotically stable if and only if $\int_0^T a(s)\, ds < 0$.

1.368 Let $f : \mathbb{R}^{n+1} \to \mathbb{R}^n$ be a C^1 function and for all $t, f(t,0) = 0$. Assume that there exist two C^1 functions $V, W : \mathbb{R}^{n+1} \to \mathbb{R}$ such that $V(t,x) \geq W(x)$ for x

from some neighborhood of 0 and $t \geq T$, $W(x) \geq 0$, 0 is a strong minimum of W, and $\partial V/\partial t + \sum_{i=1}^{n}(\partial V/\partial x_i)f_i \leq 0$. Show that $x(t) \equiv 0$ is a stable solution of the equation $x' = f(t,x)$.

1.369 Consider the differential equation $x' = f(t,x)$ with $t,x \in \mathbb{R}$, $f \in C^1$, and $f(t,0) = 0$ for all t. Assume that there exists a function $V(t,x) \geq 0$ such that for each nonzero solution $x(t)$ of this equation $dV(x(t),t)/dt < 0$. Is 0 a stable solution?

1.370 Consider the system $x' = A(t)x + f(t,x)$ where f is a Lipschitz function and $|f(t,x)| \leq C|x|^q$ for some $q > 1$ in a neighborhood of the origin. Prove the Malkin theorem: If the fundamental matrix for $y' = A(t)y$ satisfies the estimate $|Y(t,s)| \leq Ce^{a(t-s)+bs}$ and $a(q-1) + b < 0$, then the trivial solution of the full system is asymptotically stable.

1.371 Construct an example of differential equation $x' = f(t,x)$, t, $x \in \mathbb{R}$, such that 0 is asymptotically stable and is not equiasymptotically stable (0 is equiasymptotically stable if there exists $\delta > 0$ such that for each $\varepsilon > 0$ there exists $T(\varepsilon) > 0$ such that $|x(t,x_0)| < \varepsilon$ for $t > T(\varepsilon)$, where $x(t,x_0)$ is the solution which satisfies the initial condition $x(0,x_0) = x_0$, $|x_0| < \delta$).

1.372 Consider the system $x'_j = \sum_{k=1}^{n} p_{jk}f_k(y_k)$, j, $k = 1, \ldots, n$, where $f_k(y)y > 0$ for $y \neq 0$, $y_k = \sum_{m=1}^{n} a_{km}x_m$, $k = 1, \ldots, n$, a_{km} are constants, $p_{jk} = p_{jk}(x_j,t)$. Show that if the form $\sum_{k,m=1}^{n} b_{km}\xi_k\xi_m$, where $b_{km} = \frac{1}{2}\sum_{j=1}^{n}(a_{kj}p_{jm} + a_{mj}p_{jk})$, is negative definite, then the origin is stable.

1.373 Consider the differential equation $x' = f(x)$, $x \in \mathbb{R}^n$, $f \in C^1$. Assume that each eigenvalue $\lambda(x)$ of $M = (\partial f_i/\partial x_j) + (\partial f_i/\partial x_j)^T$ satisfies everywhere the inequality $\lambda(x) < -v(|x|)$, where $v(s) > 0$ is a decreasing function on $\mathbb{R}^+$ and $\int_0^\infty \exp(-\varepsilon \int_0^u v(s)\,ds)\,du < \infty$ for every $\varepsilon > 0$. Show that each solution of this equation is bounded as $t \to \infty$ and approaches the critical point.

1.374 Consider the differential equation $x' = f(x)$, $x \in \mathbb{R}^n$, $f \in C^1$. Assume that each eigenvalue of $M = (\partial f_i/\partial x_j) + (\partial f_i/\partial x_j)^T$ is negative, and assume that Tr M is bounded and $|\det M| \geq \alpha$ for some constant $\alpha > 0$. Show that each solution of this equation is bounded as $t \to \infty$ and $x(t)$ approaches the critical point.

1.375 Assume that $f: \mathbb{R}^n \to \mathbb{R}^n$ is C^1, $\partial f_i/\partial x_j = 0$ for $j < i$, $\partial f_i/\partial x_i < 0$ for $i = 1, \ldots, n$ everywhere in $\mathbb{R}^n$, and $f(x) = 0$ only at the origin. Show that each solution of the equation $x' = f(x)$ is defined on $\mathbb{R}^+$ and tends to the origin as $t \to \infty$.

1.376 Give an example of a differential equation on $\mathbb{R}$ all of whose solutions are bounded but there is an unstable solution.

1.377 Give an example of a dynamical system in the plane (x,y) such that $x = 0$, $y = 0$ is its unstable solution and each solution tends to the origin as $t \to \infty$.

1.378 Find an example of a C^1 vector field $F : \mathbb{R}^2 \to \mathbb{R}^2$ such that 0 is a stable solution of the equation $x' = F(x)$ and each other solution is unstable.

1.379 The Markus and Yamabe example. Consider the nonautonomous linear system $x' = (-b + a \cos^2 bt)x + (b - a \sin bt \cos bt)y$, $y' = (-b - a \sin bt \cos bt)x + (-b + a \sin^2 bt)y$, where $b < a < 2b$. Show that the matrix of this system has two eigenvalues with equal (and time independent) negative real parts, but the origin is unstable.

1.380 Prove the Matrosov theorem: Suppose that there exists a C^2 function V such that V is bounded from above, $V'' \geq aV + 2bV'$ for some $a > 0$ or $a = 0$ and $b \geq 0$, for every t $((b^2 + a)^{1/2} - b)V + V'$ is positive in a sufficiently small neighborhood of the origin. Then the critical point $x = 0$ of the differential equation $x' = f(t,x)$ is unstable.

1.381 Consider the system $x' = y + g_1$, $y' = z + g_2$, $z' = g_3$, where $g_1 = -xf_1$, $g_2 = -yf_2 - 2V_1x$, $f_3 = -zf_3 - 2V_2y$, and f_i are arbitrary continuous functions of x, y, z, and $V_1 = x^2$, $V_2 = x^2 + y^2$, $V_3 = x^2 + y^2 + z^2$. Using V_3 as a Liapunov function, show that for an appropriate choice of the f_i's the origin is asymptotically stable and for another choice the origin is unstable.

1.382 Show that 0 is an asymptotically stable singular point of the system $x' = y$, $y' = -x - (1 - x^2)y$, and the domain of attraction of 0 contains the unit disk.

1.383 Study stationary solutions and their stability for the system describing the Rikitake dynamo $x' + \mu x = ay + zy$, $y' + \mu y = -ax + zx$, $z' = 1 - xy$.

1.384 Show that 0 is an asymptotically stable solution of the equation $x' = -ax$, $y' = (\sin(\log t) + \cos(\log t) - 2a)y$, $\frac{1}{2} < a < \frac{1}{2}(1 + \frac{1}{2}e^{-\pi})$. Is it an asymptotically stable solution of the system $x' = -ax$, $y' = (\sin(\log t) + \cos(\log t) - 2a)\, y + x^2$?

1.385 Prove the Olech theorem: Let $f : \mathbb{R}^2 \to \mathbb{R}^2$ be a C^1 function such that $f(0) = 0$, and suppose that the Jacobian matrix $Df(y)$ is symmetric and the real parts of all its eigenvalues are negative for all $y \in \mathbb{R}^2$. If there exist positive constants ρ and r such that $|f(x)| \geq \rho$ for $|x| \geq r$, then $x = 0$ is a globally asymptotically stable solution of $x' = f(x)$.

1.386 Consider the system $x' = f(x) + by$, $y' = g(x) + dy$, where $(bg(x) - df(x))x > 0$ and $f(x)/x + d < 0$ for all $x \neq 0$, $\int_0^x (df(s) - bg(s))\, ds \to \infty$ as $|x| \to \infty$. Show that the origin is globally stable.

1.387 Study the stability of the origin for the differential system $x' = -y + rx \sin(1/r)$, $y' = x + ry \sin(1/r)$, with $r = (x + y)^{1/2}$.

1.388 Prove the stability of the origin for the system of differential equations defined by $x' = y$, $y' = -x + ry(\sin(1/r) + r^{-1} \cos(1/r))$, for $r \neq 0$, and by $x' = 0$, $y' = 0$ for $r = 0$, where $r = (x^2 + y^2)^{1/2}$.

1.389 Study the stability of $(0,0)$ for the system $x' = x(y^2 - 1)$, $y' = y(x^2 - 1)$.

1.390 Using the Ważewski inequality (see 1.132), study the stability of the system $x' = -x \sin^2 t + (a + 2b/t)y$, $y' = -ax - y \cos^2 t$.

1.391 Let x_0 be a singular point of the dynamical system $x' = f(x)$, $x \in \mathbb{R}^n$, where f is C^1 and $Df(x_0)$ has all eigenvalues with negative real parts. Show that there exists a neighborhood U of x_0 such that the boundary of U is homeomorphic to the $(n - 1)$-dimensional sphere, and for each $x \in U$, $x \neq x_0$, there exists a unique $t_0 < 0$ such that the solution $x(t)$ of this equation starting from x_0 intersects transversally the boundary of U at the moment t_0.

1.392 Consider a system $x' = f(x)$ such that div $f(x) = 0$. Prove that each bounded solution of this system which is Liapunov stable for $t \to \pm \infty$ is almost periodic.

1.393 Show that the solution $x(t) \equiv 0$ of the equation $x'' + h(t)x' = x$ is unstable if sup $h(t) < 2$.

1.394 Determine stability properties of the equation $x'' + x' + g(x) = 0$ if g is a C^1 function, $g(0) = 0$, and $xg(x) > 0$ for all $x \neq 0$.

1.395 Verify the stability of the trivial solution of the equation $x'' + g(x)x' + f(x) = 0$ under the conditions $f(x)x > 0$ and $g(x) > 0$ for all $x \neq 0$, $\int_0^x f(s) \, ds \to \infty$ as $|x| \to \infty$.

1.396 Consider the equation $x'' + g(x,x') = 0$. Assume that g is C^1 on $\mathbb{R}^2$, $g(0,0) = 0$, $g_x > 0$, $g_y > 0$. Show that each solution is defined for all $t > 0$ and $\lim_{t \to \infty} x(t) = 0$, $\lim_{t \to \infty} x'(t) = 0$.

1.397 Consider the system $x' = g(y) + xf(y)$, $y' = -h(x) + yl(x)$, for x, y in $\mathbb{R}$. Suppose that f, g, h, and l are of class C^1, defined on $\mathbb{R}$ and such that $g(y)y > 0$, $y \neq 0$, $h(x)x > 0$, $x \neq 0$, $\int_0^y g \to \infty$ as $|y| \to \infty$, $\int_0^x h \to \infty$ as $|x| \to \infty$, $f(x) < 0$, $x \neq 0$, $f(0) = 0$, $l(x) > 0$ for $x \in \mathbb{R}$. Prove that every solution tends to the origin as $t \to \infty$.

1.398 Study the stability of the trivial solution of the equation $x'' + x = \mu x'(ax' + 2bxx' + c(x')^2)$, where $a > 0$, $ac - b^2 > 0$, and μ is a parameter.

1.399 Study the stability of solutions of the equation $x'' + \frac{1}{2}(x^2 + (x^4 + 4(x')^2)^{1/2})x = 0$.

1.400 Construct a Liapunov function for the damped pendulum described by the equation $x'' + x' + \sin x = 0$ which would show the asymptotic stability of the origin.

1.401 Consider the equation $x'' + ax' + b \sin x = 0$, where a and b are positive constants. Show that for any solution $x(t)$ there is an integer k such that $x(t) \to k\pi$, $x'(t) \to 0$ as $t \to \infty$. Sketch the phase portrait.

1.402 Consider the problem of parametric excitation of a damped pendulum described by the equation $x'' + ax' = -g \sin x + A \sin(\omega t) \sin x$. Show that if $2g/a^2 < 1$ and a/ω is small, then $x = 0$ and $x = \pi$ are stable equilibrium points. If a periodic excitation $H \sin(\omega_1 t)$ is present (a new term on the right-hand side of the equation), show that there exist stable synchronized regimes of oscillations near $x = 0$ and $x = \pi$.

1.403 Study the stability of the origin for the equation $x'' = (x')^3 - \sin x$.

1.404 Show that the differential equation $x'' - x'(1 - 3x^2 - 2(x')^2) + x = 0$ has a stable periodic solution.

1.405 Discuss existence and stability of limit cycles for the system $x'' + x = \varepsilon(1 - x^2 - ay^2)x'$, $y'' + 2y = \varepsilon(1 - bx^2 - y^2)y'$, $\varepsilon \in \mathbb{R}$.

1.406 Assume that $U : \mathbb{R} \to \mathbb{R}$ has a local minimum at the origin. Is it true that zero is a stable solution of the equation $x'' = -U'(x)$?

1.407 Prove the Lagrange theorem: An equilibrium $(x_0, 0)$ of a conservative force field $-\nabla V$ is stable if the potential V has a local strict minimum at x_0. Does it suffice to assume that V has a local minimum at x_0?

1.408 Assume that a function $U : \mathbb{R}^3 \to \mathbb{R}$ has 0 as an isolated critical point which is not a minimum of U. Show that 0 is not a stable equilibrium of the equation $x'' = -\nabla U$. Is the assumption on an isolated critical point essential?

1.409 Let x_0 be an equilibrium point for $x'' = -U'(x)$ with $U \in C^1(\mathbb{R})$ bounded from below. Suppose that for x, $x \neq x_0$ close to x_0, the derivative $U''(x)$ is positive. Then x_0 is strongly stable equilibrium point; that is, for ε small enough the motions $x(t)$, $t \geq 0$, with initial data $x(0) = x_0$, $x'(0) = \varepsilon$, and described by the equation $x'' = -\lambda x' - U'(x)$, are such that $\lim_{t \to \infty} x(t) = x_0$ for all $\lambda > 0$.

1.410 Consider the Hamiltonian system with $H(p_1, p_2; q_1, q_2) = \frac{1}{2}(q_1^2 + p_1^2) - (q_2^2 + p_2^2) + \frac{1}{2}(p_1^2 p_2 - p_2 q_1^2 - q_1 q_2 p_1)$. Study the stability of the origin.

1.411 Prove the Salvadori theorem: Consider a mechanical system with dissipation such that the origin is an isolated stationary solution corresponding to a minimum of the potential energy U and for $V(q, q') = T(q, q') + U(q)$, $(V(q, q'), q') \leq -a(|q'|)$ for some continuous increasing function a, $a(0) = 0$. Then $q = q' = 0$ is asymptotically stable.

1.412 Consider a mechanical system with two degrees of freedom with potential energy described by an analytic function such that $q = 0$ is not its minimum and its Hessian at the origin does not vanish. Prove that $q = p = 0$ is unstable.

1.413 Show that the origin is unstable for the mechanical system with one degree of freedom and the potential energy $-q^8 \sin^2(1/q)$.

1.414 Consider the mechanical system on $\mathbb{R}$ with potential energy $\exp(-q^{-2}) \cos(1/q)$ for $q \neq 0$. Study the stability of the origin.

1.415 Study the stability of the origin for the mechanical system with two degrees of freedom where the potential energy is $\exp(-q_1^{-2}) \cos(1/q_1) - \exp(q_2^{-2})(\cos(1/q_2) + q_2^2)$.

1.416 Unit masses are fixed at each of 2^n vertices of a hypercube in $\mathbb{R}^n$, $n \geq 1$. A test particle which can move freely in space is attracted by the unit masses according to the Newton inverse square law of gravitation. The center of the cube is obviously a position of equilibrium for the test particle. Is this a position of stable or unstable equilibrium?

1.417 Show that if $x = 0$ is an isolated zero of a function f, then the origin is not stable for the equation $x^{(n)} = f(x)$, $n \geq 3$ (f is assumed to be sufficiently regular in order to guarantee the uniqueness of solutions).

1.418 Study the stability of position of rotation axes and that of stationary rotations of a rigid body around principal axes of inertia.

1.419 Discuss the existence and the stability of limit cycles of the system $x' = -\omega y$, $y' = \omega(x + z)$, $z' = -az + \mu(1 - b^2 x^2)x$.

1.420 Study periodic orbits and their stability for the system $x' = -y + xf(x^2 + y^2)$, $y' = x + yf(x^2 + y^2)$, where f is a (real) polynomial.

1.421 Consider the system in $\mathbb{R}^3$ $x' = f(x) - (1 + \alpha)x - z$, $y' = -\beta(f(x) - x - z)$, $z' = -\gamma(y + z)$, with some α, β, $\gamma > 0$. Show that for certain values of parameters there exist stable selfinduced oscillations.

1.422 Study periodic orbits and their stability for the Negrini-Salvadori system $x' = \mu x - y - xf(x,y) - x(x^2 + y^2)^{s+1}$, $y' = x + \mu y - yf(x,y) - y(x^2 + y^2)^{s+1}$ where $s \geq 3$ is an integer, $f(x,y) = (x^2 + y^2) \sin^2((x^2 + y^2)^{-1})$, $f(0,0) = 0$.

1.423 Prove the Poincaré criterion of stability of a periodic solution: Consider the autonomous system in the plane $x' = P(x,y)$, $y' = Q(x,y)$, with $P,Q \in C^1$ such that there exists a T-periodic solution $(\zeta(t),\eta(t))$. If $\int_0^T (P_x(\zeta(t),\eta(t)) + Q_y(\zeta(t),\eta(t)))\, dt < 0$, then the solution $(\zeta(t),\eta(t))$ is asymptotically stable.

1.424 Show that for $1/2 < \nu < 1$ the Langford system $x' = (\nu - 1)x - y + xz$, $y' = x + (\nu - 1)y + yz$, $z' = \nu z - (x^2 + y^2 + z^2)$ has periodic solutions with $z = $ const. Study their stability.

1.425 Consider the system $x' = -y + ax - ax(x^2 + y^2)^{1/2}$, $y' = x + ay - ay(x^2 + y^2)^{1/2}$, $z' = a(1 - z)(x^2 + y^2)^{1/2}$, where a is a real constant. Does it have a limit cycle?

1.426 Compute the lower Liapunov exponent of a solution of the equation $x' = (t \cos t + \sin t)x$, $y' = -(t \cos t + \sin t)y$.

1.427 Prove the Perron theorem: A linear homogeneous system in $\mathbb{R}^n$ $x' = A(t)x$ is regular; that is, the sum of its characteristic Liapunov exponents is equal to the limit $\lim_{t \to \infty} t^{-1} \int_0^t \mathrm{Re} \, \mathrm{Tr} \, A(s) \, ds$ (which does exist), if and only if for each $j = 1, \ldots, n$, $\alpha_j + \beta_j = 0$, where $\alpha_1 \leq \cdots \leq \alpha_n$ are the Liapunov exponents of the system $x' = A(t)x$ and $\beta_1 \geq \cdots \geq \beta_n$ are those of the conjugated system $y' = -A^*(t)y$.

1.428 Show that the Liapunov exponents of the linear system $x' = x$, $y' = (\sin(\log t) + \cos(\log t) - 2)y$ are not stable with respect to small (linear) perturbations of the system.

1.429 Calculate the Liapunov exponents for the system $x' = (\sin(\log(t + 1)) + \cos(\log(t + 1)))x$, $y' = (-\sin(\log(t + 1)) + \cos(\log(t + 1)))y + cx/(t + 1)$.

1.430 Give an example of a system of the form $x' = A(t)x + f(t,x)$, where $f(t,0) = 0$, $|f(t,x) - f(t,y)| \leq g(t)|x - y|$, and $\int_0^\infty g(t) \, dt < \infty$, with the Liapunov exponents different from those of the linear system $x' = A(t)x$.

1.431 It is well known that a system of n linear equations has at most n upper Liapunov exponents. Give an example of such a system with more than n lower Liapunov exponents.

1.432 Construct an example of a linear system $x' = A(t)x$ in $\mathbb{R}^2$ such that the real parts of the eigenvalues of $A(t)$ are negative but there is a solution with positive Liapunov exponent.

1.433 Study the Liapunov exponents and their stability for the system $x' = y/t$, $y' = x/t + \pi\sin(\pi t^{1/2})y$.

1.434 Give an example of a dynamical system in the plane with negative maximal Liapunov exponent and whose trivial solution is unstable.

Higher-Order Differential Equations

1.435 What is the lowest order $n > 0$ of a differential equation $x^{(n)} + a_1 x^{(n-1)} + \cdots + a_n x = 0$ having among its solutions the functions $\sin 2t$, $4t^2 e^{2t}$, $-e^{-t}$? Find the constants $a_1, \ldots, a_n$.

1.436 Show that the solution of any differential equation of the type $u^{(n)} + a_1(x)u^{(n-1)} + \cdots + a_n(x)u = f(x)$ with continuous coefficients, together with the initial conditions $u(0) = c_0$, $u'(0) = c_1$, . . . ,$u^{(n-1)}(0) = c_{n-1}$, can be reduced to the solution of a certain Volterra integral equation of the second kind $\Phi(x) + \int_0^x K(x,y)\Phi(y) \, dy = f(x)$.

1.437 Use the nonlinear integral equation $x(t) = \exp(-\int_0^t (t - s)^2 x(s) \, ds)$, $x = y''$, to establish the existence of a solution of $y^{(3)} + 2yy' = 0$ determined by $y(0) = y'(0), y''(0) = 1$, for all $t \ge 0$.

1.438 Let $W_n(y) = \det(y^{(i+j)})$, $0 \le i, j \le n$. Show that a function $y \in C^{2n}(\mathbb{R})$ satisfies the condition $W_n(y) = 0$ if and only if it solves a constant-coefficient linear homogeneous differential equation of nth order.

1.439 Show that the general solution of the equation $y^{(n)} - xy = a$, where a is a constant, is

$$y(x) = \sum_{r=0}^{n} A_r \, \omega^r \int_0^\infty \exp\left(\omega^r xt - \frac{t^{n+1}}{n+1} \right) dt$$

where $\omega^{n+1} = 1$ and the constants A_r satisfy the single relation $\sum_{r=0}^n A_r = a$.

1.440 Prove the Sansone theorem: Let $x^{(n)} + p_1 x^{(n-1)} + \lambda p_2 x^{(n-2)} + \cdots + \lambda p_{n-1} x' + \lambda p_n x = 0$ be a constant-coefficient equation, $n > 2$, $p_2 p_n \ne 0$, $a_1 < a_2 < \cdots < a_{n-1} < a_n$, and suppose that there is no nontrivial solution of the equation $p_2 x^{(n-2)} + \cdots + p_n x = 0$ vanishing at $a_2, a_3, \ldots ,a_{n-1}$ and the equation $p_2 h^{n-2} + p_3 h^{n-3} + \cdots + p_{n-1} h + p_n = 0$ does not have multiple roots. Then there exist an infinite number of real eigenvalues λ such that the considered equation has a solution $x(t) \ne 0$ vanishing at $a_1, a_2, \ldots ,a_{n-1}, a_n$.

1.441 Prove the de la Vallée-Poussin theorem: If $Lx = x^{(n)} + p_1(t) x^{(n-1)} + \cdots + p_n(t)x$, $n \ge 2$, is an ordinary differential operator with continuous coefficients in an interval (α, β), $L_j = \sup_{(\alpha,\beta)} | p_j(x) |$, h_0 is the positive root of the equation $L_n h^n/n! + L_{n-1} h^{n-1}/(n - 1)! + \cdots + L_1 h - 1 = 0$, then for all numbers $A_1, \ldots ,A_n$ and $\alpha \le a_1 < a_2 < \cdots < a_n \le \beta$ such that $a_n - a_j \le h_0$, there exists a unique solution of $Lx = 0$ satisfying $x(a_j) = A_j$. Of course, if $L_j = 0$ for all j, then the condition on h_0 is superfluous.

1.442 Prove the Bohr-Neugebauer theorem: Each bounded solution of the constant-coefficient equation $x^{(n)} + c_1 x^{(n-1)} + \cdots + c_n x = f(t)$ with a real quasiperiodic function f is quasiperiodic.

1.443 Prove the Perron theorem: If the equation $x^{(n)} + p_1(t) x^{(n-1)} + \cdots + p_{n-1}(t)x' + p_n(t)x = 0$ has continuous coefficients and $\lim_{t \to \infty} p_j(t) = a_j$, $j = 1, \ldots ,n$, and all the roots $h_1, \ldots ,h_n$ of the characteristic equation $h^n + a_1 h^{n-1} + \cdots + a_{n-1} h + a_n = 0$ are distinct real numbers, then the above equation has n linearly independent solutions such that $\lim_{t \to \infty} x_j'/x_j = h_j$, $\lim_{t \to \infty} x_j''/x_j = h_j^2, \ldots , \lim_{t \to \infty} x_j^{(n)}/x_j = h_j^n$.

1.444 Prove a generalization for nth order equations of the Dini-Hukuhara theorem: If f_j are measurable functions and $\int_0^\infty |f_j(t)| \, dt < \infty$, $j = 1, \ldots, n$, and if c_j are real numbers such that the equation $x^{(n)} + c_1 x^{(n-1)} + \cdots + c_n x = 0$ has all solutions bounded in $[0,\infty)$, then the equation $x^{(n)} + (c_1 + f_1(t))x^{(n-1)} + \cdots + (c_n + f_n(t))x = 0$ also has all solutions bounded in $[0,\infty)$.

1.445 Solve the differential equation

$$x \frac{d}{dx}\left(x \frac{d}{dx} - 1\right)\left(x \frac{d}{dx} - 2\right) \cdots \left(x \frac{d}{dx} - n + 1\right) y(x) = g(x)$$

where g is a given continuous function and $y(1) = y'(1) = \cdots = y^{(n-1)}(1) = 0$. Generalize this for the equation

$$\left(x \frac{d}{dx} + a_1\right)\left(x \frac{d}{dx} + a_2\right) \cdots \left(x \frac{d}{dx} + a_n\right) y(x) = g(x)$$

1.446 Prove the Mammana factorization theorem: A linear differential operator $L[y] = y^{(n)} + p_1(x)y^{(n-1)} + \cdots + p_n(x)y$ with continuous coefficients p_j, $j = 1, \ldots, n$, is factorizable into

$$L[y] = \left(\frac{d}{dx} + q_1(x)\right)\left(\frac{d}{dx} + q_2(x)\right) \cdots \left(\frac{d}{dx} + q_n(x)\right)$$

if and only if every nontrivial solution of the equation $L[y] = 0$ has at most $n - 1$ zeros in the considered interval.

1.447 Consider a linear differential operator of nth order, $n > 2$, $L[y] = y^{(n)} + p_1(x)y^{(n-1)} + \cdots + p_{n-1}(x)y' + p_n(x)y$ with continuous coefficients p_j, $j = 1, \ldots, n$. Suppose that there exists a system $y_1, y_2, \ldots, y_n$ of linearly independent solutions of the equation $L[y] = 0$ such that the Wronskian determinants $W_k = \det(y_j^{(i)})$, $i = 0, \ldots, k - 1$, $j = 1, \ldots, k$, $k = 1, \ldots, n - 1$, are strictly positive on an interval $[a,b]$. Then the differential inequality $L[y] \geq 0$ with the condition $y(a) = y'(a) = \cdots = y^{(n-1)}(a) = 0$ implies that $y(x) \geq 0$ in $[a,b]$. Moreover (this is the Pólya theorem generalizing the Rolle theorem), for every n times differentiable function g vanishing in $(n + 1)$ points of the interval (a,b) there is $s \in (a,b)$ such that $L[g(s)] = 0$.

Differential Equations in the Complex Domain

1.448 If $\cos z^{1/2}$ and $\sin z^{1/2}$ form a fundamental system for a linear second-order differential equation, find the equation. The same question for $z^{1/2}$, $z^{1/2} \log z$.

1.449 Let $f(z) = \sum_{k=1}^{\infty} C_k z^k$ be an analytic square matrix of z with given C_1, C_2 that satisfies the matrix differential equation $df(z)/dz = f(z)Af(z)z^{-2}$ for some given square matrix A. Show that this is possible if and only if $C_1 A C_1 = C_1$, $C_1 A C_2 = C_2 A C_1 = C_2$. If these conditions are met, show that the solution is unique and give closed-form expressions for C_k, $k \geq 3$, and $f(z)$.

1.450 Prove that if $w = w(z)$ satisfies the algebraic equation $w^n + a_2 w^{n-2} + \cdots + a_n = 0$, whose coefficients are polynomials in z, then w satisfies a linear differential equation of order $n - 1$, whose coefficients are rational functions of z.

1.451 In the equation $w' = P_0(z) + P_1(z)w + \cdots + P_n(z)w^n$ the P's are polynomials in z and $n > 1$. Show that the fixed singular points are the zeros of P_n and the point at infinity. Suppose that the P's are constants. Then show that an entire solution must be a constant. Show that there are always singular constant solutions and describe them. What happens for $n = 1$?

1.452 The solution of $w' = z^3 + w^3$, $w(0) = w_0$, has movable branch points. Find the order. If $w_0 > 0$, the branch point nearest to the origin lies on the positive real axis. How does it move if w_0 increases? What fixed singularities, if any, does this equation have?

1.453 The equation $w' = z^2 + w^2$, $w(0) = w_0$, has movable poles (of what order?). If $w_0 > 0$, there are positive poles. How does the pole nearest to the origin move when w_0 increases? Study fixed singularities of this equation.

1.454 Prove that the equation $z^3 w'' + z^2 w' + w = 0$ has no solution of the form $z^k r(z)$, where $r(z)$ is regular at $z = 0$. Prove that $z = \infty$ is a regular singular point of this equation.

1.455 Show that the general solution of $w' = 1 + z^{1/2} w^2$ is single valued on the Riemann surface of $z^{1/2}$.

1.456 A Riccati equation $y'(z) = f_0(z) + f_1(z)y(z) + f_2(z)y^2(z)$ is formally invariant under a Möbius transformation operating on the dependent variable. The Schwarzian derivative $\{w,z\} = w^{(3)}/w' - 3/2(w''/w')^2$ is a differential invariant of the group of fractional linear transformations acting on $w : \{W,z\} = \{w,z\}$ if $W = (aw + b)/(cw + d)$, $ad - bc \neq 0$. Verify these statements.

1.457 The function $z^{-1/2} \tan z^{1/2}$ satisfies the Riccati equation $y' = 1/2z - y/2z + y^2/2$. Find the general solution. Discuss the distribution of poles.

1.458 Show that the Riccati equation $y' = (z^3 - 1)^2 z^{-4} + (z^3 + 2)(z^4 - z)^{-1}y + y^2$ has a general solution of the form $y(z) = (z - z^{-2}) \tan(z^2/2 + z^{-1} + C)$.

1.459 Find a solution of the differential equation $w'' - 2zw' + 2kw = 0$, $k \geq 0$, of the form $\int_C \exp(2z\zeta) f(\zeta) \, d\zeta$, describing two possible types of contour C. Show that, if k is a positive integer, there is a solution of the form $H_k(z) = (-1)^k \exp(z^2) \, d^k(\exp(-z^2))/dz^k$.

1.460 The equation $w' = z^{-2}w$ has the origin as an irregular-singular point (consider the solution $w(z) = C \exp(-1/z)$). Show that the origin is an essential singular point of the solution which goes to zero as $z \to 0$ in the sector $|\arg z| < \pi/2$, and to infinity in $|\arg(-z)| < \pi/2$. Furthermore, $w(z)$ takes on every value $\neq 0$ and ∞ in ε-sectors centered on the imaginary axis. Show that in the disks $|z - i2^{-n}| < 2^{-n-2}$, $n = 0,1, \ldots$, the solution takes on every value except zero and infinity infinitely often.

1.461 Prove the following theorem: A point $z = a$ is a regular-singular point of the differential equation $w'' + Pw' + Qw = 0$ if and only if (i) one of the coefficients P and Q has an isolated singularity $z = a$, (ii) P either is regular or has a simple pole at $z = a$, and (iii) Q either is regular or has a pole of order ≤ 2.

1.462 Prove the Fuchs theorem: A necessary and sufficient condition for the equation $x^{(n)} + a_1(z)x^{(n-1)} + \cdots + a_{n-1}(z)x' + a_n(z)x = 0$ to have z_0 as a regular point is that the only possible singularities of the coefficients $a_k(z)$, $k = 1,2, \ldots ,n$, are poles of order at most k.

1.463 Prove the Briot and Bouquet theorem: If λ is not a positive integer then the equation $xy' - \lambda y = ax + \varphi(x,y)$, where φ is analytic at the origin, $\varphi(x,y) = \mathcal{O}(|x|^2 + |y|^2)$ as $(x,y) \to (0,0)$, has a unique analytic solution vanishing at the origin.

1.464 Find the general solution of the equation $y''(z) + \{f,z\}y/2 = 0$ where $\{f,z\} = (f''/f')' - (f''/f')^2/2$ is the Schwarzian derivative of f with respect to z.

1.465 The equation $zw' = aw + z^n$, where n is a positive integer and $a \neq n$, has a unique solution holomorphic at the origin. Find this solution explicitly. What happens when $a \to n$?

1.466 Show that, if $w(z)$ is an entire function satisfying $w' = P(z)$, where P is a polynomial of degree >2 having constant coefficients, then $w(z)$ is a constant. Specify the constants which can give solutions.

1.467 The famous Painlevé (and Gambier) transcendents (the equations with all their critical points fixed) are:
(i) $w'' = 6w^2 + z$;
(ii) $w'' = 2w^3 + zw + a$;
(iii) $w'' = (w')^2/w - w'/z + (aw^2 + b)/z + cw^3 + d/w$;
(iv) $w'' = (w')^2/2w + 3w^3/2 + 4zw^2 + 2(z^2 - a)w + b/w$;

(v) $w'' = (1/2w + 1/(w - 1))(w')^2 - w'/z + (w - 1)^2(aw + b/w)/z^2 + cw/z$
   $+ dw(w + 1)/(w - 1);$

(vi) $w'' = \frac{1}{2}(1/w + 1/(w - 1) + 1/(w - z))(w')^2 - (1/z + 1/(z - 1) +$
   $1/(w - z))w' + ((w(w - 1)(w - z))/(z^2(z - 1)^2))(a + bz/w^2 +$
   $c(z - 1)/(w - 1)^2 + dz(z - 1)/(w - z^2)).$

Verify that (vi) contains, in reality, the first five equations, which may be derived from it by a process of coalescence, that is, change of variables and passing to the limit.

Boundary Value Problems and Periodic Solutions

1.468 Let $x(a) = x(b) = 0$, $x(t) > 0$ for $t \in (a,b)$ and $x'' + x > 0$. Show that $b - a > \pi$.

1.469 Compare the (unique) solutions of the boundary value problems $x'' + 4x = 1000 \sin 100t$, $x(0) = x(\pi/4) = 1$ and $y'' + y = 1000 \sin 100t$, $y(0) = y(\pi/4) = 1$.

1.470 Determine the eigenvalues of the boundary value problem $((1 - t^2)x')'$ $+ \lambda x = 0$, $x(0) = 0$, $x(1)$ finite, associated with the problem of a rotating rope.

1.471 Consider the boundary value problem $x'' + q(t)x = r(t)$, $x(0) = x(a) = 0$, where $\sup_{[0,a]} q(t) < \pi^2/a^2$ and $r(t) \geq 0$. Show that $x(t) \leq 0$ for all $t \in [0,a]$; that is, the Green function of this problem is negative: $x(t) = \int_0^a G(t,s,a)r(s) \, ds$.

1.472 Assume that p, r, f are continuous functions on $[a,b]$ and $0 < p_0 \leq p(t) \leq p_1$, $0 < r_0 \leq r(t) \leq r_1$, $f_0 \leq f(t) \leq f_1$. Show that the eigenvalues λ_k of the problem $(x'/p(t))' + (\lambda r(t) + f(t))x = 0$, $x(a) = x(b) = 0$, satisfy the inequalities $(k^2\pi^2/p_1(b - a)^2 - f_1)/r_1 \leq \lambda_k \leq (k^2\pi^2/p_0(b - a)^2 - f_0)/r_0$.

1.473 Show that if the condition $\int_0^{1/2} t^2(f(t) + f(1 - t)) \, dt > 1$ is satisfied, then the boundary value problem $x'' + (\lambda + f(t))x = 0$, $x(0) = x(1) = 0$, has a negative eigenvalue.

1.474 Show that if p, q, $q' \in C[a,b]$ and $q^2(t)/4 - q'(t)/2 + p(t) > -(\pi/(b - a))^2$, then the boundary value problem $x'' = p(t)x + q(t)x' + r(t)$, $x(a) = A$, $x(b) = B$, has a unique solution.

1.475 Show that for every function p integrable over the interval $[a,b]$ there is $\omega_0 > 0$ such that for $\omega > \omega_0$ the trivial solution is the unique solution of the problem $x' = p(t) \cos(\omega t)x$, $x(a) = x(b) = 0$.

1.476 Consider the linear differential equation $x'' + \lambda a(t)x = 0$, $x(0) = x(1) = 0$, where $0 < a \leq a(t) \leq b < \infty$. Let v be the solution specified by the initial conditions $v(0) = 0$, $v'(0) = 1$. Show that $v(t) = t + \sum_{n=1}^\infty \lambda^n u_n(t)$, where

$u_0(t) = t$, $u_n(t) = -\int_0^t (t - s)u_{n-1}(s)a(s)\, ds$, is an expression valid for all λ for $0 \le t \le 1$. Show that the solutions of the equation $v(1) = 0$ are the required characteristic values $\lambda_1, \lambda_2, \ldots$

1.477 Show that if p is a continuous function on $[-1,1]$ and $tp(t) \ge 0$ for all t, then the boundary value problem $x^{(3)} = p(t)x$, $x(-1) = x(0) = x(1) = 0$, has a unique solution (identically equal to zero).

1.478 Show that if $p(t) \le 0$ then the boundary value problem $x^{(4)} = p(t)x$, $x(a) = x''(a) = x(b) = x''(b) = 0$, has a unique solution, the trivial one.

1.479 Find the eigenvalues and eigenfunctions of the problem $((1 + t)^2 x')' + \lambda x = 0$, $0 < t < 1$, $x(0) = x(1) = 0$.

1.480 Consider the boundary value problem $x'' + \cos^2 t(1 + \cos^2 t)^{-1} = \lambda \cos t(1 + \cos^2 t)^{-1}x$, $x'(0) = x'(\pi) = 0$. Does it have nonreal eigenvalues?

1.481 Is it possible that an eigenvalue problem for an ordinary differential equation has very close eigenfunctions for different eigenvalues?

1.482 Study the spectrum of the operator $Lu = -u''$ in $L^2(0,1)$ with the boundary conditions $u(0) = u(1)$, $u'(0) = -u'(1)$.

1.483 Determine the spectrum of the differential operator $Lu = -u'' - 2(\tanh x)u'$ in $L^p(\mathbb{R})$.

1.484 Suppose that the weight function $w(x) > 0$ is twice continuously differentiable. Consider the space $H_w(0,1)$ which is the closure of the set $\{f \in C^1[0,1], f(1) = 0\}$ in the norm induced by the scalar product $(f,g)_w = \int_0^1 w(x)f'(x)g'(x)\, dx$. Show that $H_w(0,1) \subset L^2(0,1)$ if and only if $\lim \inf_{x \to 0+} w(x)/x^2 > 0$.

1.485 Compute the kernels of the integral operators
(i) $(-d^2/dx^2 + I)^{-1}$ in $L^2(\mathbb{R})$;
(ii) $(-d^2/dx^2 + I)^{-1}$ in $L^2(0,\infty)$ with the homogeneous Dirichlet condition at the origin.

1.486 Determine real numbers h such that the operator $P_h u = du(x)/dx + u(x + h)$ defined for tempered distributions is injective and describe its kernel in another case.

1.487 Show that if $d^4u/dx^4 + ku$ is in $L^2(\mathbb{R})$ for some $k > 0$, then $d^j u/dx^j \in L^2(\mathbb{R})$ for $0 \le j \le 4$.

1.488 Given real a, b define the symmetric operator H_{ab} on $L^2(0,1)$ by $(H_{a,b}f)(x) = -d^2f/dx^2$ on $D(H_{ab}) = \{f \in C^\infty(0,1) : f'(1) + af(1) = 0, f'(0) - bf(0) = 0\}$. Show that H_{ab} is essentially selfadjoint on $D(H_{ab})$ and $(H_{ab} + I)^{-1} \le (H_{cd} + I)^{-1}$ if $a \ge c$ and $b \ge d$.

1.489 Consider a linear second-order ordinary differential operator $L = -d^2/dx^2 + b\,d/dx + c$, where c is a bounded function on $(0,1)$ and b,b' are integrable over $(0,1)$. Show that
(i) L is coercive if and only if $c + \pi^2 \geq b'/2$ and $c + \pi^2 \neq b'/2$.
(ii) L satisfies the weak maximum principle if and only if

$$\frac{b^2}{4} + c + \pi^2 \geq \frac{b'}{2} \quad \text{and} \quad \frac{b^2}{4} + c + \pi^2 \neq \frac{b'}{2}$$

1.490 Prove the Ambarzumian theorem: Denote by $\lambda_0 < \lambda_1 < \lambda_2 < \cdots$ the eigenvalues of the Sturm-Liouville problem $-u'' + q(x)u = \lambda u$, $0 \leq x \leq \pi$, $u'(0) = u'(\pi) = 0$, where q is a real continuous function. If $\lambda_n = n^2$, $n = 0,1,2,\ldots$, then $q(x) \equiv 0$.

1.491 Let $H_1 = L^2(\mathbb{R})$ and H_2 be the closure of $\mathscr{S}(\mathbb{R})$ in H_1 in the norm $\|f\|_2 = (\|xf\|^2 + \|f'\|^2)^{1/2}$. The operators $A_{\pm} : H_2 \to H_1$ (the creation and annihilation operators in quantum field theory) are defined by $A_{\pm}f = (d/dx \pm x)f$. Show that $A_{\pm}$ are Fredholm operators and compute their indexes.

1.492 Prove that the Schrödinger operator $Hu = -u'' + a(1 + |x|)^b u$ is essentially selfadjoint for $a \geq 0$ and arbitrary b or $a < 0$ and $b \leq 2$. Moreover, it is not essentially selfadjoint if $a < 0$ and $b > 2$.

1.493 Show that if $V = V_1 + V_2$ where $V_1 \in L^2(\mathbb{R})$, $V_2 \in L^\infty(\mathbb{R})$, $V(x) \leq 0$ for all x and $V(x_0) < 0$ at some continuity point x_0 of V, then the operator $-d^2/dx^2 + V$ has a negative eigenvalue.
Remark: The analogous result is true for the Schrödinger operators in two dimensions.

1.494 Give an example of a potential V in $L^2(\mathbb{R})$, with compact support, which is negative on an open set such that the operator $-d^2/dx^2 + V$ does not have negative eigenvalues.

1.495 Show that the classical formula for the distribution of eigenvalues of the Schrödinger operator $-d^2/dx^2 + V$ on $[0,\infty)$ with the homogeneous Dirichlet conditions at the origin $N(\lambda) \simeq \pi^{-1} \int_{\{V(x)<\lambda\}} (\lambda - V(x))^{1/2}\,dx$ is not valid for all potentials V which increase to ∞.

1.496 Let $\lambda_n(c)$ denote the nth eigenvalue of the operator $-d^2/dx^2 + x^2 + cx^4$, $c \geq 0$, on $\mathbb{R}$. Find a constant $c_0 > 0$ such that $\lambda_2(c) - \lambda_1(c) \geq c_0$ for all c.

1.497 Prove that the integral kernel of $\exp(-tH)$ where H is the harmonic oscillator Hamiltonian $H = \frac{1}{2}(-d^2/dx^2 + x^2 - I)$ is given by the Mehler formula

$$K_t(x,y) = (\pi(1 - e^{-2t}))^{-1/2} \exp\left(\frac{4xye^{-t} - (x^2 + y^2)(1 + e^{-2t})}{2(1 - e^{-2t})}\right)$$

1.498 Prove that the operator $B = -\frac{1}{2}d^2/dx^2 + x\,d/dx$ is positive on the probability space $L^2(\mathbb{R}, \pi^{-1/2}\exp(-x^2)\,dx)$ and essentially selfadjoint on the span of Hermite polynomials and that the semigroup generated by B is hypercontractive, that is, $\|\exp(-tB)u\|_4 \leq C\|u\|_2$ for large t.

1.499 Consider the one-dimensional Schrödinger equation described by the Lagrangian $L = \psi'\overline{\psi'} - (k^2 - u(t))\psi\,\overline{\psi}$. Change the variables $\psi = \exp(-iz)$, $\overline{\psi} = \exp(iz)$ and find the Hamiltonian form.

1.500 Consider a selfadjoint differential operator $Lx = -(px')' + qx$ with real-valued functions p, q on $[0,\infty)$. Show that if for some complex $\lambda = \lambda_0$ all solutions of the equation $Lx + \lambda x = 0$ are in $L^2[0,\infty)$, then for every complex λ all solutions are in L^2.

1.501 Let $u(t,\lambda,\alpha)$, $v(t,\lambda,\alpha)$ be the solutions of $Lx + \lambda x = 0$, $Lx = -(px')' + qx$, such that $u(0) = \sin\alpha$, $p(0)u'(0) = -\cos\alpha$, $v(0) = \cos\alpha$, $p(0)v'(0) = \sin\alpha$. Show that $x = u + mv$ with m of the form $m = -(Az + B)/(Cz + D)$, $z = \cotan\beta$, A, B, C, D certain complex numbers, satisfies the additional condition at $t = b > 0$, $(\cos\beta)x(b) + (\sin\beta)p(b)x'(b) = 0$. Then, show that the circles described by m, $0 \leq \beta \leq \pi$, tend as $b \to \infty$ either to a proper circle or to a single point. Finally, prove the Weyl limit circle–limit point theorem: The first case corresponds to the situation when all solutions are in L^2; in the second case for every complex λ there exists at most one nontrivial solution in L^2 (if Im $\lambda \neq 0$, then exactly one).

1.502 Prove that each of the cases below corresponds to the limit point in the Weyl theorem:
(i) There exists a positive function M such that $q(t) \geq -M(t)$, $\int_1^\infty (pM)^{-1/2} = \infty$, $|\,p^{1/2}(t)M'(t)M^{-3/2}(t)\,| \leq \text{const} < \infty$.
(ii) $q(t) \geq \text{const} > -\infty$, $\int_0^\infty p^{-1/2}(t)\,dt = \infty$.
(iii) $p(t) = 1$, $q(t) \geq -\text{const}\cdot t^2$.

1.503 Consider the equation $x'' = -V(x)$ with $x \in (0, \infty)$, where the potential V has locally Lipschitz derivative. Prove the Weyl theorem that
(i) there is a solution $x(t)$ which reaches the origin in finite time if and only if V is bounded from above in a neighborhood of the origin;
(ii) there is a solution $x(t)$ which escapes to infinity in finite time if and only if V is bounded from above for $x \geq 1$ and $\int_1^\infty (C - V(x))^{-1/2}\,dx < \infty$ for some $C > \sup_{x\geq 1} V(x)$.

1.504 Prove the Weyl criterion: The operator $H = -d^2/dx^2 + V(x)$ where V is a continuous real function on $(0,\infty)$ is essentially selfadjoint on $C_0^\infty(0,\infty)$ if and only if there exist a complex number λ and a solution u of the eigenvalue problem $-u'' + Vu = \lambda u$ which is not in L^2 in a neighborhood of infinity and there exists a solution which is not in L^2 in a neighborhood of the origin (one then says that V corresponds to the limit point case).

1.505 Show that if a continuous real function V on $(0,\infty)$ satisfies the following condition: there is a positive differentiable function M such that $V(x) \geq -M(x)$, $\int_1^\infty M(x)^{-1/2} dx = \infty$, $M'(x)M(x)^{-3/2}$ is bounded for large x, then at least one solution of the equation $u'' = Vu$ is not in L^2 in a neighborhood of infinity.

1.506 Give examples of continuous potentials V on $(0,\infty)$ such that
(i) V is complete at infinity in the sense of quantum mechanics (at least one solution of the equation $u'' = Vu$ is not in L^2 at ∞) but not classically complete at infinity (there is a solution $x(t)$ of the equation $x'' = -V(x)$ which goes to infinity in a finite time);
(ii) V is complete at infinity in the classical sense but not in the quantum sense.

1.507 Discuss solvability of the boundary value problem $x''(t) + |x(t)| = 0$, $x(0) = 0$, $x(b) = B$.

1.508 Estimate the number of solutions of the boundary value problem for the Duffing equation $x'' + c \sin x = 0$, $x(0) = x(1) = 0$, with a positive parameter c.

1.509 Show that the problem $x'' + x(1 + (x')^2)^{1/3} + h(t) = 0$, $x(a) = x(b) = 0$, has infinitely many solutions.

1.510 Prove that the problem $x'' = -x^2$, $x(-c) = x(c) = 0$, admits exactly two solutions for each real number c.

1.511 Consider the boundary value problem $x'' = x^2$, $x(0) = 0$, $x(\tau) = B$. Show that there exists a constant $c > 0$ such that for $B > -c\tau^2$ this problem has two solutions, for $B = -c\tau^2$ one, and for $B < -c\tau^2$ there is no solution.

1.512 Study the structure of the set of solutions of the boundary value problem $x'' + \lambda x^2 = 0$, $x(0) = x(1) = 0$, with a real parameter λ.

1.513 Consider the boundary value problem $x''(t) + f(t,x(t)) = 0$, $x(a) = A$, $x(b) = B$, where $|f(t,x) - f(t,y)| \leq K|x - y|$ and f is continuous with respect to (t,x). Prove that this problem has a unique solution whenever $K(b - a)^2 < \pi^2$.

1.514 Show that the problem $x'' + \delta \sin x + h(t) = 0$, $x(a) = x(b) = 0$, has a solution. This solution is unique if $|\delta| < \pi^2/(b - a)^2$.

1.515 Prove the Scorza-Dragoni theorem: If a continuous function f is uniformly bounded on $[a,b] \times \mathbb{R} \times \mathbb{R}$, then the boundary value problem $x'' = f(t,x,x')$, $x(a) = A$, $x(b) = B$, has a solution for arbitrary A and B.

1.516 Let $f(t,x,x')$ be a continuous bounded function, $|f(t,x,x')| \le m$ for all $0 \le t \le a$ and all $x, x' \in \mathbb{R}^n$. Show that solutions of the problem $x'' = f(t,x,x')$, $x(0) = x(a) = 0$ (which exist), satisfy the estimates $|x(t)| \le ma^2/8$, $|x'(t)| \le ma/2$.

1.517 Give an example of a boundary value problem of the form $x'' = f(t,x,x')$ with a genuinely nonlinear function f, $x(a) = x(b) = 0$, which has infinitely many solutions.

1.518 Prove that the boundary value problem (arising in certain radiation problems and in electrohydrodynamics) $x'' = \exp x$, $x(a) = x(b) = 0$, has a solution on any interval $[a,b]$.

1.519 Study the structure of solutions of the boundary value problem $x'' + \frac{1}{2}\lambda \exp x = 0$, $x(0) = x(1) = 0$.

1.520 Show that the problem $(r^2 u')' + \sigma r^2 \exp u / \int_a^1 r^2 \exp(u(r)) \, dr = 0$, $r \in [a,1]$, $u(a) = u(1) = 0$, has a solution for all σ.

1.521 Let $u = u(x)$ be a solution of the problem $(xu')' = xf(u)$, $u'(a) = -\sigma$, $u(\infty) = 0$, $\sigma > 0$, $a > 0$, and f be a strictly increasing, twice continuously differentiable function, vanishing at zero. Prove that $u \simeq x^{-1/2} \exp(-\lambda x)$ where $f'(0) = \lambda^2$, $\lambda > 0$.

1.522 Assume that f is a continuous function on $\mathbb{R}$. Let $x(t)$ be a periodic solution of the equation $x' = f(x)$. Show that $x(t)$ is constant.

1.523 Discuss existence of periodic solutions of the equation $x' + Cx = f(t)$, where f is a periodic function. Do the same for the right-hand side of the form $\exp(ht)f(t)$ with periodic f and solutions of the form $\exp(ht)g(t)$, g periodic.

1.524 Assume that $f : \mathbb{R}^2 \to \mathbb{R}$ is a C^1 function and $f(t + T,x) = f(t,x)$. Let x be a solution of $x' = f(t,x)$ defined on $[0,\infty)$. Show that the sequence $x(nT)$ is monotone.

1.525 Let $A(t)$ be a continuous matrix on $[0,T]$. Assume that if $x(t)$ is a solution of the equation $x' = A(t)x$ such that $x(0) = x(T)$ then $x(t) = 0$. Prove that for each continuous $b(t)$ there exists a unique solution $y(t)$ of the equation $y' = A(t)y + b(t)$ such that $y(0) = y(T)$. Moreover, there exists a constant $C > 0$, independent of $b(t)$, such that $\sup_t |y(t)| \le C \sup_t |b(t)|$.

1.526 Let $A(t)$ be a continuous $n \times n$ T-periodic matrix. Assume that $\varphi \equiv 0$ is the unique T-periodic solution of $x' = A(t)x$. Prove that there exists $\delta > 0$ such that for each function $f : \mathbb{R} \times \mathbb{R}^n \to \mathbb{R}^n$ satisfying $f(t,x) = f(t + T,x)$, $|D_2 f(t,x)| < \delta$, the equation $x' = A(t)x + f(t,x)$ has a unique T-periodic solution φ_f. Prove also that $\varphi_f \to 0$ if $f \to 0$ uniformly.

1.527 Prove the Massera theorem for linear equations: If the linear inhomogeneous system $y' = A(t)y + f(t)$ with T-periodic continuous functions A and f has a solution bounded for $t \geq 0$, then it has a T-periodic solution.

1.528 Consider the scalar differential equation $x' = f(t,x)$, where $f : \mathbb{R}^2 \to \mathbb{R}$ is T-periodic with respect to t, locally Lipschitz with respect to x, and continuous. Show that if this equation has a solution y defined and bounded on $\mathbb{R}^+$, then it has a T-periodic solution.
Remark: This result is not true if $x \in \mathbb{R}^2$ (the examples are rather complicated).

1.529 Prove that if $f : \mathbb{R}^2 \to \mathbb{R}$ is periodic in t and decreasing in x, then $x' = f(t,x)$ has a periodic solution if and only if it has a bounded solution.

1.530 Prove that if f is a periodic function with period T in t, then x is a T-periodic solution of the equation $x' = f(t,x)$ if and only if

$$x(t) = x(0) + T^{-1} \int_0^T f(s,x(s))\, ds$$

$$+ \int_0^T (f(s,x(s)) - T^{-1} \int_0^T f(\tau,x(\tau))\, d\tau)\, ds.$$

1.531 Let $f : \mathbb{R} \times \mathbb{R}^n \to \mathbb{R}^n$ be a C^1 function and $f(t + T,x) = f(t,x)$ for some $T > 0$. Assuming that the differential equation $x' = f(t,x)$ has finitely many periodic solutions, show that the periods have the form kT, $k \in \mathbb{N}$.

1.532 Show by means of an example that a T-periodic differential equation may have a periodic solution whose period T' is incommensurable with T.

1.533 Assume that $f : \mathbb{R} \times \mathbb{R}^n \to \mathbb{R}^n$ is a C^1 function and $f(t + T,x) = f(t,x)$ for some $T > 0$. Let x be a periodic solution of the equation $x' = f(t,x)$ with the period T'. Show that if T/T' is irrational, then for each t_0, $f(t,x(t_0))$ is constant.

1.534 Prove the Mawhin-Walter theorem: Consider the differential equation $x' = f(t,x)$, where $f : \mathbb{R} \times [0,1] \to \mathbb{R}$ is a continuous function such that $f(t,0) = f(t,1)$, f being locally Lipschitz in x, uniformly with respect to t. Assume that there exists $a \in L^1$, $a(t) \geq 0$, such that $f(t,x) \leq a(t)$ for $x \geq 0$, $f(t,x) \geq -a(t)$ for $x \leq 0$ and let $F^+(t) = \lim \sup_{x \to +\infty} f(t,x)$, $F^-(t) = \lim \inf_{x \to -\infty} f(t,x)$. If $\int_0^1 F^+(t)\, dt < 0 < \int_0^1 F^-(t)\, dt$, then there exists at least one solution of the equation satisfying $x(0) = x(1)$.

1.535 Let $f : \mathbb{R} \times \mathbb{R}^n \times \mathbb{R}^m \to \mathbb{R}^n$ be a C^1 function and $f(t + T, x, \lambda) = f(t,x,\lambda)$ for all t and some $T > 0$. Assume that p is a T-periodic solution of $x' = f(t,x,0)$ such that the solution of $y' = D_2 f(t,p(t),0)y$, $y(T) = y(0)$, is $y(t) \equiv 0$. Show that there exist ε, $\delta > 0$ such that for $|\lambda| < \delta$ the equation $x' = f(t,x,\lambda)$ has a unique T-periodic solution $p(t,\lambda)$ such that $|p(t,\lambda) - p(t)| < \varepsilon$.

1.536 Consider the equation $x' = f(x) + g(t)$ where $f'(x) \geq k > 0$ and g is a continuous almost periodic function. Show that this equation has exactly one almost periodic solution.

1.537 Let $g : \mathbb{R} \to \mathbb{R}$ be a C^1 function and $b : \mathbb{R} \to \mathbb{R}$ be a T-periodic even continuous function. Show that the solution of the problem $x'' + g(x) = b(t)$, $x'(0) = x'(T) = 0$, is T-periodic.

1.538 Let $g, b : \mathbb{R} \to \mathbb{R}$ be C^1 odd functions and b is $2T$-periodic. Show that the solution x of the problem $x'' + g(x) = b(t)$, $x(0) = x(T) = 0$, is $2T$-periodic.

1.539 Let $\alpha(t,z)$, $\beta(t,z)$ be continuous functions for t, $z \in \mathbb{R}$ with the following properties: α, β are periodic of period $T > 0$ in t for fixed z, $\alpha > 0$, $|\beta(t,z)| \to \infty$ and $|\alpha(t,z)/\beta(t,z)| \to 0$ as $|z| \to \infty$ uniformly in t. Show that the equation $x'' = x\alpha(t,x') + \beta(t,x')$ has at most one solution of period T.

1.540 Prove the Knobloch theorem: Let $f : [a,\infty) \times \mathbb{R} \to \mathbb{R}$ be a C^1 function (this assumption may be relaxed to the continuity) that is T-periodic in t. The equation $x'' = f(t,x,x')$ has a T-periodic solution if and only if it has a subsolution α (that is, $\alpha'' \geq f(t,\alpha,\alpha')$) and a supersolution β (that is, $\beta'' \leq f(t,\beta,\beta')$) that are T-periodic functions satisfying the condition $\alpha \leq \beta$.

1.541 Show that the equation $x'' = 3x^3 + \cos t$ has 2π-periodic solutions.

1.542 Consider the forced "negative stiffness" Duffing equation $x'' + \alpha x' + x + x^3 = \beta \cos t$, $\alpha > 0$, $\beta \geq 0$. Show that solutions remain bounded for all time and that for $0 < \beta \ll \alpha \ll 1$ there are precisely three periodic orbits of period 2π, one a saddle and the other two attractors.

1.543 Consider the equation of the forced oscillations of a simple pendulum $x'' + (mg/\ell) \sin x = F(t)$ with an odd, T-periodic function (for instance, harmonic forcing $F(t) = A \sin(2\pi t/T)$). Prove the existence of a unique T-periodic solution of this equation if $T < (\pi/2)(\ell/mg)^{1/2}$.

1.544 Suppose that $f \in C^1(\mathbb{R}^2)$, $f(0,0) < 0$ and $f(x,y) > 0$ for large $x^2 + y^2$. Show that the equation $x'' + f(x,x')x = 0$ has a nontrivial periodic solution.

1.545 Consider the nonautonomous Van der Pol equation $y'' + \varepsilon(y^2 - 1)y' + y = \varepsilon c \sin \omega t$, where $\varepsilon \geq 0$, $c \in \mathbb{R}$, $\omega^2 = 1 + \varepsilon\beta$, $\beta \in \mathbb{R}$. Show that this equation has at least one simple solution for each β, which assures the existence of a solution of period $T = 2\pi/\omega$ for ε sufficiently small. Note that if $c = 0$, the equation has a periodic solution of period $T = 2\pi + \mathbb{O}(\varepsilon^2)$. Thus, the application of an external excitation of small amplitude and frequency near that of the limit cycle of the autonomous equation gives rise to a periodic oscillation whose amplitude "recalls" that of the limit cycle but whose period is that of the excitation. This nonlinear phenomenon is called "synchronization."

1.546 Study periodic orbits of the special Liénard equation $x'' + (x')^3 - ax' + x = 0$.

1.547 Consider the Liénard differential equation $x'' + f(x)x' + g(x) = 0$, where f, g are odd functions, $xg(x) > 0$ for $x \neq 0$ and $f(x) \not\equiv 0$ in a neighborhood of the origin. Prove the Opial theorem: If $F(x) = \int_0^x f(s)\, ds$ and $\int_0^x g(s)|F(s)|^{-1}\, ds \geq (1/4 + \varepsilon)|F(x)|$ for $0 \leq x \leq x_1$ and some $\varepsilon > 0$, then all sufficiently small solutions are periodic. Moreover, if $\int_0^x g(s)|F(s)|^{-1}\, ds \leq |F(x)|/4$, then there exist arbitrarily small nonperiodic solutions.

1.548 Prove the Seifert theorem: If in the differential equation $x'' + g(x) = f(t)$, $xg(x) > 0$ for $x \neq 0$, $|g(x)| \geq \max |f(t)|$ for sufficiently large $|x|$, $f(t + T) = f(t)$, and if T is sufficiently small, then there exists a T-periodic solution.

1.549 Prove the Opial theorem: Suppose that for the equation $x'' + g(x) = f(t)$, $|g(x)| \to \infty$, $xg(x) > 0$ for $x \neq 0$, $f(t + T) = f(t)$, $\liminf_{|x| \to \infty} \tau(x) > T/4 > 0$, where $\tau(x) = 2^{-1/2}\, \text{sgn}\, x \int_0^x (G(x) - G(s))^{-1/2}\, ds$, $G' = g$. Then there exists a T-periodic solution of this forced equation.

1.550 Prove the Levinson-Smith-Sansone theorem: If $f(x) < 0$ for $|x| < \delta$ and $f(x) > 0$ for $|x| > \delta$, $\lim_{x \to \infty} F(x) = \infty$ ($F' = f$), then the equation $x'' + f(x)x' + x = 0$ has a unique (nontrivial) periodic solution.

1.551 Show that the Rayleigh equation $x'' - \mu(x' - (x')^3/3) + x = 0$ possesses for $\mu > 0$ a unique periodic solution close to $2 \cos t$. This solution is (orbitally asymptotically) stable.

1.552 Discuss the existence of solutions of smallest period 2π of the Reuter equation $x'' + x = \varepsilon(3\nu \cos 2t - x^2) - \varepsilon^2(\lambda x' + \mu x) - \varepsilon^2 \mu x^2$.

1.553 Show that the equation $x'' + \mu \sin(x') + x = 0$ has an infinite number of limit cycles.

1.554 For which values of the parameter μ does the equation $x'' + (\mu + \cos x - (x')^2/2)x' + \sin x = 0$ have a periodic solution?

1.555 Give an asymptotic expansion of periodic solutions of the equation $x'' + x = 4\mu x^3$, $\mu \in \mathbb{R}$.

1.556 Study periodic solutions of the system $x'' + x - y^3 = 0$, $y'' + y + x^3 = 0$.

1.557 Prove that the nonautonomous system $x'' = -\nabla U(x,t)$, where $x \in \mathbb{R}^n$, U is a C^1 real-valued function, T-periodic in t such that $U(x,t) \to \infty$ as $|x| \to \infty$ uniformly in t, has a T-periodic solution.

1.558 Consider a system in $\mathbb{R}^n$ of the form $x'' + Ax + f(x) = 0$, where $f(x) = \nabla F(x) = o(|x|)$ near the origin, A is a selfadjoint matrix with the eigenvalues

$0 < \lambda_1^2 \leq \lambda_2^2 \leq \cdots \leq \lambda_k^2$, $k \leq n$. Then this system has a family of (nontrivial) periodic solutions $x = x_\varepsilon(t)$ with minimal period T_ε, such that $(x_\varepsilon, T_\varepsilon) \to (0, 2\pi/\lambda_k)$ as $\varepsilon \to 0$. If F is real analytic, this family depends analytically on ε and $x_\varepsilon(t) = \sum_{j=1}^\infty \alpha_j(t)\varepsilon^j$, $T_\varepsilon(t) = 2\pi/\lambda_k + \sum_{j=1}^\infty \beta_j\varepsilon^j$, where $\alpha_1(t)$ is $2\pi/\lambda_k$ periodic and satisfies $x'' + Ax = 0$.

1.559 Assume that $g(t,x,u)$ is a uniformly continuous function and $\lim_{|x|+|u|\to\infty} |g(t,x,u)| /(|x| + |u|) = 0$. Show that the problem $x''(t) + g(t,x,x') = 0$, $x(a) = x(b) = 0$, has a solution.

1.560 Let $f(t,x,x')$ be a continuous function for $0 \leq t \leq a$ and all $(x,x') \in \mathbb{R}^{2n}$ and satisfy the Lipschitz condition with respect to x, x' of the form $|f(t,x_1,x_1') - f(t,x_2,x_2')| \leq \alpha |x_1 - x_2| + \beta |x_1' - x_2'|$ and $\alpha a^2/8 + \beta a/2 < 1$. Show that the problem $x'' = f(t,x,x')$, $x(0) = 0$, $x(a) = 0$, has a unique solution.

1.561 Let $f(t,x,x')$ be a continuous function on $\mathbb{R}^3$ and strictly increasing in x for fixed (t,x'). Prove that the problem $x'' = f(t,x,x')$, $x(0) = x_0$, $x(a) = x_a$, can have at most one solution. Show by example that "strictly increasing" cannot be replaced by "nondecreasing."

1.562 Assume that the problem $x''(t) + p(t)x'(t) + q(t)x(t) = 0$, (i) $x(a) \sin \alpha + x'(a) \cos \alpha = 0$, $x(b) \sin \beta + x'(b) \cos \beta = 0$, has a unique solution $x(t) \equiv 0$. Show that the problem $x'' + g(t,x,x') = 0$, (i), has a solution if

$$\lim_{|x|+|u|\to\infty} \frac{|g(t,x,u) - p(t)u - q(t)x|}{|x| + |u|} = 0$$

uniformly with respect to t.

1.563 Prove that the problem $x'' + g(t,x,x') = 0$, $x(a) = x(b) = 0$, has a solution if $|g(t,x,u)| \leq A |u|^2 + B$ and $\partial g(t,x,u)/\partial x \leq -\varepsilon$, where A, B, and $\varepsilon > 0$ are some constants.

1.564 Consider the two-point boundary value problem $x'' = f(t,x,x')$, $x(0) = x(1) = 0$, where f is a continuous function, $yf(t,y,0) > 0$ for large $|y|$, $|f(t,y,p)| \leq A(t,y)p^2 + B(t,y)$ for some functions A and B bounded on each compact subset of $[0,1] \times \mathbb{R}$. Prove that there exists a solution of this problem.
Remark: The classical Bernstein theorem (where one supposes $f_y \geq k > 0$ for some constant k) is a consequence of this result.

1.565 Prove that the boundary value problem $x'' = f(t,x,x')$, $x(a) = A$, $x(b) = B$, where f satisfies the Carathéodory condition (that is, f is measurable with respect to t, continuous with respect to x and x', and f locally has a t-integrable majorant) cannot have two solutions if f is nondecreasing with respect to x and monotone in x'.

1.566 Let $f(t,x,u)$ be a C^1 function of $t \in [0,1]$, x, $u \in \mathbb{R}$ and let x be a solution of the problem $x'' = f(t,x,x')$, $x(0) = a$, $x(1) = b$. Assuming $\partial f((t,x,u)/\partial x > 0$, prove that there exists $\varepsilon > 0$ such that if $\mid \beta - b \mid < \varepsilon$, then there exists a solution y of the problem $y'' = f(t,y,y')$, $y(0) = a$, $y(1) = \beta$.

1.567 Show that the boundary value problem $x'' + 2x^3 = 0$, $x(0) = x(A) = 0$, has a countably infinite number of distinct solutions.

1.568 Consider the boundary value problem $x'' - 2xx' + \lambda x = 0$, $x(0) = x(\pi) = 0$, with the parameter $\lambda > 0$. Prove that the points $(\lambda,x) = (n^2,0)$, $n = 1,2, \ldots$, are bifurcation points and for $n^2 < \lambda < \mu$ the solutions x_λ, x_μ which remain on the nth bifurcating branch are monotone with respect to the parameter: $\mid x_\lambda(t) \mid \leq \mid x_\mu(t) \mid$.

1.569 Consider a Hamiltonian system $x' = y$, $y' = -V'(x)$ in the plane, where V is a smooth potential function with a nondegenerate relative minimum at the origin (hence the origin is a center for this system). Let T denote the period function which assigns to the amplitude of each periodic orbit its minimum period. Define

$$N(x) = 6V(x)(V''(x))^2 - 3(V'(x))^2V''(x) - 2V(x)V'(x)V^{(3)}(x)$$

$$= (V'(x))^4 \; (V(x)/(V'(x))^2)''$$

Prove that if $N(x) \geq 0$, then T is monotone increasing, and if $N(x) \leq 0$, then T is monotone decreasing. Apply this result to determine existence and uniqueness of solutions for the Neumann boundary value problem $x'' = a(x')^2 + bx + c$, $x'(0) = x'(1) = 0$.

1.570 Consider the boundary value problem $u'' + f(u) = 0$, $-L \leq x \leq L$, $u(-L) = u(L) = 0$, where $f(u) = -u(u - a)(u - b)$ for $0 < a < b/2$. Show that there exists L_0 such that if $L < L_0$ then there are no nonconstant solutions of this problem and if $L \geq L_0$ then such solutions exist.

1.571 Prove that the boundary value problem $x'' = f(t,x)$, $x(a) = A$, $x(b) = B$, with a continuous function f, has a solution if and only if there exist a subsolution and a supersolution of this problem: g and h in $C^2(a,b)$ such that $g(t) \leq h(t)$, $g''(t) \geq f(t,g(t))$, $h''(t) \leq f(t,h(t))$, $g(a) \leq A \leq h(a)$, $g(b) \leq B \leq h(b)$.

1.572 Determine all pairs (A,B) such that the boundary value problem $x'' = \mid x' \mid^n \operatorname{sgn} x'$, $x(0) = A$, $x(\tau) = B$, $n \geq 0$, has a solution.

1.573 Determine some (sufficient) solvability conditions of the problem $x' = x^k y^l$, $y' = x^m y^n$, $\int_0^1 x^p(t) \, dt = A$, $\int_0^1 y^q(t) \, dt = B$, where the powers are meant in the following sense: $x^k = \mid x \mid^k \operatorname{sgn} x$.

1.574 Consider the equation $x'' = p/(q^2 + x^2)$ with the boundary conditions $x(0) = A$, $x(1) = B$, $p, q \in \mathbb{R}$. Show that for sufficiently large q this problem has a unique solution for every pair (A,B), while for small q (relative to p) the problem may have three solutions. Find a method to determine q such that this problem has exactly two solutions.

1.575 Prove that the problem $x' = tx^{1/3}$, $\ell(x) = c$, $-1 \le t \le 1$, cannot have unique solutions for all $c \in \mathbb{R}$ whenever ℓ is a continuous functional.

1.576 Prove the existence of solutions of the Thomas-Fermi equation $t^{1/2}x'' = x^{3/2}$ with the limit conditions $x(0) = 1$, $\lim_{t \to \infty} x(t) = 0$.

1.577 Prove the existence of solutions to the boundary value problem for the Thomas-Fermi equation $t^{1/2}x'' = x^{3/2}$, $0 < t \le 1$, $x(0) = 1$, $x(1) = 0$.

1.578 Show that the boundary value problem $x'' = 2x(x')^3(x^2 + \pi^2)^{-2}$, $x(0) = A$, $x(\tau) = B$, has a solution whenever $\tau \ge 1$ and it does not have any solution for certain A and B if $\tau < 1$.

1.579 Consider the Euler boundary value problem $x'' + px(1 - (x')^2)^{3/2} = 0$, $x(0) = x(1) = 0$, describing the action of an axial thrust on a uniform elastic rod. Prove that the rod deflects out of its plane whenever p exceeds the smallest eigenvalue of the associated linear problem $x'' + px = 0$.

1.580 Let f be a continuous function on $[0,1]$ and $g : \mathbb{R} \to \mathbb{R}$ satisfy the Lipschitz condition with a constant ≤ 40. Show that the problem $-x^{(3)} + g(x) = f(t)$, $x(0) = x'(0) = x(1) = 0$, has a unique solution.

1.581 Show that the boundary value problem $((x')^3)' = 1$, $0 < t < 1$, and $x(0) = x(1) = 0$, does not have classical solution. It has, however, a suitable weak solution $x \in H_0^1(0,1)$.

1.582 Assume that the function $f : \mathbb{R}^{n+1} \to \mathbb{R}$ is continuous and bounded. Show that the Picard problem $x^{(n)} = f(t,x,x', \dots ,x^{(n-1)})$, $x(t_1) = y_1, \dots ,x(t_n) = y_n$, $t_i \ne t_j$, $i \ne j$, admits at least one solution. Study uniqueness of solutions to the boundary value problem $x^{(3)} = (q + x^2)^{-1}$, $x(a) = A$, $x'(a) = B$, $x(b) = C$. Answer the same question for the Picard problem $x(t_j) = y_j$, $j = 1,2,3$.

1.583 Assume that $f : \mathbb{R}^{n+1} \to \mathbb{R}$ is a continuous bounded function. Show that the Nicoletti problem $x^{(n)} = f(t,x,x', \dots ,x^{(n-1)})$, $x(t_1) = y_1, \dots ,x^{(n-1)}(t_n) = y_n$, has at least one solution.

1.584 Show that the (singular) boundary value problem $x'' + x + x^{-3} = 0$, $x(0) = x(\pi) = 0$, does not have any solution.

1.585 Study the singular boundary value problem for the equation (occurring in the theory of elasticity) $(1 - x')x'' - cx = 0$, $0 \le t \le 1$, $c \ge 0$, with the following conditions: (i) $x(0) = x_0$, $x(1) = 1$; (ii) $x'(0) = 0$, $x'(1) = d < 1$.

1.586 Let u_R be the solution of the boundary value problem for the Poisson-Boltzmann equation $(ru')' = rf(u)$, $u(r) = u$, $r \in [a,R]$, $u'(a) = -\sigma/a$, $u(R) = 0$, where $f(u) = \sigma\mu_0 \exp(\alpha u) + N\mu_+ \exp(\beta u) - N\mu_- \exp(-\beta u)$, $\mu_\pm = (\int_a^R r \exp(\pm\beta u) \, dr)^{-1}$, $\mu_0 = (\int_a^R r \exp(\alpha u) \, dr)^{-1}$. Show that if $R \to \infty$ and $N \to \infty$ so that $NR^{-2} \to \rho > 0$, then u_R tends uniformly on each interval to the function u which is the solution of the problem $(ru')' = 4\rho r \sinh u$, $u'(a) = -\sigma/a$, $u(\infty) = 0$. Moreover, if $NR^{-2} \to 0$ then $u_R \to \infty$, and if $NR^{-2} \to \infty$ then $u_R \to 0$.

Singular Perturbations, Asymptotic Theory, Averaging

1.587 Study the asymptotic behavior as $\varepsilon \to 0$ of solutions of the equation $\varepsilon^2 u'' + u = F(t)$ with the initial conditions $u(0) = u'(0) = 0$.

1.588 Determine the asymptotic expansion of solutions of the singular perturbation problem for the beam equation

$$\varepsilon^2 \frac{d^4u}{dx^4} - \frac{d^2u}{dx^2} = p(x), \qquad u(0) = u(1) = \frac{du}{dx}(0) = \frac{du}{dx}(1), 0 < \varepsilon \ll 1.$$

1.589 Show that the equation $\varepsilon u' + e^{2x}u = e^{3x}$, $x \in [0,a]$, does not have solutions analytic with respect to ε for $\varepsilon = 0$.

1.590 Consider the problem of beats for a linear oscillator in which the driver frequency ω is close to the natural frequency $\omega_0 : x'' + \omega_0^2 x = F \cos \omega t$. Solve this equation for arbitrary initial data and $\varepsilon = (\omega_0 - \omega)/\omega_0$ the small parameter. Compare this with the result of the formal application of the two-variable method of asymptotic expansion.

1.591 Find an asymptotic expansion in μ up to μ^2 for the solution of $x' = -x^2 + 6\mu/t$, $x(1) = 1 + 3\mu$.

1.592 Study the asymptotic behavior as $\varepsilon \to 0$ ("boundary layer approximation") for the solutions of the boundary value problems
(i) $\varepsilon^2 u'' - u = 1, 0 < x < 1, u(0) = u(1) = 0$.
(ii) $\varepsilon u'' + u' = x, 0 < x < 1, u(0) = u(1) = 0$.

1.593 Find the derivative of the solution with respect to the parameter $\partial x/\partial \mu$ for $\mu = 0$ of the problem $x' = x + \mu(t + x^2)$, $x(0) = 1 + \mu^2$.

1.594 Obtain an asymptotic formula with a remainder $\mathcal{O}(\varepsilon^2)$ for the solution of the Cauchy problem for the Riccati equation $x' = 2tx + \exp(-t^2) + \varepsilon tx^2$, $x(0) = 0$.

1.595 Consider the singular perturbation problem $\varepsilon u'' + u - u^3 = 0$, $u(0) = u(1) = 0$, $\varepsilon \to 0$. Derive the first-order approximation for the solution.

1.596 Study the asymptotic behavior as $\varepsilon \to 0$ of the solutions of the singular perturbation problem $\varepsilon x'' = x - x^3$, $0 < t < 1$, $x(0) = A$, $x(1) = B$.

1.597 Study the behavior of solutions of the singular perturbation problem $x' + x = \varepsilon(x'')^2$ as ε tends to 0.

1.598 Solve the following nonlinear problem (which is a model of the shock structure for an isothermal shock for large Mach number): $u'' + \varepsilon u'/u^2 - u' = 0$, $x \in \mathbb{R}$, $0 < \varepsilon \ll 1$, $u(-\infty) = 1$, $u(\infty) = \varepsilon$. Compare the exact solution with the solution obtained by perturbation methods.

1.599 Consider the initial value problem for the equation $x'' + (1 - \mu)/x^2 = \mu/(x - 1)^2$, $0 \le x \le 1$, $\mu \ll 1$, $x(0) = 1 - \mu^2$, $x'(0) = (2(2 - c^2))^{1/2}\mu^{1/4}$, $0 \le c^2 \le 2$. This is a limiting case of the equation of motion in the problem of two fixed force centers, which corresponds to starting near the equilibrium point with a small positive velocity. Find the half-period of the periodic orbit and compare with the expression obtained using matched asymptotic expansions.

1.600 Calculate the solution of $x'' + x + \varepsilon x \mid x \mid = 0$, $x(0) = 0$, $x'(0) = v$, to $\mathcal{O}(\varepsilon)$.

1.601 Study the equation $x' = \varepsilon x \cos t$ by the method of averaging. Does it have a hyperbolic limit set? Compare the averaged and the exact solutions.

1.602 Study the nonlinear systems $x' = \varepsilon(x - x^2) \sin^2 t$ and $x' = \varepsilon(x \sin^2 t - x^2/2)$ by the method of averaging.

1.603 Consider the equation $x'' + 4\varepsilon(\cos^2 t)x' + x = 0$ and its (crude) averaging $z'' + 2\varepsilon z' + z = 0$ with initial conditions $z(0) = 0$, $z'(0) = 1$. Compare their solutions.

1.604 Consider the equation $x' = \varepsilon a(x)f(t)$, $x(0) = x_0$, where a is a sufficiently smooth function in a domain containing x_0 and $f(t) = \sum_{n=1}^{\infty} 2^{-n} \cos(2^{-n}t)$ is the almost periodic function. Show that the error made by replacing this equation with its averaged form is larger than $\mathcal{O}(\varepsilon)$ as $\varepsilon \to 0$.

1.605 Consider the equation of the Einstein pendulum $x'' + \omega^2(\varepsilon t)x = 0$ with slowly varying frequency $0 < m \le \omega(\varepsilon t) \le M < \infty$. Establish the relation $\omega(\varepsilon t)x^2 + (x')^2/\omega(\varepsilon t) = \text{const} + \mathcal{O}(\varepsilon)$ on the time scale $1/\varepsilon$ (which is a well-known adiabatic invariant of the system).

1.606 Determine the asymptotic solution of the Mathieu equation (for a linear oscillator with frequency modulation) $x'' + (1 + 2\varepsilon \cos(2t))x = 0$ with initial values $x(0) = \alpha$ and $x'(0) = \beta$, $\varepsilon \to 0$.

Equations in Banach Spaces

1.607 Let C be a contraction in a Banach space X, $\lambda > 0$, $x \in X$. Then, if one denotes by $u_\lambda(t,x)$ the unique solution of the initial value problem $u' = \lambda^{-1}(Cu - u)$, $t \geq 0$, $u(0) = x$, one has the Chernoff estimate $\| u_\lambda(t,x) - C^n x \| \leq ((n - t/\lambda)^2 + t/\lambda)^{1/2} \| x - Cx \|$ for every $n \geq 1$ and $t > 0$. Prove it.

1.608 Prove that for bounded operators A and B in a Banach space

$$\exp(t(A + B)) = \exp(tA) +$$

$$\sum_{n=1}^{\infty} \int_0^t \int_0^{s_1} \cdots \int_0^{s_{n-1}} \exp((t - s_1)A)B \exp((s_1 - s_2)A)$$

$$\cdots B \exp((s_{n-1} - s_n)A) B \exp(s_n A) \, ds_n \cdots ds_1$$

1.609 Give an example of the equation $x' = Ax$ with a bounded operator A in a Banach space such that $\lim_{t \to \infty} x(t) = 0$ for all solutions but $\| \exp(tA) \|$ does not tend to zero.

1.610 Give an example of a linear differential equation $x' = A(t)x$ in a Banach space X such that the subspace $\{x(0) : x(t) \in E(X)\}$ is dense in X but not equal to the whole space X, when $E(X)$ is
(i) $L^1((0,\infty);X)$;
(ii) $L^\infty((0,\infty);X)$;
(iii) $M((0,\infty);X) = \{x : (0,\infty) \to X : \sup_t \int_t^{t+1} | x(s) | \, ds < \infty\}$.

1.611 Consider a 1-periodic linear differential equation $x' = A(t)x$ with x from a Hilbert space X and the associated operator equation $U' = A(t)U$ with the initial condition $U(0) = I$. The equation is said to have the Floquet representation of order m if $U(t) = P(t) \exp(-tB)$ for some operator B and m-periodic operator function P. The well-known Floquet theorem says that in the finite-dimensional case (X complex, respectively real) there exists such a representation (of order 1 or 2 respectively). Give examples of 1-periodic differential equations such that
(i) $X = \mathbb{R}^2$ and there is no Floquet representation of order 1;
(ii) X infinite dimensional and there is no Floquet representation of any order.

1.612 Show that the equation $x' = -Sx$, where S is the shift in the space $\ell^2(\mathbb{Z})$, has no nontrivial solutions which are almost periodic.

1.613 Consider two linear systems in a Banach space E, $x' = A(t)x + u(t)$ and $y' = A(t)y$. Suppose that for every function $u \in L^p((0,\infty);E)$, $1 < p < \infty$, the first problem has a uniformly bounded solution. Then there exist $\alpha > 0$ and a positive function $N(t)$ satisfying the inequality $|y(t)| \leq N(t_0) \exp(-\alpha(t - t_0)^{1/q}) |y(t_0)|$ for $q^{-1} = 1 - p^{-1}$ and all $t \geq t_0$.

1.614 Solve the infinite system of ordinary differential equations $dp_n/dt = \mu p_{n+1}(t) - (\lambda + \mu)p_n(t) + \lambda p_{n-1}(t)$, $n = 1,2,3, \ldots$, $dp_0/dt = \mu p_1(t) - \lambda p_0(t)$, which describes the number of customers in a single-server queueing system with Poisson arrivals and exponential service times; p_n is the probability that the system size is n.

1.615 Show by constructing an example that the Peano existence theorem for ordinary differential equations is not true in Banach spaces.

1.616 Give an example of a differential equation in an infinite-dimensional Banach space such that uniqueness of solutions of the Cauchy problem does not imply continuous dependence on the initial values (as holds in finite-dimensional spaces, compare with 1.9).

Functional-Differential Equations

1.617 Show that the maximum m_n of $|P_n(u)|$, $|u| \leq 1$, where P_n are Legendre polynomials defined by the relation $d(1 - u^2)P_n'(u))/du + n(n + 1)P_n(u) = 0$, satisfies the difference equation $m_{n+1} = (2n + 1)(n + 1)^{-1}m_{n-1}$, $m_0 = m_1 = 1$.

1.618 Determine the solutions of the exponential form $\exp(ax)$, $\exp(ax + by)$ respectively, $a, b \in \mathbb{C}$, of the following difference and functional equations:
(i) $u(x) = \frac{1}{2}(u(x + 1) + u(x - 1))$;
(ii) $u(x) = \frac{1}{2}\int_{x-1}^{x+1} u(t)\, dt$;
(iii) $u(x,y) = (2\pi)^{-1} \int_0^{2\pi} u(x + \cos\vartheta, y + \sin\vartheta)\, d\vartheta$;
(iv) $u(x,y) = \pi^{-1} \int_0^1 r\, dr \int_0^{2\pi} u(x + r\cos\vartheta, y + r\sin\vartheta)\, d\vartheta$.
Do there exist solutions which grow faster in x, y than such an exponential function?

1.619 Consider the difference equation $ay(n + 2) - by(n + 1) + cy(n) = g(n) = 1$ if $n \in \mathbb{N}$ and $=0$ if $n \notin \mathbb{N}$, $n \geq 0$. Give necessary and sufficient conditions for boundedness of solutions.

1.620 Study the Hutchison-Wright differential-difference equation $x'(t) = -ax(t - 1)(1 + x(t))$.

1.621 Find special solutions (of the exponential form $\exp(zt)$) of the differential-delay equations
(i) $x'(t) + ax(t - 1) = 0$;
(ii) $x'(t) + ax'(t - 1) + bx(t - 1) = 0$.

1.622 Study stability of the trivial solution of the differential-delay equation $x'(t) + 3\sin x(t) + 2x(t - \tau) = 0$.

1.623 Find periodic solutions of the differential-delay equation $x'(t) + ax(t) + bx(t - \tau) = f(t)$, where f is a periodic function.

1.624 It is well known that a solution of a linear homogeneous nonautonomous differential equation with bounded coefficients which converges to zero faster than any exponential function $\exp(-ct)$, $c > 0$, vanishes identically. Show that this is no longer true for linear differential-delay equations.

1.625 Give an example of an autonomous (nonlinear) differential-delay equation with nonunique backward continuation of solutions on $(-\infty, 0]$.

1.626 Show that for an n-dimensional autonomous linear delay differential equation the set of all solutions with the initial values from $C([-1, 0]; \mathbb{R}^n)$ for a moment t can have dimension $< n$.

1.627 Show that the delay differential equation $x'(t) = (\sin t)x(t - 2\pi)$ cannot be reduced to an autonomous equation by a 2π-periodic change of variables.

1.628 Show that the zero solution of the neutral delay equation $x'(t) - x'(t - 1) + cx(t) = 0$, where $c > 0$ is a constant, is stable but not uniformly asymptotically stable.

1.629 Suppose that $f'_k = f_{k+1}$ for a doubly infinite sequence (f_k), $k \in \mathbb{Z}$, of functions on $\mathbb{R}$. Prove the Roe theorem: If $|f_k|_\infty$ are uniformly bounded, then $f_0(x) = c \cos x + d \sin x$ for some constants c and d.

1.630 Suppose that $f : \mathbb{R}^+ \to \mathbb{R}$ satisfies the conditions $f'(x) = f(cx)$ for some $c > 1$ and $f(0) = 0$. Must it be constant?

Integral Equations

1.631 Show that the integral equation $u(x) = A + Bx + \int_0^x (C + D(x - t))u(t)\,dt$ has a solution of the form $u(x) = K \exp(mx) + L \exp(nx)$ with some constants K, L, m, n depending on A, B, C, D.

1.632 Study solutions of the Lalesco-Picard integral equation $u(x) = \lambda \int_{-\infty}^\infty \exp(-|x - t|)u(t)\,dt$.

1.633 Prove that for $\lambda < 1/2$ the unique solution in $L^1(\mathbb{R})$ of the integral equation $u(x) = \lambda \int_{-\infty}^\infty \exp(-|x - t|)u(t)\,dt + g(x)$ can be written as

$$u(x) = g(x) + \lambda(1 - 2\lambda)^{-1/2} \int_{-\infty}^\infty \exp(-(1 - 2\lambda)^{-1/2}|x - t|)g(t)\,dt.$$

1.634 Prove that for $|\lambda| < 1$ the Milne integral equation $u(x) = (\lambda/2) \int_0^\infty (\int_{|x-y|}^\infty t^{-1}e^{-t}\,dt)u(y)\,dy$ has the unique solution $u \equiv 0$ in the class $L^\infty[0, \infty)$.

1.635 Prove the existence and uniqueness of the solution of the integral equation $u(x) = \int_0^x y^{x-y}u(y)\,dy$. Show that $u(x) = Cx^{x-1}$ satisfies this equation and that $\int_0^1 u^2(x)\,dx$ does not exist.

1.636 Solve the equation $u(x) - \lambda \int_0^x a(x)b(y)u(y)\, dy = f(x)$.

1.637 Show that the equation $u(x) - \lambda \int_0^1 (x^2 + y^2) \sin u(y)\, dy = 0$ has non-trivial solutions in $L^2(0,1)$ for suitably restricted values of $|\lambda|$.

1.638 Consider the Weyl integral equation $u(x) = \lambda \int_0^\infty \sin(xs)u(s)\, ds$. Show that $\lambda = (2/\pi)^{1/2}$ is an eigenvalue with infinite-dimensional eigenspace.

1.639 Determine the equations satisfied by the functions u and v minimizing the integral $\int_a^b \int_a^b (K(x,y) - u(x)v(y))^2\, dx\, dy$, where K is a (nonsymmetric) kernel.

1.640 Give an example of a Volterra equation $u(x) = \int_0^x K(x,t)u(t)\, dt$ with a continuous kernel K such that $u \equiv 0$ is the only integrable solution, but there are also nonintegrable solutions.

1.641 Verify that the integral operator with the kernel $K(s,t) = \sum_{n=1}^\infty n^{-2} \sin ns \sin(n + 1)t$, $0 \le s, t \le 2\pi$, has nonzero solutions.

1.642 Let G be a convex bounded domain in $\mathbb{R}^3$. Show that the boundary value problem for the stationary transport equation $(s,\, \mathrm{grad}\ \psi) + \alpha\psi = \lambda h(x)(4\pi)^{-1} \int_S \psi(x,s')\, ds'$, $\psi(x,s) = 0$ for $x \in \partial G$, $(s,v_x) < 0$ (s is an element of the unit sphere S) is equivalent to the Peierls integral equation

$$\varphi(x) = \int_G \exp(-\alpha\, |\, x - y\, |\,)(4\pi\, |\, x - y\, |^2)^{-1} (\lambda h(y)\varphi(y) + f(y))\, dy$$

where

$$\varphi(x) = (4\pi)^{-1} \int_S \psi(x,s')\, ds' \quad \text{and} \quad f, h \in C(\overline{G}), \ h(x) > 0 \ \alpha > 0$$

1.643 For the Peierls integral equation in a region $G \subset \mathbb{R}^3$ $\varphi(x) = \lambda(4\pi)^{-1} \int_G \exp(-\alpha\, |\, x - y\, |)\, |\, x - y\, |^{-2}\varphi(y)\, dy$, $\alpha > 0$, prove the following estimate for the first characteristic value: $\lambda_1 \ge \alpha/(1 - \exp(-\alpha\, \mathrm{diam}(G)))$.

1.644 Solve the integral equation $(1 - h)u(t) + \pi^{-1}(t-i) \int_{-\infty}^\infty u(\tau)(\tau - t)^{-1}\, d\tau = f(t)$, where $t \in \mathbb{R}$, $h \in \mathbb{R}\backslash\{0,2\}$, in the class of functions such that
(i) $u(t) = o(|\, t\, |^{-a})$ for $|\, t\, | \to \infty$ and some $a > 0$;
(ii) $|\, v(t) - v(\infty)\, | = \mathbb{O}(|\, t\, |^{-a})$.

1.645 Show that the Hammerstein integral equation $u(x) = \frac{1}{2} \int_0^1 a(x)a(t)(1 + u^2(t))\, dt$, where $a(x) > 0$ for all $0 \le x \le 1$ and $\int_0^1 a^2(x)\, dx > 1$, has no real solutions.

1.646 Prove the global existence and uniqueness of the solution of the nonlinear Volterra equation $u(t) = f(t) + \int_0^t K(t,s,u(s))\, ds$, where $K : [0,T] \times [0,T] \times \mathbb{R} \to \mathbb{R}$ is a continuous function, $|\, K(t,s,x) - K(t,s,y)\, | \le L\, |\, x - y\, |$, and f is a continuous function.

1.647 Suppose that $f \in C[0,\infty)$ and $k \in L^1_{loc}$, g is continuous and satisfies a bound $|g(u)| \le a + b|u|$ for some constants a and b. Show that all solutions of the integral equation $u(x) = f(x) + \int_0^x k(x - s)g(u(s))\, ds$ exist for all $x \ge 0$.

1.648 Suppose that k is nonnegative and continuous, $\int_0^\infty k(s)\, ds < 1$. Let u solve the integral equation $u(x) = 1 + \int_0^x k(x - s)u(s)\, ds$, $x \in [0,\infty)$.
(i)   Show that $u(x)$ is bounded on $[0,\infty)$.
(ii) Show that $\lim_{x\to\infty} u(x)$ exists and is equal to $(1 - \int_0^\infty k(s)\, ds)^{-1}$.

1.649 Let $g : \mathbb{R} \to \mathbb{R}$ be a continuous function such that $\lim\inf_{x\to\infty} (g(x)/x^{1+a}) > 0$ for some $a > 0$. Let $k \in L^1(0,B)$ satisfy $k(x) \ge bx$ a.e. on $[0,B]$ for some constant $b > 0$. Show that for any $x_0 \in (0,B]$ there exists a constant u_0 such that the solution u of $u(x) = u_0 + \int_0^x k(x - s)g(u(s))\, ds$ cannot be continued to the entire interval $[0,x_0]$.

1.650 Consider the integral equation $u(x) = u_0 + K \int_0^x (x - s)(u(s))^{1+a}\, ds$, where u_0, K, and a are positive constants. Show that the solution $u = u(x)$ tends to ∞ as x approaches $u_0^{-a/2}(2K/(2 + a))^{-1/2} \int_1^\infty (s^{2+a} - 1)^{-1/2}\, ds$.

1.651 Show that the integral equation $u(x) = \int_0^x (\pi(x - s))^{-1/2}g(u(s))\, ds$, where $g(u) = 1 - u$ for $u < 1/2$, $(1 - (2u - 1)^{1/3})/2$ for $u \in [1/2,1]$, and $1 - u$ for $u > 1$, has a unique continuous solution u for $x \ge 0$. Show that the Picard iterates $u_0(x) \equiv 0$, $u_{n+1}(x) = \int_0^x (\pi(x - s))^{-1/2}g(u_n(s))\, ds$, $n = 0,1,2, \ldots$ do not converge to the unique solution of this equation.

1.652 Let u be the unique solution of $u(x) = \int_0^x (\pi^{1/2}(x - s))^{-1/2}(1 - u(s))\, ds$. Show that $u = u(x)$ exists for all $x \ge 0$, it is positive and increasing for $x > 0$, and it tends to 1 as x tends to ∞.

1.653 Let u solve the integral equation $u'(x) = -x \int_0^x \exp(-(x - s)(x + s)^2) u(s)\, ds$, $u(0) = 1$. Show that $u = u(x)$ exists for all $x \ge 0$, u is decreasing, and $\lim_{x\to\infty} u(x) > 0$.

1.654 Let $k : [0,1] \times [0,1] \to [0,\infty)$ be a nonnegative continuous function and $p \in (0,1)$. Prove that if there exists a continuous function q such that $q(x) > 0$ for $x \in (0,1]$ and $mq(x) \le \int_0^x k(x,s)(q(s))^p\, ds \le Mq(x)$, $0 < m \le M < \infty$, then the equation $u(x) = \int_0^x k(x,s)(u(s))^p\, ds$, $x \in [0,1]$, has a unique solution u such that $\inf_{x\in[0,1]} u(x)/q(x) > 0$ and $\sup_{x\in[0,1]} u(x)/q(x) < \infty$.

1.655 Show that the equation $u(x) = \int_0^x (x - s)^a(u(s))^p\, ds$, $a > -1$, $p \in (0,1)$, has a unique continuous solution u such that $u(x) > 0$ for $x > 0$.

1.656 Let k be a nonnegative function such that $\int_0^1 k(s)\, ds < \infty$. Let $g : [0,\infty) \to [0,\infty)$ be a continuous nondecreasing function. Show that the equation

$u(x) = \int_0^x k(x - s)g(u(s))\, ds$ has a nondecreasing continuous solution on an interval $[0,\delta]$ for some $\delta \in (0,1]$.

1.657 Let $g : [0,\infty) \to [0,\infty)$ be a nondecreasing continuous function such that $g(0) = 0$ and $g(x) > 0$ for $x > 0$. Prove that the initial value problem $u^{(n)}(x) = g(u(x))$, $u(0) = u'(0) = \cdots = u^{(n-1)}(0) = 0$ has a nontrivial solution on $[0,\varepsilon)$ for some $\varepsilon > 0$ if and only if the generalized Osgood condition $\int_0^\varepsilon (s/g(s))^{1/n} s^{-1}\, ds < \infty$ holds.

Applications to Mechanics, Physics, Biology, and Geometry

1.658 Find the motion of a particle falling freely under the influence of gravity and the resistance R of the form $R = mg(A + B \log |v|)$, $0 < A < 1$, $B > 0$, m is the mass of the particle, g is the gravity constant, and v is the velocity of the particle.

1.659 Consider a particle of mass m falling freely under the influence of gravity and the resistance of the form $mg(a + bv^n)$, where a, b, n are positive constants, $a < 1$, and v is the velocity. Show that v tends to $((1 - a)/b)^{1/n}$ as t tends to infinity.

1.660 Show that the motion of total energy E verifying $x'' = -x^{a-1}$ has the period $T(E)$ proportional to $E^{-1/a-1/2}$. Show that the proportionality coefficient is $2 \int_{-1}^{1} (1 - x^a)^{-1/2}\, dx$.

1.661 Show that the period of motion $T(E)$ with energy E verifying $mx'' = -V'(x)$ with V such that $V(0) = 0$, $V'(0) = 0$, $V''(0) > 0$, is such that $\lim_{E \to 0^+} T(E) = 2\pi(m/V''(0))^{1/2}$.

1.662 Let $V(x)$ be a C^∞ convex even function vanishing at the origin. Consider the motion associated to the potential energy V, having total energy E. Show that its period is larger than the period of the motion with potential energy $W(x) = \frac{1}{2}(\sup_\zeta V''(\zeta))x^2$.

1.663 Find the limit as $E \to \infty$ of the period of the motion with energy E developing with potential energy V such that $V(x) = V(-x)$, $\lim_{x \to \infty} V(x)/x^2 = +\infty$.

1.664 Find the limit as $E \to \infty$ of the period of the motion with energy E developing with potential energy V such that $V(x) = V(-x)$, $\lim_{x \to \infty} V(x)/x^2 = 0$, $\lim_{x \to \infty} V(x) = \infty$.

1.665 Solve the Huygens isochronism problem: Determine a curve in a vertical plane such that a material point gliding along this curve reaches the lowest point (the global minimum) in the same time independently of the initial condition.

1.666 Prove that if all the motions under the action of a force with potential energy V verifying:
(i) $V(x) = V(-x)$,
(ii) $V'(x) > 0$ for $x > 0$,
(iii) $\lim_{x \to \infty} V(x) = \infty$,
have the same period, then there is a $k > 0$ such that $V(x) = \frac{1}{2}kx^2 + V(0)$. Prove also that there exist infinitely many smooth functions verifying (ii) and (iii) and originating motions with period independent of energy.

1.667 Suppose that the force F on $\mathbb{R}$ is directed toward some fixed point, which is taken as the origin, and is a function of the distance from this point and the velocity v, $F = F(x,v)$. Moreover, assume that the motion is isochronic, that is, for each $x_0 \in \mathbb{R}$ the motion starting from x_0 with velocity 0 reaches the origin at the time T which does not depend on x_0. Find the form of the force F.

1.668 Let $T(E)$ be the period of the motion with energy E developing under the action of a potential V verifying: $V(x) = V(-x)$, $V'(x) \neq 0$ for $x \neq 0$, $\lim_{x \to \infty} V(x) = \infty$ and $V(0) = 0$. Prove that V is given by $\int_0^{V(x)} T(E)(V(x) - E)^{-1/2} \, dE = 4\pi(m/2)^{1/2}x$.

1.669 Let S^1 act on $\mathbb{R}^2 \setminus \{0\}$ by rotations and let V be a potential on $\mathbb{R}^2$ invariant under S^1, so V is a function of the radial coordinate r only. If $T(v) = \frac{1}{2}m|v|^2$, show that the amended potential is given by $V_\ell(r) = V(r) + \ell^2/2mr^2$. Show that the motion in the plane under the potential V is governed by motion on the half-line $r > 0$ under the potential V_ℓ.

1.670 Find the trajectories of motion in the central field of force F of the form $F(x) = a/|x|^2 + b/|x|^3$, where a and b are positive constants.

1.671 Find the trajectories (in an analytic form) of motion in the central field of force F of the form $F(x) = -\mu/|x|^5$, $\mu > 0$.

1.672 Prove that for an arbitrary field of central forces there exist initial data such that the trajectory is a circle of a given radius r. Show that the velocity along this trajectory is constant.

1.673 Prove the Binet formulas: If $r = r(\varphi)$ is the equation of the trajectory of a material point in a field of central forces, then the projection of the force on the radius is $-mc^2r^{-2}(d^2(1/r)/d\varphi^2 + 1/r)$ and the square of the velocity is $c^2((d(1/r)/d\varphi)^2 + 1/r^2)$, where $c^2 = r^2\varphi'$, is the double of the sectorial velocity. Determine the central force if the trajectories are of the type $r = p/(1 + e \cos(\omega(\varphi - \varphi_0)))$.

1.674 Solve the Suslov problem: Show that if the trajectory of a material point is a plane curve $\omega(x,y) = 0$, then the potential energy is of the form

$U = \frac{1}{2} M^2(x,y) | \nabla \omega |^2 + \Phi(x,y,\omega) - \text{const}$, where M, Φ are differentiable functions, $\Phi(x,y,0) = 0$.

1.675 Prove the Bertrand theorem: If all the trajectories of a material point are conical sections, then the potential is inverse proportional to the distance from the center which is in the focus of the conic.

1.676 Assume that all bounded solutions of the equation $x'' = U'(| x |)x/ | x |$, $x \in \mathbb{R}^2$, are periodic. Show that $U(r) = ar^2$, $a \geq 0$, or $U(r) = -k/r$, $k \geq 0$.

1.677 Suppose that the force F is such that the law of areas is satisfied; that is, the radius vector of each motion traces out equal areas in equal times. Find the force F.

1.678 Consider the motion in the field of central force F on $\mathbb{R}^2$ which in polar coordinates has the form $F(r,\varphi) = (a + b \cos 2\varphi)/r^2$ where $a > 0$ and $b > 0$. Show that each trajectory of motion is an algebraic curve.

1.679 Find all central forces on $\mathbb{R}^2$ such that each trajectory is an algebraic curve.

1.680 Show that the Kepler laws imply the inverse square law of gravitation force.

1.681 Prove that in the Kepler two-body problem: If $x \times v \neq 0$, $E > 0$, then $| x | / | t |$ tends to $(2E)^{1/2}$ as $| t | \rightarrow \infty$. (The hypothesis $x \times v \neq 0$ rules out the possibility of a collision with the origin in a finite time.)

1.682 Consider a classical particle moving in $\mathbb{R}^n$ under the influence of a time-dependent force $f(x(t),t)$, whose trajectories satisfy the equation $x'' = f(x(t),t)$ in $\mathbb{R}^n$. Assume that:
(i) $f(x,t)$ is bounded, continuous in (x,t), and locally Lipschitz in x with the Lipschitz constant uniform for t in compact sets;
(ii)   there exist R_0, τ_0, C_0 such that $| f(x,t) | \leq C_0 | x |^{-\alpha} | t |^{-\beta}$, for any (x,t) with $| x | \geq R_0$, $| t | \geq \tau_0$, where α and β are nonnegative and $\alpha + \beta > 2$;
(iii) there exist nonnegative constants D, γ, and δ with $\gamma + \delta > 2$ such that $| f(x,t) - f(y,t) | \leq Dr^{-\gamma} | t |^{-\delta} | x - y |$, for any x, y, t satisfying $| x |$, $| y | \geq r \geq R_0$ and $| t | \geq \tau_0$.
Prove the following scattering result: If a, $b \in \mathbb{R}^n$, $b \neq 0$, then there exists a unique solution $x(t)$ of the above equation which is asymptotic to $a + bt$, that is, $\lim_{t \to \infty} (x'(t) - b) = 0$, $\lim_{t \to \infty} (x(t) - bt - a) = 0$.

1.683 Suppose that the law of attraction is linear: $F(r) = \text{const} \cdot r$, rather than the inverse square law. Show that in the n-body problem the origin can be moved to the center of the mass and that the resulting equations become independent and can be solved completely.

1.684 Show that in the n-body problem there is no arrangement of the attracting particles so that they all remain at rest.

1.685 Prove the Sundman total collapse theorem: All the particles in the n-body problem cannot come together at the same time (finite or infinite) unless the angular momentum is zero.

1.686 Prove the virial theorem in a sharp form due to Pollard: In the n-body problem $\lim_{t \to \infty} t^{-1} \int_0^t T = -E$ is true if and only if $\lim_{t \to \infty} t^{-2}I(t) = 0$, where I is the moment of inertia and T is the kinetic energy. In the classical theorem it is assumed that I and T remain bounded for $t \geq 0$ and the conclusion reads: $\lim_{t \to \infty} t^{-1} \int_0^t T = \lim_{t \to \infty} \frac{1}{2} t^{-1} \int_0^t U$.

1.687 Assuming that the particles in the n-body problem move for all time $t > 0$ without obstruction, show that if $E > 0$, then $I \to \infty$ as $t \to \infty$. Conclude that at least one distance between the particles cannot remain bounded. This does not say that some distance becomes infinite.

1.688 In the classical n-body problem, let $r_{ij}(t)$ denote the distance between particles p_i and p_j at time t. Let $q(t) = \min(r_{ij})$, $s(t) = \min(r'_{ij})$, m_i is the mass of p_i, $M_1 = \sum_{i=1}^n m_i m_j$, $M_2 = \min(m_i m_j/(m_i + m_j))$, $M = M_1/M_2$. Units are chosen so that the constant of gravitation is 1. Prove that if $s(0) > 0$ and $s^2(0)q(0) > 8M$, then $q(t) \to \infty$ as $t \to \infty$.

1.689 Assume that in a bounded domain Ω of $\mathbb{R}^3$ there is certain distribution of mass that creates a smooth gravitational potential. Show that for every point $Q \in \mathbb{R}^3$ and positive energy $E \geq 0$ there exists a direction v such that the material point with energy E moving in the direction v escapes to infinity.

1.690 Show that if a system in $\mathbb{R}^2$ $q' = Q(q,p,t)$, $p' = P(q,p,t)$, conserves the Lebesgue measure in the phase space, then this system is Hamiltonian; that is, $Q = \partial H/\partial p$, $P = -\partial H/\partial q$, for some function H. Is it true in a higher-dimensional situation?

1.691 Show that the Hamiltonian $H = \frac{1}{2}(p^2 + k^2 x^2) - xpA(t)$, where the function A satisfies the Riccati equation $A' - A^2 + u(t) = 0$, generates the one-dimensional Schrödinger equation $x'' + (k^2 - u(t))x = 0$.

1.692 Find two Hamiltonian forms of the equation $x' = F(x,t)$ such that the evolution coordinate would be t and x respectively.

1.693 Find the Hamiltonian form of the system describing the interaction of three waves $c'_1 = -ik_1 c_3 \bar{c}_2 \exp(-i\gamma(t))$, $c'_2 = -ik_2 \bar{c}_1 c_3 \exp(-i\gamma(t))$, $c'_3 = -ik_3 c_2 c_1 \exp(i\gamma(t))$.

1.694 Find first integrals of the three-particle Toda lattice described by the Hamiltonian $H(q,p) = (p_1^2 + p_2^2 + p_3^2)/2 + \exp(q_3 - q_1) + \exp(q_1 - q_2) + \exp(q_2 - q_3)$.

1.695 Suppose that the Hamiltonian $H(q_i,p_i,t)$, $i = 1, \ldots ,n$, of a mechanical system satisfies the condition $\det(\partial^2 H/\partial p_i \partial p_j) \neq 0$. Show that this system does not admit nontrivial first integrals of the form $f(q_i,t) = $ const. Determine the form of a Hamiltonian if the corresponding system of the canonical equations admits a set of n first integrals $f_1(q_i,t) = \alpha_1, \ldots ,f_n(q_i,t) = \alpha_n$.

1.696 A system of differential equations $q_i' = A_i(q_j,p_j,t)$, $p_i' = B_i(q_j,p_j,t)$, $i, j = 1, \ldots , n$, admits $2n$ independent first integrals $w_k(q_j,p_j,t) = $ const. Show that this system is Hamiltonian if and only if all the Poisson brackets of w_k, w_m, $k, m = 1, \ldots , 2n$, are also first integrals.

1.697 Give an example of a Hamiltonian system on $T^*\mathbb{R}^2$ that has an equilibrium at the origin, one direction is attracting, one is repelling, and there is a two-dimensional manifold of closed orbits.

1.698 Prove that if a system with the Lagrangian $L = \frac{1}{2}\sum_{i=1}^{n}(q_i')^2 - V(q_i)$ admits a first integral obtained via the Noether theorem, then there exist constants a, c_i and a skew symmetric matrix (b_{ij}) such that

$$\sum_{i=1}^{n}(aq_i + \sum_{j=1}^{n} b_{ij}q_j + c_i)\frac{\partial V}{\partial q_i} + 2aV = 0$$

Compute the corresponding integral.

1.699 Find the general solution of the stationary Hamilton-Jacobi equation $H(\partial W/\partial q,q) = E$ for the Hamiltonian of the two-dimensional oscillator $H(p,q) = \frac{1}{2}(p_1^2 + p_2^2) + \frac{1}{2}(\omega_1^2 q_1^2 + \omega_2^2 q_2^2)$.

1.700 Solve the Cauchy problem for the evolution of the Hamilton-Jacobi equation for one-dimensional flow of noninteracting particles

$$\frac{\partial S(q,t)}{\partial t} + \frac{1}{2}\left(\frac{\partial S(q,t)}{\partial q}\right)^2 = 0$$

with the initial data $S(q,0) = cq$, $c \in \mathbb{R}$, $S(q,0) = -q^2/2$, $S(q,0) = -q^3/3$, $S(q,0) = -q \arctan q + \frac{1}{2}\log(1 + q^2)$.

1.701 Show that for the one-dimensional oscillator the time mean $\bar{f}(q_0,p_0) = \lim_{T\to\infty} T^{-1}\int_0^T \varphi(q(q_0,p_0,t), p(q_0,p_0,t))\, dt$ for all φ is a function of the Hamiltonian $H = \frac{1}{2}(p^2 + \omega^2 q^2)$ only.

1.702 Discuss possibilities of stabilization of a simple pendulum with rapidly oscillating suspension (vertically and/or horizontally)—the Bogolubov-Kapitsa problem. In particular, consider the vertical unstable equilibrium and an arbitrary position of the pendulum.

1.703 Study the motion of a simple pendulum with rotating (in the horizontal plane) suspension point described by the equation $x'' + \sin x - \alpha \sin 2x - \beta \cos x = 0$.

1.704 The suspension point of a double pendulum (of lengths ℓ and mass m) oscillates in a horizontal line harmonically: $a \sin \omega t$. Find the oscillations of this pendulum (in the linear approximation) if $\omega^2 = g/\ell$.

1.705 Describe small oscillations of the composed pendulum consisting of n simple pendulums of length ℓ and mass m suspended in a single file.

1.706 A material point moves inside the paraboloid $z = \alpha x^2 + \beta y^2$ rotating with a constant angular velocity Ω around the z-axis. Determine, in coordinates connected with the paraboloid, small oscillations around the stable equilibrium.

1.707 Show that the period of the simple pendulum described by the differential equation $\varphi'' + g(\sin \varphi)/\ell = 0$ with the maximum displacement ω from its equilibrium position is $P(k) = 2\pi(l/g)^{1/2}(1 + (1/2)^2 k^2 + (1 \cdot 3/2 \cdot 4)^2 k^4 + (1 \cdot 3 \cdot 5/2 \cdot 4 \cdot 6)^2 k^6 + \cdots)$, where $k = \sin(\omega/2)$.

1.708 A simple pendulum with variable string length and subject to viscous damping is described by $\theta'' + (2r'/r + c/m)\theta' + g \sin \theta/r = 0$, where $\theta(t)$ is the angle the string makes with the vertical, m the mass of the bob, c the damping coefficient, and g the acceleration due to gravity. Determine the exact expression for $r(t)$ such that the bob moves horizontally for all time, the initial conditions being $\theta(0) = \theta_0$, $\theta'(0) = 0$, where $0 < \theta_0 < \pi/2$. (When r is constant, no exact elementary solution of this differential equation is known; see the preceding problem.)

1.709 Study the evolution of energy of a simple pendulum with slowly varying length $\ell = \ell(\varepsilon t)$, $0 < \varepsilon \ll 1$.

1.710 Find three independent first integrals of the system describing motion of a particle under the gravitation force of a center and a constant force (the Cellerier and Saint-Germain problem).

1.711 Consider the motion of a particle on a line under a potential $V(x)$. Assume that the line rotates in a plane with constant angular velocity ω. Show that the correct motion of the particle is that of a particle on the line in the effective potential $W(x) = V(x) - \frac{1}{2}m\omega^2/x^2$.

1.712 If a ball were dropped in a hole bored through the center of the Earth, it would be attracted toward the center with a force directly proportional to the distance of the body from the center. Find the equation of the motion, the amplitude of the motion, and the period.

1.713 A cubical block of wood is 2 ft on a side. When depressed and released in water, it oscillates with a period of 1 sec. What is the specific density of the wood?

1.714 Determine the frequencies of small oscillations of a molecule consisting of three atoms, two of them of mass m and the middle one of mass M, connected by elastic forces.

1.715 Four equal uniform rods, each of mass m and length $2a$, are smoothly joined to form a rhombus $ABCD$, which is hung from a fixed point A. A and C are connected by a light elastic string of natural length a and modulus of elasticity $2mg$. Determine the energy equation for that motion of the system in which C remains vertically below A. What is the length of the equivalent simple pendulum, for small oscillations about the position as the angle BAD equals $2\pi/3$?

1.716 The attracting force between two balls of mass m and radius r depends only on the distance between their centers. In the initial moment the balls do not move and the distance between their centers is equal to $a > 2r$. Consider two cases:
(i)   one of the balls is fixed,
(ii)   both are movable.
Show that the quotient of the times elapsed before the collision is equal to $2^{1/2}$.

1.717 The ends of a rod move along two perpendicular lines $0x$, $0y$. The velocity of the end sliding on the $0y$ axis is constant and directed toward the origin. Show that the acceleration of each point of the rod is perpendicular to the $0y$ axis and is inversely proportional to the cube of the distance of this point to the $0y$ axis.

1.718 A particle falls in a vertical plane from rest under the influence of gravity and a force perpendicular and proportional to its velocity. Derive the equation of the trajectory and identify the curve.

1.719 Find the trajectory of a material point on a two-dimensional sphere if the meridians are crossed at equal angles.

1.720 A particle moves under gravity from the highest point of a sphere of radius r. Prove that it cannot clear the sphere unless its initial velocity exceeds $(gr/2)^{1/2}$.

1.721 A smooth plane inclined at an angle α to the horizontal is rigidly connected with a fixed vertical axis about which it revolves with uniform angular velocity ω. A heavy particle moves under gravity on the plane. Prove that if x is the displacement of the particle from the point of the plane on the axis, measured upward along the line of greatest slope, then $d^4x/dt^4 + \omega^2(3\cos^2\alpha - 1)\,d^2x/dt^2 + x\omega^4\cos^2\alpha = g\omega^2\sin\alpha$.

1.722 A uniform, flexible cable weighing w_1 lb/ft supports a horizontal bridge. The weight of the bridge is uniformly distributed and is w_2 lb/ft. The weight of the supporting rods is negligible. Show that the differential equation of the curve of the cable is given by $Hy'' = w_1(1 + (y')^2)^{1/2} + w_2$.

1.723 An arched bridge is to be build of stone of uniform density. The weight of the stone is w lb for each square foot of surface. The bridge is so constructed that at each point the resultant tension due to the weight of the stones above it acts in a direction tangent to the arch of the bridge. Find the equation of the arch. (Take the x-axis at the top of the bridge, the y-axis through the highest point of the arch, and the positive direction downward.)

1.724 It is well known that if a uniform thin flexible cord is suspended freely from its endpoint in a uniform gravitational field, the shape of the cord will be an arc of catenary. Determine the shape of the cord if a very long one is used which requires replacement of the uniform gravitational field approximation by the inverse square field.

1.725 A particle is projected with velocity less than that from infinity under a force tending to a fixed point and varying inversely as the nth power of the distance. Prove that if $n \geq 3$ the particle will ultimately fall into the center of force.

1.726 A particle falls from rest under gravity through a stationary cloud. The mass of the particle increases by accretion from the cloud at a rate which is mkv, where m is the mass and v the velocity of the particle and k is a constant. Find the distance the particle has fallen after time t.

1.727 A particle whose mass increases through condensation of moisture at a constant time rate m_0/τ, where m_0 is the initial mass of the particle and τ is a constant, moves freely under gravity. The particle is projected from the origin of coordinates with a velocity whose horizontal and vertical components are u and v respectively. Prove that the coordinates of the particle when its mass is m are given by $x = u\tau\log(m/m_0)$, $y = \frac{1}{4}g\tau^2(1 - m^2/m_0^2) + (v\tau + \frac{1}{2}g\tau^2)\log(m/m_0)$, the y-axis being vertically upward. Show also that, if $v > 0$, the greatest height attained by the particle is $\frac{1}{2}(v\tau + \frac{1}{2}g\tau^2)\log(1 + 2v/g\tau) - \frac{1}{2}v\tau$.

1.728 A comet approaches the sun with velocity v from a great distance in such a manner that it would miss the sun by a distance b were it not deflected. Let the actual orbit be given by $u(\varphi)$, where $1/u$ is the distance from the sun and φ is the angle measured relative to the point of closest approach. Show that conservation of angular momentum yields $\varphi' = vbu^2$ and that conservation of energy yields $(u')^2/u^4 + v^2b^2u^2 - 2\gamma Mu = v^2$ where γ is the gravitational constant and M the mass of the sun. Hence obtain $u(\varphi)$ in the form $u = (\sin \Phi + \cos \Phi)/b \cos \Phi$ where Φ is a constant angle whose tangent is $\gamma M/v^2 b$. Show also that the comet will eventually recede to infinity having its direction of motion turned through an angle 2Φ.

1.729 Discuss the Zhukowskiĭ equations of motion of a glider: $y' = -\sin \varphi - ay^2$, $\varphi' = y - y^{-1} \cos \varphi$, where y is the velocity and φ is the angle between the trajectory of the glider and the x-axis.

1.730 The equation $x'' + x = kx^2$, where k is a small positive parameter, appears in the theory of equatorial satellite orbits of an oblate spheroid: x describes the variations from a constant in $1/r$, where r is the distance from the center of the spheroid to the satellite, and t is an angular variable. Discuss the phase plane trajectories of this equation and express the solutions in terms of elliptic functions. Obtain an upper bound for the perturbations due to the nonlinear term. Give a power series expansion in k of the solutions (the Poincaré expansion).

1.731 Consider a model for repeated impacts of a ball with a massive, sinusoidally vibrating plate given by the mappings $\varphi_{j+1} = \varphi_j + v_j$, $v_{j+1} = \alpha v_j - \gamma \cos(\varphi_j + v_j)$. Here v_j are velocities, $\varphi_j = t_j$ (rescaled) times of impacts, $0 < \alpha \leq 1$ is the coefficient of restitution. Prove that in the Hamiltonian case, $\alpha = 1$, there exist unbounded orbits of the above discrete dynamical system. Construct an attracting set in the dissipative case $\alpha < 1$ (when all orbits remain bounded). Show that for γ sufficiently large and $\alpha = 1/2$ there exists a hyperbolic horseshoe for the above mapping.
Remark: This bouncing ball problem provides a highly nontrivial example of a discrete dynamical system with a direct physical interpretation.

1.732 Determine moments M_1, M_2, and M_3 in the equations of the motion of a gyroscope with (self)excitation $Ax_1' = (B - C)x_2x_3 + M_1$, $Bx_2' = (C - A)x_3x_1 + M_2$, $Cx_3' = (A - B)x_1x_2 + M_3$, where A, B, C are the principal moments of inertia and x_1, x_2, x_3 are the projections of the angular velocity, so that $\exp(2\lambda t)T = \text{const}$ and $\exp(2\lambda t)G^2 = \text{const}$ for $T = \frac{1}{2}(Ax_1^2 + Bx_2^2 + Cx_3^2)$, $G^2 = A^2x_1^2 + B^2x_2^2 + C^2x_3^2$ (the kinetic energy and the square of the kinetic momentum, respectively).

1.733 Integrate the equations of motion of a symmetric gyroscope with Klein-Sommerfeld damping $Ap' = -(C - A)rq - A\lambda^2 p$, $Aq' = (C - A)rp - A\lambda^2 q$, $Cr' = -C\mu^2 r$.

1.734 Consider the Euler equations for the (instantaneous) rotation of a solid suspended at its center of inertia (A, B, C are the central moments of inertia of the solid and x, y, z are the components of instantaneous rotation in the central axes of inertia). Show that the stationary rotations of a solid around its axes of greatest and smallest inertia are orbitally stable but not Liapunov stable.

1.735 A long smooth straight wire is inclined at an acute angle α to the upward vertical and is rotated about the upward vertical through one end O with constant angular velocity ω. A small bead of unit mass is free to slide on the wire and is initially at rest relative to the wire and at a distance ℓ from O. Show that the bead will move continuously outward from the axis of rotation if $\ell > (g/\omega^2) \cotan \alpha \cdot \cosec \alpha$. If $\ell = (2g/\omega^2) \cotan \alpha \cdot \cosec \alpha$, show that the system of forces required to maintain the system at constant angular velocity has, after time t, a moment $(8g^2/\omega^2) \cotan \alpha \cdot \cos \alpha \cdot \cosh^3(\tfrac{1}{2}\omega t \sin \alpha) \cdot \sinh(\tfrac{1}{2}\omega t \sin \alpha)$ about the axis of rotation.

1.736 A bead of mass m is threaded on a smooth circular wire of radius a which is made to rotate about a fixed vertical diameter with constant angular velocity ω. At time t the bead is at an angular distance θ from the lowest point of the wire, and $\theta = \pi/2$, $d\theta/dt = 0$ when $t = 0$. Express $d\theta/dt$ as a function of $\cos \theta$. If $a\omega^2 = 4g$, describe the motion of the bead, and show that the couple required to rotate the wire is $2ma^2\omega^2 \sin \theta \cdot \cos \theta(\cos \theta(1/2 - \cos\theta))^{1/2}$.

1.737 A bead of mass M slides on a smooth wire in the form of a circle, of radius a, with its plane vertical. One end of a light elastic string, of modulus of elasticity λ and unstretched length ℓ, $0 < \ell < 2a$, is tied to the bead, and the other end of the string is attached to the highest point of the wire. Show that a position of equilibrium exists in which the string is inclined to the vertical at an angle θ given by $\cos \theta = \tfrac{1}{2}\lambda\ell/(\lambda a - Mg\ell)$, provided that this fraction lies between 0 and 1. Show that if this position of equilibrium exists it is stable.

1.738 C is the center of a thin uniform circular plate of radius a and mass M. A and B are the corners of an equilateral triangle of sides $4a$ with vertex at C. The triangle lies in a plane normal to that of the plate with AB parallel to the plate. Attracting particles, each of mass m, are placed at A, B respectively. If a particle of unit mass is in equilibrium at the circumcenter of the triangle ABC, prove that $m/M = 32(1 - 4/(19)^{1/2})/3$.

1.739 Let there be a series of cups of equal capacity full of water and arranged one below another. Pour into the first cup an equal quantity of wine at constant rate and let the overflow in each cup go into the cup just below it. Assuming that complete mixture takes place instantaneously, find the amount of wine in each cup at any time t and the limit as t goes to infinity.

1.740 An infinite earthed cylindrical conductor of radius a is enclosed by a concentric infinite cylindrical conductor of radius b whose potential is V. A uniform magnetic field H is applied parallel to the common axis of the conductors. An electron of mass m whose charge is $-e$ ($e > 0$) leaves the inner conductor with zero velocity. Show that it cannot reach the outer conductor if $V < eH^2((b^2 - a^2)/b)^2/(8mc^2)$.

1.741 A submarine was spotted from a torpedo boat. After that the submarine submerged and went straight away at constant velocity v_1 in an unknown direction. The velocity, v_2, of the torpedo boat is greater than v_1, and the destroyer should appear just over the submarine in order to destroy it. Determine the strategy of pursuit.

1.742 The Carathéodory problem: A wire of a given length ℓ is hung on the glass of a window with its ends fixed. Determine the form of the wire such that its projection by sunlight on the floor will be (i) minimal, (ii) maximal.

1.743 Consider the flow in $\mathbb{R}^2$ associated with a reflecting particle: for $t > 0$, set $F_t(q,p) = q + tp$ if $q > 0$, $q + tp > 0$ and $F_t(q,p) = -q - tp$ if $q > 0$, $q + tp < 0$, and $F_t(-q,p) = -F_t(q,p)$, $F_{-t} = F_t^{-1}$. What is the exact generator of the induced unitary flow on $L^2(\mathbb{R}^2)$? Is it essentially selfadjoint on the C^∞ functions with compact support away from the line $q = 0$?

1.744 A man initially at $(0,0)$ walks along the straight shore $0x$ of a lake towing a rowboat initially at $(0,a)$, by means of a rope of length a, which is always held taut. Show that the boat moves in path (called a tractrix) with parametric equations $x = a \log(\cotan(\theta/2) - \cos \theta)$, $y = a \sin \theta$.

1.745 A fighter plane whose velocity is v is chasing a bomber plane whose velocity is w. The nose of the fighter plane is always pointed toward the bomber, which is flying in a direction making an angle β with the horizontal. Find the path traced by the fighter as plotted by an observer in the bomber. The same question for an observer on the ground.

1.746 It has been recommended by highway safety officials that for reasonable safety in a lane of traffic with car A_1 following A_2, A_1 should remain at least t_0 seconds behind A_2. Thus starting at any given time, A_1 should take a time interval of at least t_0 seconds to reach the location that A_2 occupied at that time. Now suppose that the velocities and spacing are initially constant and optimal and that A_2 accelerates. What should be the acceleration of A_1 so as to maintain the minimum safe separation?

1.747 Escape velocity with drag. If one assumes that the frictional force (or drag) retarding a missile is proportional to the density of the air $\rho(x)$, at altitude x (above the earth), and to the square of velocity, then the differential equation of the motion of the missile can be written in the form $x'' + \rho(x)(x')^2 + (x + 1)^{-2} = 0$, $x(0) = 0$, $x'(0) = v_0$ (after a proper normalization of the constants).

(i) Show that escape is not always possible (for instance, if $\rho(x) \geq (2x + 2)^{-1}$).

(ii) Find the necessary and sufficient conditions on $\rho(x)$ in order to allow escape.

(iii) Give an explicit formula for the escape velocity when it exists.

1.748 Determine the minimal time path of a jet plane from takeoff to a given point in space. Assume the highly idealized situation in which the total energy of the jet (kinetic + potential + fuel) is constant and, what is reasonable, that the jet burns fuel at its maximum rate, which is also constant. Assume also that $v = y = x = t = 0$ at takeoff, which gives $v^2 + 2gy = at$ at the energy equation.

1.749 Towns A and B are directly opposite each other on the banks of a river of width D which flows with constant velocity U. A boat leaving town A travels with constant velocity V always aimed toward town B. Show that its path is given by

$$x = \frac{D}{2}(t^{1-U/V} - t^{1+U/V}), \qquad y = Dt$$

and it will not arrive at town B unless $V > U$.

1.750 A cat pursues a mouse. The cat's velocity equals 1 and the mouse's velocity is $1/d$, where d is the distance between the cat and the mouse (with no restriction on the direction). Show that if $d(0) > 1$ then the mouse can escape so that $d(t) > 1$ for all t.

1.751 Let $x(t)$ denote the population of a given species at time t. According to the logistic law of population growth $x(t)$ satisfies the equation $x' = ax - bx^2$, where a, b are the constants called the vital coefficients of the population. For the human population $a = 0.029$ and $b = 2.695 \times 10^{-12}$. Show that, according to this model, the human population of the earth will tend to the limiting value of 10.76×10^9.

1.752 In a certain community the number of individuals is increasing at a constant rate r. Suppose that at time $t = 0$ there is a number N_0 of infected individuals and that the population at this time is N. Show that the initial value problem for the number of infected individuals M in the community at any time $t > 0$ is given by $M' - k(N + rt)M = -kM^2$, $M(0) = N_0$, where k is a constant. Solve this equation.

1.753 The initial value problem for the Lotka-Volterra system $x' = ax - bxy$, $y' = -cy + dxy$, $x(0) = x_0 > 0$, $y(0) = y_0 > 0$, describes the predator-prey system; $x(t)$, $y(t)$ are the prey and predator population at time t, respectively, a, b, c, d some positive constants. Show that the solution $x(t)$, $y(t)$ of this problem is periodic and the average values X and Y of $x(t)$ and $y(t)$ are

$$X = \lim_{t \to \infty} t^{-1} \int_0^t x(s)\, ds = \frac{c}{d}, \qquad Y = \lim_{t \to \infty} t^{-1} \int_0^t y(s)\, ds = \frac{a}{b}$$

1.754 The Newton problem. In seeking the form of a solid of revolution which experiences a minimum resistance when it moves through a fluid in the direction of the axis of revolution, one is led to minimize the following integral: $\int_a^b y(y')^3/(1 + y'^2)\, dx$. Show that the first integral of the Euler equation is

$y = C(1 + p^2)^2/p^3$, where $p = y'$ and C is an arbitrary constant. Observe that $x = C(\log p + p^{-2} + \frac{3}{4}p^4) + C_1$, where C_1 is an arbitrary constant. Note that the above equation provides a solution of the Euler equation in parametric form.

1.755 According to the Fermat principle light propagates in an inhomogeneous medium described by the refraction coefficient n so that $\delta[\int_s^t n \, d\tau] = 0$. By the optical-mechanical analogy the trajectories of a particle in the potential field $V = -\frac{1}{2}n^2 +$ const coincide with the rays in that medium. Describe these rays if $n(x,y,z) = \alpha(x^2 + y^2 + z^2)^{-1/4}$.

1.756 A map $A : H \to H$ is said to be derived from a variational principle if there is a function $L : H \to \mathbb{R}$ such that $dL(x)v = (A(x),v)$, where $(\cdot,\cdot)$ is an inner product on H. Prove the Vainberg theorem: A comes from a variational principle if and only if $DA(x)$ is a symmetric linear operator.

1.757 Discuss the solutions of the variational problem

$$\min \int_a^b x^c (1 + |y'|^2)^{1/2} \, dx, \quad c > 0.$$

1.758 Study the Newton problem: Minimize the functional $\int_a^b x\varphi(|y'|) \, dx$, where $\varphi(r) > 0$, φ decreases, $\varphi(\infty) = 0$.

1.759 Consider the problem of minimizing the scalar-valued function $g : \mathbb{R}^n \to \mathbb{R}$. One step of the steepest descent method $x_{k+1} = x_k - a_k \nabla g(x_k)$ may be regarded as one explicit Euler step with step length a_k applied to the differential equation $dx/dt = -\nabla g(x)$ with the initial condition $x(0) = x_0$. Discuss the behavior of the trajectories of this system in the neighborhood of local minimizers and maximizers of the function g.

1.760 Consider the problem of solving the nonlinear equation $f(x) = 0$ where $f : \mathbb{R}^n \to \mathbb{R}^n$ is a sufficiently regular function. The classical Newton iteration $x_{k+1} = x_k - (f'(x_k))^{-1}f(x_k)$ may be regarded as one explicit Euler step with unit step length applied to the differential equation $x' = -(f'(x))^{-1}$ with the condition $x(0) = x_0$. Solve this problem. What can be said about convergence of the Newton method?

1.761 Consider two parallel lines ℓ_1, ℓ_2 in the plane, a fixed point O, and a curve γ. Denote by T the intersection of the tangent to γ at a point M and ℓ_1, by S the intersection of ℓ_2 and the line containing O and M. Determine the curve γ if it is known that the triangle MST has the constant area $a^2/2$.

1.762 Write a differential equation of all circles in the plane.

1.763 Write a differential equation satisfied by all conic sections which do not have an asymptote parallel to the y-axis.

1.764 Determine a differential equation for the curves of second degree.

2
PARTIAL DIFFERENTIAL EQUATIONS

General questions and some linear equations

2.1 Show that, in general, the solution of the Cauchy problem does not depend continuously on the initial data.

2.2 Verify that there is no analytic solution of the Cauchy problem for the heat equation $u_t = u_{xx}$ with the initial condition $u(x,0) = (1 + x^2)^{-1}$.

2.3 Most of the proofs of the Cauchy-Kowalevskaya theorem use analytic majorant series. Prove this theorem using soft analysis methods, that is, simple a priori estimates and the Schauder fixed-point theorem.

2.4 Prove the local solvability of the partial differential equation with constant coefficients $P(D) u = f, f \in \mathcal{D}'(\mathbb{R}^n)$, along the scheme of M. E. Taylor and J. Dadok. The statement to prove is: For almost everywhere (a.e.) α, $0 \le \alpha_j \le 1$, $j = 1, \ldots, n$, the operators $P(D + \alpha) : \mathcal{D}'(\mathbb{T}^n) \to \mathcal{D}'(\mathbb{T}^n)$ and $P(D + \alpha) : C^\infty(\mathbb{T}^n) \to C^\infty(\mathbb{T}^n)$ are isomorphisms. The crucial lemma is: For a.e. α as above there exist C, N satisfying the inequality $\mid P(k + \alpha) \mid^{-1} \le C(1 + \mid k \mid)^N, k \in \mathbb{Z}^n$.

2.5 The Garabedian-Grushin example. Prove that the equation $u_x + ixu_y = f$ is not (locally) solvable in $\mathcal{D}'$ for certain smooth functions f.

2.6 Show that examples of locally not solvable equations can be pasted together; that is, if $P(x,D)$ is a (pseudo) differential operator, $\Omega \subset \mathbb{R}^n$, and for every open subset $\omega \subset \Omega$ there exists $\varphi \in C_0^\infty(\omega)$ such that $Pu = \varphi$ has no solution of the class $\mathscr{E}'\,(\Omega)$ in ω, then there is a function $f \in \mathscr{S}(\Omega)$ for which $Pu = f$ is not solvable in any open subset of Ω. Moreover, the complement of the set of such functions f is of the first category in $\mathscr{S}(\Omega)$.

2.7 Show that the operator $\partial/\partial\bar{z} + c(z)$, $z = x + iy$, where c is an analytic function in the plane, is analytic-hypoelliptic.

2.8 Let $u_0 = u_0\,(z)$ be an entire function in the complex plane. Show that if $u = u(t,z)$ is an entire function in $\mathbb{C}^2$, satisfying the equation $u_t = z^2\,u_z$ in $\mathbb{C}^2$, $u(0,z) = u_0(z)$, then $u(t,z) = u_0(z/(1-zt))$. Derive from this that the Cauchy problem does not have, in general, an entire function as its solution.

2.9 It may be proved that the operators $I + |\,x\,|^{2\nu}(-\Delta)^\mu$ are hypoelliptic for $0 < \mu < \nu$, $\nu \in \mathbb{N}$ (so in particular $I - |\,x\,|^{2\nu}\Delta$ for $\nu \geq 2$). Show that $I - |\,x\,|^2\Delta$ is not hypoelliptic.

2.10 Consider the following operator in $\mathbb{C}^2$: $L = \frac{1}{2}(\partial/\partial z_1 + i\,\partial/\partial z_2)$. Describe all the solutions $h = h(z_1,z_2)$ of the homogeneous equation $Lh = 0$ which are holomorphic in the polydisk $\{(z_1,z_2) \in \mathbb{C}^2 : |\,z_1\,| < r_1, |\,z_2\,| < r_2\}$, $r_1,r_2 > 0$. If h is such a solution, prove that $h_0(z_1) = h(z_1,0)$ can be extended as a holomorphic function in the disk $\{z_1 \in \mathbb{C} : |\,z_1\,| < r_1 + r_2\}$. Derive from this that the Cauchy problem for the operator L with the initial condition $h = h_0(z_1)$ when $|\,z_1\,| < r_1, z_2 = 0$, does not have, in general, a holomorphic solution in the above polydisk.

2.11 Let $f_1, \ldots, f_n$ be n functions belonging to $C_0^\infty(\mathbb{R}^{2n})$ and assume $n > 1$. Assume that the following compatibility conditions are satisfied: $\partial f_j/\partial\bar{z}_k = \partial f_k/\partial\bar{z}_j$ for $j, k = 1, \ldots, n$. Show that there is a C^∞ function u in $\mathbb{R}^{2n}$ which satisfies the (overdetermined) system of equations $\partial u/\partial\bar{z}_j = f_j$, $j = 1, \ldots, n$, in the whole of $\mathbb{R}^{2n}$ and which, furthermore, has compact support.

2.12 Let K be a compact subset of $\mathbb{C}^n$, $n > 1$, whose complement is connected and let h be a holomorphic function of $z = (z_1, \ldots, z_n)$ is $\mathbb{C}^n\backslash K$. Derive from the result of the preceding problem that there is an entire function H in $\mathbb{C}^n$ which is equal to h in $\mathbb{C}^n\backslash K$.

2.13 Let $P(D)$ be a differential operator with constant coefficients in $\mathbb{R}^n$. Prove that every distribution u such that $P(D)u = 0$ in $\mathbb{R}^n$, with its support contained in a half-space whose boundary (which is a hyperplane in $\mathbb{R}^n$) is noncharacteristic with respect to $P(D)$, must vanish identically in $\mathbb{R}^n$.

2.14 Construct a fundamental solution for the operator

$$-\frac{\partial^2}{\partial x^2} + 2\frac{\partial^2}{\partial x\partial y} + y^2\frac{\partial^2}{\partial z^2}$$

2.15 Prove the following generalization of the Harnack theorem: Let $P(D)$ be a differential operator with constant coefficients in $\mathbb{R}^n$, having a fundamental solution which is C^∞ in the complement of the origin. Let N_Ω be the space of distribution solutions of the homogeneous equation $P(D)h = 0$ on an open subset $\Omega \subset \mathbb{R}^n(N_\Omega) \subset C^\infty(\Omega)$ since P is hypoelliptic here. The following topologies on N_Ω are identical:
  (i)   the C^∞ topology (uniform convergence of the functions and all their derivatives on every compact subset of Ω),
  (ii)   the C^0 topology,
  (iii) the topology induced by $\mathscr{D}'(\Omega)$ (the functions $f_\alpha \in N_\Omega$ converge if the integrals $\int f_\alpha \varphi$ converge for any test function $\varphi \in \mathscr{D}(\Omega)$, uniformly on bounded subsets of $\mathscr{D}(\Omega)$; it suffices to consider sequences f_α and thus require the convergence of integrals for each individual φ).

2.16 Suppose that a C^1 complex function w defined in a plane region G of diameter $R < R_0$ satisfies the inequality $| \partial w / \partial \bar z | \leq K | w |$ and w is continuous up to the boundary Γ of G. Show that $\sup_G | w | \leq k \max_\Gamma | w |$, where R_0 and $k > 0$ depend only on K.

2.17 Prove the Carleman unique continuation theorem: If a C^1 complex function w defined on a plane region G containing the origin satisfies the inequality $| \partial w / \partial \bar z | \leq K | w |$ and $w(z) = \mathcal{O}(| z |^N)$ as z tends to 0 for every $N > 0$, then w is identically zero.

2.18 Suppose that $s > s'$, $s \geq -n/2$, are real numbers, Ω is an open subset of $\mathbb{R}^n$, $u \in C_0^\infty(\Omega)$. Prove that for every $\varepsilon > 0$ there exists $\eta > 0$ such that diam $\Omega \leq \eta$ implies the inequality $\| u \|_{s'} \leq \varepsilon \| u \|_s$ where $\| . \|_s$ denotes the norm in the Sobolev space $H^s(\mathbb{R}^n)$. Give an example showing that the hypothesis $s \geq -n/2$ is important.

2.19 Consider the Beltrami equation $\partial w / \partial \bar z = \mu(z) \, \partial w / \partial z$ with a bounded measurable function μ, $| \mu(z) | \leq k < 1$, in a domain $G \subset \{ | z | < 1 \}$. Prove that $w(z) = f(\varphi(z))$, where φ is a homeomorphism of the unit disk, $\varphi(0) = 0$, $\varphi(1) = 1$, φ, φ^{-1} Hölder continuous, f is analytic on $\varphi(G)$.

2.20 Solve the partial differential equation $pf + \sum_{j=1}^n x_j \, \partial f / \partial x_j = h$ on a star-shaped open set in $\mathbb{R}^n$, where $p \in \mathbb{R}$, $f = f(x)$, $h = h(x)$.

2.21 Show that there is no continuously differentiable solution of the Günter equation $u_x + u_y = w(x - y)$, where w is any nowhere differentiable function (for instance, the Weierstrass one).

2.22 Prove the Picone theorem: Let u be a solution of the equation $a(x,y)u_x + b(x,y)u_y = -u$ of class C^1 in the closed unit disk D in the plane. If $a(x,y)x + b(x,y) \, y > 0$ on the boundary of D, then u vanishes identically.

2.23 Give necessary and sufficient conditions on the function u_0 defined and C^∞ in the open interval $(-1,1)$, in order that there be a solution u to the initial value problem $u_t = iu_x$ for $x^2 + t^2 < 1$, $u(x,0) = u_0(x)$ for $x \in (-1,1)$.

2.24 Consider the equation in the plane $xu_y - yu_x = f(x^2 + y^2)$, where f is a C^∞ function such that $f(t) = 0$ if $t < 1$ or $t > 2$, $f(3/2) = 1$. Show that this equation has no distribution solution in $\mathbb{R}^2 \backslash \{0\}$.

2.25 Prove that there exist functions $u, a \in C^\infty(\mathbb{R}^2)$ such that $u_y + au_x = 0$ and supp $u = \{(x,y) : y \geq 0\} \supset$ supp a.

2.26 Explain why there are no solutions of the linear equation $u_x + u_y = u$ which pass through the line $x = t$, $y = t$, $u = 1$.

2.27 Solve the system $u_t = u_x + v_x$, $v_t = v_x$, $x \in \mathbb{R}$, $t \geq 0$, with the initial conditions $u(x,0) = u_0(x)$, $v(x,0) = v_0(x)$. Show that the solution operator $S(t)$ $: < u_0,v_0 > \mapsto <u(t),v(t)>$ is not continuous in $L^2(\mathbb{R})$.

2.28 Consider the following lower-order perturbation of the system in the preceding problem: $u_t = u_x + v_x$, $v_t = v_x + u$, $x \in \mathbb{R}$, $t \geq 0$, with the same initial conditions. Solve this system.

2.29 Let the operators L_1, L_2 be defined by $L_1u = au_x + bu_y + cu$, $L_2u = du_x + eu_y + fu$, where a, b, c, d, e, f are constants with $ae - bd \neq 0$. Prove that
(i)   the equations $L_1u = w_1$, $L_2u = w_2$ have a common solution, if $L_1w_2 = L_2w_1$;
(ii)   the general solution of $L_1L_2u = 0$ has the form $u = u_1 + u_2$, where $L_1u_1 = 0$, $L_2u_2 = 0$.

2.30 Let L be the linear differential operator $Lf(x) = \sum_\alpha a_\alpha(x)D^\alpha f(x)$, $x = (x_1, \ldots, x_n)$. Prove that $f(x) = S_1(x)A(S_2(x))$, where S_1, S_2 are specified functions and A is an arbitrary suitably differentiable function of $S_2(x)$, satisfies the linear partial differential equation $Lf(x) = 0$ in a domain, if and only if $S_1(x)$ $\exp(cS_2(x))$ is a particular solution of this equation in that domain, for every real c in some interval.
Example: $\exp(c(x + iy))$ and $A(x + iy)$ satisfy the Laplace equation and $R^{-1} \exp(c(R - vt))$ and $R^{-1}A(R - vt)$, $R^2 = x^2 + y^2 + z^2$, satisfy the wave equation for an arbitrary suitably differentiable function A.

Elliptic equations

2.31 Show that if the problem $\Delta u = 0$ for $|x| <, |y| < 1$, $u = 0$ for $|x| = 1$, $|\nabla u| = 1$ for $|y| = 1$, has any solution at all, then it has at least two solutions.

2.32 Find a nontrivial solution of the homogeneous mixed boundary value problem $\Delta u = 0$ for $x^2 + y^2 < 1$, $\partial u/\partial v + u = 0$ for $x^2 + y^2 = 1$.

2.33 Solve the mixed boundary value problem $\Delta u = 0$ in $(0,a) \times (0,b)$, $u(a,y) = u(x,0) = u(x,b) = 0$ for $x \in [0,a]$ and $y \in (0,b)$, $(\partial u/\partial x)(0,y) + \lambda u(0,y) = g(y)$ for $y \in (0,b)$, with a bounded continuous function g and a real λ.

2.34 Solve the Cauchy problem for the Laplace equation $\Delta u = 0$, $u = u(x,y,z)$, $u(x,y,0) = f(x,y)$, $u_z(x,y,0) = g(x,y)$ with analytic functions f and g.

2.35 Prove that for every smooth function u defined on the ball $B = \{x : |x - x_0| \le R\} \subset \mathbb{R}^3$, $u(x_0) = (4\pi R^2)^{-1} \int_{\partial B} u \, dS - (4\pi)^{-1} \int_B \Delta u(x)(|x - x_0|^{-1} - R^{-1}) \, dx$. Generalize to the n-dimensional case.

2.36 Show that if $u \in C^2$ then

$$-\Delta u(x_0) = \lim_{r \to 0} 2nr^{-2} \left(u(x_0) - \sigma_n^{-1} \int_{S^{n-1}} u(x_0 + ry) \, dy \right)$$

Draw as a corollary inequalities between the spherical averages and the values at the point x_0 for super- and subharmonic functions.

2.37 Derive the Poisson formula for the values of a harmonic function in a disk using the mean value property of harmonic functions and the conformal map that transforms the disk with a given point onto the unit disk and its center.

2.38 Show that the harmonicity of functions in $\mathbb{R}^n$, $n \ge 3$, is preserved only by the similarity transformations.

2.39 Let $\Delta u = f$ in $G \subset \mathbb{R}^n$. Show that the Kelvin transform of u, defined by $v(x) = |x|^{2-n}u(x/|x|^2)$ for $x/|x|^2 \in G$, satisfies $\Delta v(x) = |x|^{-n-2} f(x/|x|^2)$.

2.40 Let u be a harmonic function in an open bounded set $G \subset \mathbb{R}^n$. Prove the interior gradient bound

$$|\nabla u(x_0)| \le n(\text{dist}(x_0, \partial G))^{-1} (\sup u - u(x_0))$$

If $u \ge 0$ in G infer that $|\nabla u(x_0)| \le n(\text{dist}(x_0, \partial G))^{-1}u(x_0)$.

2.41 Prove that for every harmonic function u defined in a convex open set $D \subset \mathbb{R}^n$ and satisfying the condition $u = g$ on ∂D for some function g twice differentiable in $D \cup \partial D$ the following estimate holds: $|\nabla u| \le \max_{\partial D} |\nabla g| + (\text{diam } D/2) \sup_D |\Delta g|$.

2.42 Solution of the boundary value problem for the Laplace equation by the Schwarz alternating procedure. Suppose that $\Omega \subset \mathbb{R}^n$ is the union of two overlapping domains G, G' with piecewise smooth boundaries. Assume further that the boundaries Γ, Γ' of G, G' intersect at an angle different from zero and not at vertices or at along edges of Γ, $\Gamma'(\Omega$ may be a multiply connected domain, of course). Assuming solvability of the boundary value problem for arbitrary continuous boundary data for G, G', prove that the solution of the Dirichlet problem for $\Omega(u = f$ on $\partial\Omega)$ can be obtained as the limit of functions defined below. Let $\Gamma = a \cup b$, $\Gamma' = a' \cup b'$, so that $b = \Gamma \cap \overline{G'}$, $b' = \Gamma' \cap \overline{G}$, hence $\partial\Omega = \cup a'$. Solve the boundary value problem for G with given data f on a and $(\sup_{\partial\Omega} f)$ on b obtaining a function u_1. Then solve the boundary value problem for G' with f on a' and u_1 on b' obtaining the function u_1'. Then u_2 is the solution of the boundary value problem on G with the data f on a and u_1' on b and so on. Prove that the functions u_n in G and u_n' in G' converge uniformly to harmonic functions u, u' which are identical in the intersection of G and G'.

2.43 Prove the Kellogg theorem: Let U be a bounded open set in $\mathbb{R}^n$, $n \geq 2$, such that the Dirichlet problem for the Laplacian is solvable in U, and let a continuous real function f on U satisfy the restricted mean value property; that is, for every $a \in U$ there is $r = r(a)$ such that $\{x : |x - a| \leq r\} \subset U$ and

$$f(a) = \frac{\int_{\{x : |x-a| = r\}} f(x)\, dS(x)}{\int_{\{x : |x - a| = r\}} dS(x)}$$

where dS is the Lebesgue measure on the $(n - 1)$-dimensional sphere. Prove that f is a harmonic function on U.

2.44 f is a continuous real function on $\mathbb{R}^n$ whose mean

$$M(x,r) = \frac{\int_{\{y : |y - x| \leq r\}} f(y)\, dy}{\int_{\{y : |y - x| \leq r\}} dy}, \qquad r > 0,\ x \in \mathbb{R}^n$$

satisfies the condition $\forall x \in \mathbb{R}^n \lim_{r \to 0} r^{-2}(M(x,r) - f(x)) = 0$. Prove that f is a harmonic function. Is the continuity assumption essential?

2.45 Show that the surfaces $(x^2 + y^2)^2 - 2a^2(x^2 - y^2) + a^4 = c$ can form a family of equipotential surfaces, and find the general form of the corresponding potential function.

2.46 Consider the family of surfaces Σ defined by the equations

$$\frac{x^2}{a^2 + s} + \frac{y^2}{b^2 + s} + \frac{z^2}{c^2 + s} = 1$$

where $a > b > c$. If $-c^2 < s < \infty$, then these surfaces are ellipsoids, if $-b^2 < s < -c^2$ or $-a^2 < s < -b^2$ they are hyperboloids. If $s = \infty$, this is a sphere of infinite radius, if $s = -c^2$ the surface degenerates to an elliptic disk in the xy plane. Show that these surfaces can be equipotential and this potential equals $V = A \int_0^s ((a^2 + s)(b^2 + s)(c^2 + s))^{-1/2} ds + B$, where A and B are determined by the conditions at ∞ and on Σ.

2.47 Show that a necessary and sufficient condition that a family of surfaces $F(x,y,z) = c$, where F has continuous partial derivatives of the second order, may be the equipotential surfaces of a Newtonian potential (solutions of the Laplace equation) is the $\Delta F/|\nabla F|^2$ is a function $\varphi(F)$ of F only. Show that if this condition is fulfilled, the potential is $U = c_1 \int^F \exp(-\int^t \varphi) dt + c_2$ with some constants c_1 and c_2.

2.48 The Liapunov example. Show that the Hölder condition satisfied by the density of the double layer potential is not sufficient to guarantee finiteness of the normal derivative of this potential.

2.49 Let $u_1, \ldots, u_k$ denote the harmonic measures of the boundary components of a domain G of connectivity k. Prove that $u_1 + \cdots + u_k \equiv 1$, but that no nontrivial linear relationship can exist among any smaller number of these functions.

2.50 Does there exist a discontinuous solution of the Laplace equation $\Delta u = 0$?

2.51 Prove that if $\Delta u = 0$ in a bounded open set $G \subset \mathbb{R}^n$ and $u = \partial u/\partial v = 0$ on an open, smooth portion of the boundary ∂G, then u is identically zero.

2.52 Explain the following "paradox": The function $u(x,y) = (x^2 + y^2 - 2x)/(x^2 + y^2)$ is harmonic and vanishes on the circle $x^2 + y^2 - 2x = 0$. Does it mean that the Dirichlet problem in this circle with the homogeneous boundary data has a nonunique solution?

2.53 Consider the function

$$u(x,y) = \exp\left(\frac{x}{x^2 + (y-1)^2}\right) \cos\left(\frac{y-1}{x^2 + (y-1)^2}\right)$$

which is harmonic in the unit disk in the plane. Prove that its trace on the unit circle is not a distribution (but only an analytic functional).

2.54 Construct an example of a positive harmonic function in the unit disk such that its radial limits are infinite on a subset of measure zero of the unit circle. This would show sharpness of the Fatou theorem.

2.55 Prove that a function u harmonic in the unit disk in the plane has the limit in the mean on the unit circle if and only if
(i) $\int_{|x|=r} u^2(x) ds$, $0 < r < 1$, is bounded.
This condition is equivalent to
(ii) $\int_{|x|<1} |\nabla u(x)|^2 (1 - |x|) dx < \infty$, or

(iii) $\int_0^{2\pi} \int_{\Omega(\vartheta)} |\nabla u|^2 \, dx \, d\vartheta < \infty$ (finiteness of the Luzin area integral), where $\Omega(\vartheta)$ is the region obtained by the rotation by angle ϑ, $0 \le \vartheta < 2\pi$, of the region bounded by two tangents to the circle $\{|x| = 1/2\}$ and passing through $(1,0)$ and the larger arc of this circle.

2.56 The A. V. Bitsadze example. Consider the boundary value problem in the unit ball in $\mathbb{R}^3$ for the Laplace equation $\Delta u = 0$ with the condition $\partial u/\partial x = f$ on the unit sphere. Show that this problem has an infinite-dimensional kernel. Give a supplementary condition on a part of the boundary which would guarantee the unique solvability.

2.57 Prove the Bôcher theorem: If u is a positive harmonic function in the punctured ball $\{x : 0 < |x| < 1\} \subset \mathbb{R}^3$, then there exist a constant $b \ge 0$ and a function h harmonic in the unit ball such that $u(x) = b/|x| + h(x)$.

2.58 Prove that for every positive harmonic function u in $\{x : |x| > 1\} \subset \mathbb{R}^3$ there exists $\lim_{x \to \infty} u(x)$.

2.59 Prove the Zaremba theorem: If u is a harmonic function in a domain $D \subset \mathbb{R}^2$ and u vanishes on the boundary of D except for a finite number of points A_j, $j = 1, \ldots, k$, where $u(z)/\log |z - A_j|$ tends to zero as $z \to A_j$, then u vanishes identically in D.

2.60 Show that any function harmonic in the region bounded by two concentric spheres is the sum of a function which is harmonic in the interior of the outer sphere and a function which is harmonic outside the inner sphere.

2.61 Consider a harmonic function u in the unit disk D in $\mathbb{R}^2$, continuous up to the boundary. Show that $u \in C^1(\overline{D})$ if and only if $u_{|\partial D}$ and its Hilbert transform are in $C^1(\partial D)$. Generalize to domains in $\mathbb{R}^n$ with smooth boundaries.

2.62 For $f \in C^\alpha(\mathbb{T}^1)$, $0 < \alpha < 1$, consider the Poisson integral u of f in the unit disk D in $\mathbb{R}^2$. Show that
(i) $|\nabla u| \le C(1 - r)^{\alpha-1}$ and $u \in C^\alpha(\overline{D})$.
(ii)   The same result as in (i) for the harmonic conjugate of u.
(iii)   Conclude that the Hilbert transformation acts in $C^\alpha(\mathbb{T}^1)$ for $0 < \alpha < 1$.

2.63 The Lebesgue example of a domain in $\mathbb{R}^3$ homeomorphic to a ball, with its boundary smooth except for one point and such that the Dirichlet problem for the Laplacian has no classical solution for certain continuous boundary data. Let $v(x) = (x_1^2 + x_2^2 + x_3^2)^{1/2} + x_1 \log((x_1^2 + x_2^2 + x_3^2)^{1/2} - x_1)$. Consider the domain bounded by the level set $\{v = 1\}$ (which is tangent to the x_1-axis at the origin) and the boundary condition equal to v off the origin and -1 at the origin. Show that the Dirichlet problem with this continuous boundary condition does not have a classical solution.

2.64 Prove the Liouville theorem: If u is a harmonic function in $\mathbb{R}^n$ and $|u(x)| \leq C(1 + |x|^m)$, then u is a (harmonic) polynomial of degree less or equal to m.

Remark: The same assertion remains valid under the weaker hypothesis $u(x) \geq -c(1 + |x|^m)$ for some $c > 0$.

2.65 Prove the following version of the Phragmén-Lindelöf theorem: Let u be a harmonic function of x and y in the upper half-plane $\{y > 0\}$, and suppose that $u = 0$ at all finite points of the x-axis. If in addition u fulfills the requirement $\lim \sup_{r \to \infty} u/r = 0$, then it must vanish identically.

2.66 Prove the Phragmén-Lindelöf theorem for harmonic functions: If u is a harmonic function in the strip $\{x : 0 \leq x_n \leq \pi h\}$, $h > 0$, and $u \leq 0$ for $x_n = 0$ and $x_n = \pi h$, and moreover $u(x) \leq c \, \exp(h^{-1}(a_1 |x_1| + \cdots + a_{n-1} |x_{n-1}|))$ for some $c > 0$ and $\sum_{j=1}^{n-1} a_j^2 \leq 1$, then $u \leq 0$ everywhere.

2.67 Solve the Poisson equation $\Delta u = A$, $A \in \mathbb{R}$, in the disk of radius $r = a$ with the boundary condition $\partial u/\partial v \,|_{r=a} = B$, choosing B in such a way that this problem has a solution.

2.68 Show that the Poisson equation $\Delta u = (x_2^2 - x_1^2)(2 |x|^2)^{-1} ((n + 2) \times (-\log |x|)^{-1/2} + 2(-\log |x|)^{-3/2})$ with continuous right-hand side has no classical solution in the ball $\{|x| < 1/2\} \subset \mathbb{R}^n$.

2.69 Prove that for a subharmonic function u, $M(r) = \sup_{|x| \leq r} u(x)$ is a convex function of $\log r$ if $n = 2$ and of r^{2-n} if $n > 2$.

2.70 Prove the Hadamard three-circle theorem: Let u be a subharmonic function in the annulus $\{r_1 < |x| < r_2\}$ in the plane and $m(r) = \max_{|x| = r} u(x)$. Then $m(r)$ satisfies the inequality

$$m(r) \leq \frac{m(r_1) \log(r_2/r) + m(r_2) \log(r/r_1)}{\log (r_2/r_1)}$$

for all $r_1 < r < r_2$. Generalize to the n-dimensional case.

2.71 Assume that $u \in C^2 (\Omega)$ is strictly positive and $w = u \exp(\sum_{j=1}^n a_j x_j)$ is subharmonic for every choice of constants $a_1, \ldots, a_n$. Show that $\log u$ is a subharmonic function in Ω.

2.72 For $f \in C^2(\mathbb{R})$, $f'(t) > 0$, and $v \in C^2(\Omega)$ let the function $f(\sum_{j=1}^{n} a_j x_j + v)$ be subharmonic for every choice of constants $a_1, \ldots, a_n$. Show that f is a convex function.

2.73 Show that the equation $\Delta u = -1$ in $\{ |x| < 1, |y| < 1 \}$, with the boundary conditions $u = 0$ if $|x| = 1$ and $u_x = u_y$ if $|y| = 1$, has at most one solution.

2.74 Suppose that u solves the overdetermined problem (of Pompeiu type) $\Delta u = -1$ in a bounded smooth domain $\Omega \subset \mathbb{R}^n$, $u = 0$ and $\partial u/\partial v = c = \text{const}$ on $\partial \Omega$. Prove that Ω is a ball and $u(x) = (n^2 c^2 - |x|^2)/(2n)$.

2.75 Show that if $|\Delta u| \leq M$ and $u \geq 0$ in the disk $x^2 + y^2 < R^2$, then $(R - r)(R + r)^{-1}(u(0,0) - \frac{1}{4}M (R^2 + (R + r)^2)) \leq u(x,y) \leq (R + r) \times (R - r)^{-1}(u(0,0) + \frac{1}{4}M (R^2 + (R - r)^2))$.

2.76 Let $u \in C^2(G)$ satisfy the Poisson equation $\Delta u = f$ in G. Then

$$\sup_G \text{dist}(x, \partial G) | \nabla u(x) | \leq C(\sup_G | u | + \sup_G \text{dist}(x, \partial G)^2 | f(x) |)$$

and for all x, y in G, $x \neq y$,

$$\frac{\min (\text{dist}(x, \partial G), \text{dist} (y, \partial G))^2 | \nabla u(x) - \nabla u(y) |}{| x - y |} \leq C(\sup_G | u | $$

$$+ \sup_G \text{dist}(x, \partial G)^2 | f(x) |) \left(\log \frac{\min(\text{dist}(x, \partial G), \text{dist}(y, \partial G))}{| x - y |} + 1 \right)$$

2.77 Prove that a radial solution of the Poisson equation $\Delta u = f$ in the ball $B(0, r_0) \subset \mathbb{R}^n$, where f is a bounded measurable function, has locally bounded second derivatives $\partial^2 u/\partial x_i \, \partial x_j$, $i, j = 1, \ldots, n$.
Remark: This result is not true for general (nonradial) functions; see the next problem.

2.78 Give examples of continuously differentiable functions u in $\mathbb{R}^2$ such that
(i) u_{xx} and u_{yy} are continuous but the mixed derivative u_{xy} is not continuous.
(ii) Δu is a continuous function but u_{xx} is not continuous.

2.79 It is well known that solutions of the Poisson equation $\Delta u = f$ with Hölder continuous functions f are twice continuously differentiable. Show that the assumption on f can be weakened to the Dini condition: If $|f(x) - f(x_0)| \leq \varepsilon(|x - x_0|)$ for every x in a neighborhood of x_0 and $\int_0^1 \varepsilon(r)r^{-1} \, dr < \infty$, then u is C^2. Prove it and show that this condition is also necessary: given an increasing positive function ε such that $\int_0^1 \varepsilon(r)r^{-1} \, dr = \infty$, there exists a continuous function f with compact support satisfying $|f(x)| \leq \varepsilon(|x|)$ with its Newtonian potential u not having second derivatives at the origin.

2.80 How can one explain the fact that, given a bounded open subset Ω of $\mathbb{R}^n$, for suitable values of $\lambda > 0$ the homogeneous equation $(\Delta + \lambda)u = 0$ has nontrivial solutions in $H_0^1(\Omega)$, although no linear partial differential equation with constant coefficients can have compactly supported distribution solutions different from zero?

Remark: This last assertion can be established by means of the Fourier transform, applying the Paley-Wiener theorem.

2.81 Let G be the square $\{\, |x| < \pi/2, \ |y| < \pi/2 \,\}$. Show that any solution of the problem $\Delta u + u = 0$ in G, $u = g(x,y)$ on ∂G is unique. Is the same result true for the square $\{\, |x| < \pi 2^{-1/2}, \ |y| < \pi 2^{-1/2} \,\}$?

2.82 Let B be the ball of radius π centered at the origin in $\mathbb{R}^3$. Show that a solution of the equation $\Delta u + u = w(x)$ with vanishing boundary values can exist only if $\int_B w(x)\, |x|^{-1} \sin |x| \, dx = 0$.

2.83 Prove the mean value theorem for C^2 solutions of the equation $-\Delta u + k^2 u = 0$ in $\mathbb{R}^n$ (the so-called metaharmonic functions) $(\sigma_n r^{n-1})^{-1} \int_{S(x_0,r)} u \, dS = C(n,r,k)u(x_0)$, where

$$ C(n,r,k) = 2\pi^{-1/2}\Gamma\left(\frac{n}{2}\right)\Gamma\left(\frac{n-1}{2}\right)^{-1} \int_0^1 \cosh(krt)(1-t^2)^{(n-3)/2}\, dt $$

Moreover, for $n = 3$ show that $u(x_0) = k^3(4\pi(kr\cosh(kr) - \sinh(kr)))^{-1} \times \int_{B(x_0,r)} u \, dx$.

2.84 Prove that each C^4 function of three variables such that its spherical mean $(4\pi R^2)^{-1} \int_{|x-x_0|\,=\,R} u \, dS$ is equal to $u(x_0)g(R)$ for some function g is an eigenfunction of the Laplacian $\Delta u + cu = 0$, where c is a constant and $g(R) = \sin(Rc^{1/2})/(Rc^{1/2})$.

2.85 Denote by u a complex solution of the reduced wave equation $\Delta u + k^2 u = 0$ in some unbounded planar region, and assume that it satisfies the radiation condition $\lim_{r\to\infty} r^{1/2}(\partial u/\partial r - iku) = 0$, $\lim \sup_{r\to\infty} r^{1/2}u < \infty$. Prove the Rellich theorem: If $\lim_{R\to\infty} \int_{r=R} |u|^2 r \, d\vartheta = 0$, then u vanishes identically. Establish the identity

$$ u(\xi,\eta) = \frac{1}{4}i \int_{\partial D}\left(\frac{u(x,y)\partial H_0^{(1)}(kr)}{\partial v} - H_0^{(1)}(kr)\frac{\partial u(x,y)}{\partial v}\right) ds $$

for arbitrary solution of this equation, satisfying the radiation condition in the exterior of any simple closed curve ∂D. Here $H_0^{(1)}(kr) = \pi^{-1} \int_{-\pi/2+i\infty}^{\pi/2-i\infty} \exp(ikr\cos z)\, dz$ is the first Hankel function.

2.86 Show that if G is a bounded domain in $\mathbb{R}^3$, $G_1 = \mathbb{R}^3\backslash G$, $\mu \in L^1(G) \cap L^\infty(G)$ with its support in G, then

$$V(x) = \int_G \exp(ik \mid x - y \mid) \mid x - y \mid^{-1} \mu(y) \, dy$$

$$\overline{V}(x) = \int_G \exp(-ik \mid x - y \mid) \mid x - y \mid^{-1} \mu(y) \, dy$$

$$V_0(x) = \int_G \exp(-k \mid x - y \mid) \mid x - y \mid^{-1} \mu(y) \, dy$$

satisfy
(i) $V, \overline{V}, V_0 \in C^1(\mathbb{R}^3) \cap C^\infty(G_1)$;
(ii) V and $\overline{V}$ satisfy in G $\Delta u + k^2 u = 0$, V_0 satisfies $\Delta u - k^2 u = 0$, $V_0(x) \rightarrow$ 0 as $\mid x \mid \rightarrow \infty$;
(iii) V and $\overline{V}$ satisfy the Sommerfeld condition $u(x) = \mathcal{O}(\mid x \mid^{-1})$, $\partial u(x)/\partial \mid x \mid \mp iku(x) = \mathcal{O}(\mid x \mid^{-1})$ as $\mid x \mid \rightarrow \infty$.

2.87 Consider the partial differential equation $u_{xx} + u_{yy} \pm a^2(x,y)u = 0$. Show that by a change of variables $(x,y) \mapsto (s,t)$ this equation may be reduced to the equation $u_{ss} + u_{tt} \pm u = 0$ if and only if $\log a$ is a harmonic function. Find explicit solutions of $u_{xx} + u_{yy} \pm \exp(2kxy)u = 0$.

2.88 Find some solutions of the equation $(x^2 + y^2)(u_{xx} + u_{yy}) - au = 0$ which are not regular at the origin.

2.89 Sketch the nodal lines of the eigenfunctions in the unit square membrane corresponding to the eigenvalues $5\pi^2$, $10\pi^2$, $13\pi^2$, $17\pi^2$.

2.90 Given a membrane D (a plane region) of area A, show that when $\varepsilon > 0$ is small enough, each square inside D of area ε must contain nodes of every eigenfunction u_n, $\Delta u_n + \lambda_n u_n = 0$, such that $n > \pi A/\varepsilon$.

2.91 Prove that under the Dirichlet boundary condition $u = 0$, the nth eigenvalue of an elliptic selfadjoint operator in a domain G never exceeds the nth eigenvalue of the same operator in any subdomain of G.

2.92 Let λ_n be the nth eigenvalue of the elliptic selfadjoint operator in the domain G under the boundary condition $u = 0$, and let μ_n be the nth eigenvalue for the condition $\partial u/\partial v + bu = 0$ on part of the boundary of G and $u = 0$ on the remaining part of this boundary. Show that $\mu_n \leq \lambda_n$.

2.93 If, in the above problem, in the boundary condition $\partial u/\partial v + bu = 0$ the function b is either increased or diminished at every point, then each individual eigenvalue can change only in the same sense.

2.94 If, in the elliptic selfadjoint differential equation $Lu + \lambda cu = 0$, the coefficient c varies at every point in the same sense, then, for every boundary condition, the nth eigenvalue changes in the opposite sense.

2.95 Prove the Faber-Krahn inequality: $\lambda_1 \geq \pi k_0^2/A$ for the first eigenvalue of $-\Delta$ on a plane region of area A, where k_0 is the smallest positive root of the Bessel function J_0.

Remark: The three-dimensional version of this inequality reads: $\lambda_1 \geq \pi^2(4\pi/3V)^{2/3}$, where V is the volume of the considered domain.

2.96 Let $\lambda_1 < \lambda_2 \leq \cdots$ and $0 = \mu_1 < \mu_2 \leq \cdots$ denote the eigenvalues of the Dirichlet and Neumann problems respectively for the Laplacian in a domain $D \subset \mathbb{R}^n$. It is an immediate consequence of the variational formulation of the eigenvalue problems that $\mu_k \leq \lambda_k$ for $k = 1,2, \ldots$. Prove that if D is any nonempty bounded convex domain in $\mathbb{R}^n$, then $\mu_{k+n} \leq \lambda_k$ for $k = 1,2, \ldots$.

2.97 The difference $K(x,y) = N(x,y) - G(x,y)$ between the Neumann function and the Green function of the linear elliptic partial differential equation $\Delta u = Pu$ (where $P(x) \geq 0$ in a region D) is called the kernel function. Show that K has the reproducing property $u(y) = \langle u(x),K(x,y) \rangle =: \int (\nabla u \cdot \nabla K + PuK)\, dx$. Prove that $K(x,y)$ is an extremal function for the minimum problem $\min\langle u,u \rangle / u(y)^2 = K(y,y)^{-1}$ where $u \neq 0$ is subject to the restriction $\Delta u = Pu$. Show that the reproducing property is a consequence of the extremal property of K.

2.98 Show formally that the Green function G is a Lagrange multiplier for the minimum problem in the preceding exercise when it is treated as a Bolza problem in the calculus of variations with the equation $\Delta u = Pu$ imposed as a side condition.

2.99 Give an example of a harmonic function in the unit disk such that its boundary values on the unit circle are continuous but its Dirichlet integral is infinite.

2.100 Show that not every minimizing sequence of continuous, piecewise smooth functions for the Dirichlet integral tends to the solution of the corresponding Dirichlet problem.

2.101 Show that, whatever the complex number λ, the skew linear form $a(u,v) = \int_\Omega (\lambda u \bar{v} + (\Delta u)(\Delta \bar{v}))\, dx$ is not coercive on $H^2(\Omega)$, where Ω is a bounded open set in $\mathbb{R}^n$, $n \geq 2$. Describe the boundary value problem formally associated with the equation $a(u,v) = \int_\Omega f\bar{v}\, dx$ for all $v \in H^2(\Omega)$, where $f \in L^2(\Omega)$.

2.102 Consider the open set $G = \{(x,y,z) : z > (1 + \frac{1}{2}(x^2 + y^2))^{1/2}\}$ in $\mathbb{R}^3$. Prove that

(i)   the skew linear form $\int_G \nabla u \cdot \nabla \bar{v}\, dx$ is not coercive on $H_0^1(G)$,

(ii)   there are no nontrivial harmonic functions in G belonging to $H_0^1(G)$,

(iii)   there is a nontrivial harmonic function in G, continuous in $\bar{G}$, vanishing on the boundary of G (the points at infinity are not included in this boundary).

2.103 Let $w = z \exp(-(\log(1/|z|))^{1/2})$. By considering the relation $\partial w/\partial \bar{z} = v(z) \, \partial w/\partial z$, where $\partial/\partial z = \frac{1}{2}(\partial/\partial x - i\partial/\partial y)$, $\partial/\partial \bar{z} = \frac{1}{2}(\partial/\partial x + i\,\partial/\partial y)$, show that $u = \mathrm{Re}\, w = x \exp(-(\log(1/r))^{1/2})$, $r = |z|$, satisfies a uniformly elliptic equation of divergence form $(au_x + bu_y)_x + (bu_x + cu_y)_y = 0$, in which the coefficients $a \to 1$, $b \to 0$, $c \to 1$ at the origin and are regular in $0 < r < 1$. Observe that $u(0,0) = 0$, $u(x,y) > 0$ for $x > 0$ and $u_x(x,0) > 0$ for $x > 0$ and $u_x(0,0) = 0$. Compare with the boundary point lemma of Hopf.

2.104 Consider the equation $u_{xx} + y^2 u_{yy} = 0$ in the rectangle $R = \{0 < x < \pi,\ 0 < y < Y\}$, such that $u \in C(R) \cap C^2(R)$ satisfies the boundary conditions $u(0,y) = u(\pi,y) = 0$, $0 \le y \le Y$. Show that the Dirichlet problem for this equation is not solvable for general boundary data.

2.105 Show that the Dirichlet problem in the unit disk for the Bitsadze (elliptic but not strongly elliptic) equation $\frac{1}{4}(\partial^2/\partial \bar{z}^2)u = u_{xx} + 2iu_{xy} - u_{yy} = 0$ with homogeneous boundary conditions has infinitely many solutions.

2.106 Show that the Dirichlet problem for the system

$$\frac{\partial^2 u_1}{\partial x^2} - \frac{\partial^2 u_1}{\partial y^2} - 2\frac{\partial^2 u_2}{\partial x \partial y} = 0, \qquad 2\frac{\partial^2 u_1}{\partial x \partial y} + \frac{\partial^2 u_2}{\partial x^2} - \frac{\partial^2 u_2}{\partial y^2} = 0$$

is neither of Fredholm nor of Noether solvability type.

2.107 Verify that $u_1 + iu_2 = \alpha(1 - z\bar{z}) + z\bar{z}(2\pi i)^{-1} \int_C f(t)(t - z)^{-1}\, dt - (2\pi i)^{-1} \int_C \overline{f(t)}(t - z)^{-1}\, dt$, where α is an arbitrary constant, solves the elliptic system $x\,\Delta u_1 + y\,\Delta u_2 - 2(u_{1x} + u_{2y}) = 0$, $y\,\Delta u_1 - x\,\Delta u_2 + 2(u_{2x} - u_{1y}) = 0$ with the boundary condition $u_1 + iu_2 = f$, f a Hölder function on the unit circle C.

2.108 If the transformation of coordinates $\xi = \xi(x,y)$, $\eta = \eta(x,y)$ brings a linear elliptic equation $au_{xx} + 2bu_{xy} + cu_{yy} + du_x + eu_y + fu = 0$ with analytic coefficients a, b, c, d, e, and f into canonical form, prove that the quantity $\Gamma = (\xi - \xi_0)^2 + (\eta - \eta_0)^2$ satisfies the first-order partial differential equation

$$a\Gamma_x^2 + 2\, b\Gamma_x\Gamma_y + c\Gamma_y^2 = (4ac - b^2)^{1/2}\frac{D(\xi,\eta)}{D(x,y)}$$

Show that the fundamental solution of the above elliptic equation can be found in the form $A(x,y;x_0,y_0) \log \Gamma + B(x,y;x_0,y_0)$ with coefficients $A \ne 0$ and B that are regular at the point (x_0,y_0). Use this result to establish that all solutions of this equation are real analytic functions of the variables x and y.

2.109 Let $Lu = au_{xx} + 2bu_{xy} + cu_{yy} = 0$ in an exterior domain $\{r > r_0\}$, L being uniformly elliptic. Prove that if u is bounded on one side then u has a limit (possibly infinite) as $r \to \infty$.

2.110 Construct for arbitrary integer $m \geq 1$ distributions E_k in $\mathbb{R}^n$, $k = 1, \ldots, m$, such that $\Delta E_{k+1} = E_k$, $k = 1, \ldots, m - 1$, $\Delta E_1 = \delta$, E_k is a linear combination of the functions r^{2k-n}, $r^{2k-n} \log r$, $r = |x|$, $k = 1, \ldots, m$. Show that, for a suitable choice of the constants $C_{m,n}$, the following is true:
(i)   if $2m - n$ is not an even positive integer, $E_m = C_{m,n} r^{2m-n}$ is a fundamental solution of Δ^m,
(ii)   if $2m - n \geq 0$ is an even integer, $E_m = C_{m,n} r^{2m-n} \log r$ is a fundamental solution of Δ^m.
Derive from this that, if n is odd and $\lambda > 0$, a fundamental solution of $\lambda - \Delta$ is given by $-r^{2-n} \sum_{m=0}^{\infty} C_{m+1,n} \lambda^m r^{2m}$, whereas if n is even $-r^{2-n} \sum_{m=0}^{p-2} C_{m+1,n} \lambda^m r^{2m} - r^{2-n} (\log r) \sum_{m=p+1}^{\infty} C_{m+1,n} \lambda^m r^{2m}$, where $p = n/2$. Compare this with $(4\pi r)^{-1} \exp(-\lambda^{1/2} r)$ when $n = 3$.

2.111 Consider an elliptic differential operator $Lu = \sum_{j,k=1}^{n} a_{jk}(y) \, \partial^2 u / \partial y_j \, \partial y_k$. Let A_{jk} denote the quotient of the cofactor of a_{jk} and $A(y) = \det(a_{jk}(y))$, and $\sigma_n = 2\pi^{n/2}/\Gamma(n/2)$ is the area of the sphere S^{n-1}. Verify that the E. E. Levi function $H(x,y) = ((n-2)\sigma_n A(y)^{1/2})^{-1} (\sum_{j,k=1}^{n} A_{jk}(y)(x_j - y_j)(x_k - y_k))^{1-n/2}$ for $n > 2$ and $H(x,y) = (2\pi A(y)^{1/2})^{-1} \log (\sum_{j,k=1}^{2} A_{jk}(y)(x_j - y_j)(x_k - y_k))^{-1/2}$ for $n = 2$ satisfies the identity $\sum_{j,k=1}^{n} a_{jk}(y) \, \partial^2 H / \partial x_j \partial x_k = 0$. Moreover, $H = \mathbb{O}(r^{2-n})$, $\nabla H = \mathbb{O}(r^{1-n})$, $D^2 H = \mathbb{O}(r^n)$, $n > 2$.
Remark: If L has constant coefficients, H is a fundamental solution of L.

2.112 Study solutions of the equation

$$(\sin \theta)^{-1} \frac{\partial(\sin \theta \, \partial u / \partial \theta)}{\partial \theta} + (\sin \theta)^{-2} \frac{\partial^2 u}{\partial \varphi^2} + n(n+1)u = 0$$

2.113 Find a fundamental solution of the elliptic equation $u_{zz} + u_{rr} + \lambda r^{-1} u_r = 0$. Discuss the limiting case in which λ and r both become infinite while their ratio remains bounded.

2.114 Find a fundamental solution for the system of equations of elastic equilibrium

$$\mu \, \Delta u_j + (\lambda + \mu) \frac{\partial(\operatorname{div} u)}{\partial x_j} = 0, \qquad j = 1,2,3$$

2.115 Show that for a differential operator P on $\mathbb{R}^n$, which is invariant under isometries, the following conditions are equivalent:

(i) $P = c\Delta$ for some $c > 0$.

(ii) $P \neq 0$ and there exists an open subset Ω of $\mathbb{R}^n$ such that for all smooth u
       $u(x) = \max u$ implies $Pu(x) \leq 0$.

(iii) For every open bounded subset Ω of $\mathbb{R}^n$ P satisfies the weak maximum
       principle on Ω.

(iv)   For every connected open subset Ω of $\mathbb{R}^n$ P satisfies the strong maximum
       principle on Ω.

2.116 Parabolic maximum principle and dissipativity. Let P be a partial differential operator of order m with continuous coefficients in an open subset Ω of $\mathbb{R}^n$ and $c \in \mathbb{R}$. Prove that the following conditions are equivalent:

(i) $m \leq 2$ and $P = \sum_{j,k} a_{jk}(x) \, \partial^2/\partial x_j \, \partial x_k + \sum_j a_j(x) \, \partial/\partial x_j + a_0(x)$ is an
       elliptic operator, $a_0(x) \leq c$.

(ii)   For every $u \in C^m(\Omega)$ and $x \in \Omega$ such that $u(x) = \max u \geq 0$, $Pu(x) \leq cu(x)$.

(iii) For every $u \in C^m(\Omega) \cap C_0(\Omega)$, $\lambda > 0$ with $\lambda c < 1$, $\|u\|_\infty \leq (1 - \lambda c)^{-1} \|u - \lambda Pu\|_\infty$.

(iv)   For every $u \in C^{m,1}(\Omega \times (0,T)) \cap C([0,T];C_0(\Omega))$, $\|u(T)\|_\infty \leq e^{cT} \|u(0)\|_\infty + \int_0^T e^{c(T-t)} \|u_t(t) - Pu(t)\|_\infty \, dt$.

(v) $a_0 = P1$ satisfies $a_0(x) \leq c$ and for every $u \in C^{m,1}(\Omega \times (0,T))$, $u_t \leq Pu$,
       $\max(u,0) \in C_0(\Omega \times (0,T])$ implies $u \leq 0$ on $\Omega \times (0,T)$.

(vi)   For every bounded open subset Ω_0 of Ω and $u \in C^{m,1}(\Omega_0 \times (0,T)) \cap C(\overline{\Omega}_0 \times [0,T])$, $u_t \leq Pu$ implies $\max u(T) = \max_{\Omega_0 \times 0 i \cup \partial\Omega_0 \times [0,T)} (e^{c(T-t)} \max(u(x,t),0))$.

2.117 Consider the equation

$$\sum_{i,j=1}^n \left(\delta_{ij} + \frac{g(r)x_i x_j}{r^2} \right) \frac{\partial^2 u}{\partial x_i \, \partial x_j} = 0$$

Show that it has a radially symmetric solution $u = u(r)$, $r \neq 0$, satisfying the ordinary differential equation

$$\frac{u''}{u'} = \frac{1 - n}{r(1 + g)}$$

(i)   If $n = 2$ and $g(r) = -2/(2 + \log r)$, show that the equation is uniformly
       elliptic in the disk of radius e^{-3}, with continuous coefficients at the origin,
       and has bounded solutions $a + b/\log r$ in the punctured disk that do not
       satisfy the estimate $\lim \sup_{x \to 0} u(x) \leq \sup_{|x| = r} u(x)$.

(ii)   If $n \geq 3$ and $g(r) = -(1 + (n - 1)\log r)^{-1}$ the equation is uniformly
       elliptic in the disk of radius e^{-1} and has continuous coefficients at the

origin. Show that the corresponding solution $u = u(r)$ satisfies the condition $u = \mathcal{O}(r^{2-n})$ as $r \to 0$ but does not satisfy the estimate in (i).

(iii) If $n \geq 3$, determine a function $g(r)$ such that the equation is elliptic and has a bounded solution $u = u(x)$ in the punctured disk, continuous at $r = 0$, that does not satisfy the estimate in (i).

2.118 Show that in the Gilbarg-Serrin theorem on the maximum principle for functions with possible singularities a smoothness condition for the coefficients of the considered elliptic operator is essential.

2.119 Show that for solutions of the three-dimensional biharmonic equation $\Delta^2 u = 0$, $(4\pi R^2)^{-1} \int_{|x-x_0|=R} u \, dS = u(x_0) + \frac{1}{6}R^2 \Delta u(x_0)$.

2.120 Suppose that Ω is a domain such that each line parallel to the x_1 axis intersects $\partial\Omega$ in at most two points. Prove that each solution of the biharmonic equation $\Delta^2 u = 0$ can be represented in the form $u = x_1 u_1 + u_2$, where u_1 and u_2 are harmonic functions.

2.121 Discuss analytic solutions of the equation $\Delta^2 u + a_1 \Delta u + a_2 u = 0$.

2.122 Prove the Pizetti formula: If φ is a real analytic function in a neighborhood of the origin, then

$$\sigma_n^{-1} \int_{S^{n-1}} \varphi(r\omega) \, d\omega = \Gamma\left(\frac{n}{2}\right) \sum_{k=0}^{\infty} \frac{(r/2)^{2k} \Delta^k \varphi(0)}{k! \, \Gamma(k+n/2)}$$

2.123 Show that the solutions of the polyharmonic equation $\Delta^{m+1} u = 0$ in a simply connected domain containing the origin are given by the formula $u(x) = \sum_{k=1}^{m} |x|^{2k} u_k(x)$, where u_k are arbitrary harmonic functions.

2.124 Show that if $\Delta u \geq \Delta^2 u$ (in $\mathscr{D}'(\mathbb{R}^n)$) for some $u \in C^2(\mathbb{R}^n) \cap \mathscr{S}'(\mathbb{R}^n)$ and u is not a constant, then $u(x) < \max_{|y-x|=r} u(y)$ for all $x \in \mathbb{R}^n$ and $r > 0$.

2.125 Imitating the Hadamard counterexample concerning the Cauchy problem for the Laplace equation, piece together exponential and trigonometric functions to construct an infinitely differentiable coefficient $p = p(x,y)$ such that the elliptic partial differential equation $(\partial/\partial x + i \, \partial/\partial y)^6 u - i x^6 \, \partial^5 u/\partial y^5 - \partial^4 u/\partial y^4 + pu = 0$ has an infinitely differentiable solution u of the general form $u(x,y) = \varphi(x) \exp(ik(x)y) + \psi(x) \exp(i\ell(x)y)$ which vanishes identically in the left half-plane $x \leq 0$ but differs from zero somewhere in the right half-plane $x > 0$.

2.126 Show that a nontrivial solution of the equation $\Delta u = u^2$ in a plane domain Ω cannot assume its maximum in an interior point of Ω.

2.127 Show that the boundary value problem $\Delta u = u^3$ in the unit disk in $\mathbb{R}^2$, $u = 0$ on the unit circle, has the unique solution $u \equiv 0$.

2.128 Prove that there exists a nonconstant positive solution of the equation $\Delta u + au(1 - u) = 0$ on the square $[-L,L] \times [-L,L]$ with sufficiently large L, $u = 0$ on the boundary.

2.129 Consider the problem $\Delta u + \mu u - u^3 = 0$ in a bounded open subset Ω of $\mathbb{R}^n$, $u = 0$ on $\partial\Omega$. Show that if $\mu > 0$ is sufficiently small, then $u \equiv 0$ is the unique solution of this problem.

2.130 Study solvability of the Dirichlet problem for the equation $\Delta u - u^{-1} \mid \nabla u \mid^2 = 0$ in the unit disk D in the plane, $u = f$ on the unit circle C.

2.131 Study the Dirichlet and the Newmann problem for the equation $\Delta u - \mid \nabla u \mid^2 = 0$.

2.132 Consider a nonlinear Poisson equation $\Delta u = f(x,u,\nabla u)$ in a bounded region Ω in $\mathbb{R}^n$ with smooth boundary where $u = 0$, such that $u \mapsto f(x,u,\nabla u)$ determines a continuous map $F : C^{1+\alpha} \to C^\alpha$ and $\|Fu\|_\alpha \le M$ for all $u \in C^{1+\alpha}$ and some $M < \infty$. Prove that this boundary value problem has at least one solution $u \in C_0^{2+\alpha}(\Omega)$.

2.133 Show that the (supercritical) elliptic problem $\Delta u - u + \mid u \mid^p u = 0$ for the $p \ge 4/(n - 2)$ has no nontrivial smooth solutions defined on $\mathbb{R}^n$, $n > 2$, with $\mid u \mid \to 0$ as $\mid x \mid \to \infty$.

2.134 Prove that if u is a solution of the Dirichlet problem $\Delta u + f(u) = 0$ in $D \subset \mathbb{R}^n$, $u = 0$ on ∂D, then

$$\int_{\partial D} \frac{1}{2} (x,\nu) \left(\frac{\partial u}{\partial \nu}\right)^2 dS + F(0)\int_{\partial D} (x,\nu) \, dS + \left(\frac{n}{2} - 1\right) \int_D \mid \nabla u \mid^2 dx$$
$$- n \int_D F(u) \, dx = 0$$

where $F'(t) = f(t)$ and ν is the exterior normal vector.

2.135 Show that the real solutions of the Liouville equation $\Delta u + 2Ke^u = 0$, $u : D \subset \mathbb{R}^2 \to \mathbb{R}$, are of the form

$$u = 2 \log\left(\frac{\mid f' \mid}{1 + K\mid f\mid^2/4}\right)$$

where $f(z)$ is, apart from simple poles, any complex analytic function. In particular, prove that for negative values of K, the Liouville equation $\Delta u + 2Ke^u = 0$ cannot have solutions in the entire plane.

2.136 Consider the problem $\Delta u + \sigma \exp u/\int_\Omega \exp(u(x)) \, dx = 0$, $u = 0$ on $\partial\Omega$, in a domain $\Omega \subset \mathbb{R}^3$. Assume that Ω is starlike, $\int_{\partial\Omega} ds/(x,\nu) = A < \infty$, where v is the exterior unit normal to $\partial\Omega$. Prove that this problem does not have classical solutions of $\sigma > 3/A$.

2.137 Prove that for sufficiently small σ the problem $\Delta u + \sigma \exp u/ \int_\Omega \exp(u(x)) \, dx = 0$ in $\Omega \subset \mathbb{R}^3$, $u = 0$ on $\partial\Omega$, has a solution.

2.138 Let $D \subset \mathbb{R}^2$ be a bounded, simply connected domain, the boundary of which is divided into two piecewise analytic arcs Γ_0 and Γ_1, and let u, v be classical solutions of the problems $\Delta u + f(x) = 0$ in D, $u = 0$ on Γ_0, $\partial u/\partial v + u = 0$ on Γ_1, $\Delta v + c = 0$ in D, $v = 0$ on Γ_0, $\partial v/\partial v = 0$ on Γ_1, where c is a positive constant such that $f(x) \le c$. Prove that $u(x) \le v(x)$ in D.

2.139 Obtain a gradient bound for solutions of the semilinear equation $\Delta u + f(u) = 0$ in a bounded region $\Omega \subset \mathbb{R}^n$, where f is a positive C^1 function.

2.140 Prove that the equation $u_{xx} + u_{yy} = F(x,y,u,u_x,u_y)$ has only analytic solutions when F is analytic.

2.141 Give an example of an elliptic nonlinear boundary value problem with smooth coefficients with no classical solution.

2.142 Establish the uniqueness of the solution of the Dirichlet problem for the (quasilinear) minimal surface equation $(1 + u_y^2)u_{xx} - 2u_x u_y u_{xy} + (1 + u_x^2) u_{yy} = 0$ over a convex region.

2.143 Prove the Bernstein minimal graph theorem: Let $u(x,y)$ solve the minimal surface equation

$$\frac{\partial(u_x(1 + u_x^2 + u_y^2)^{-1/2})}{\partial x} + \frac{\partial(u_y (1 + u_x^2 + u_y^2)^{-1/2})}{\partial y} = 0$$

for all $(x,y) \in \mathbb{R}^2$. Then the graph of u is a plane. The following lemmas may be useful:
(i) The minimal surface equation is equivalent to $(1 + u_y^2)u_{xx} - 2u_x u_y u_{xy} + (1 + u_x^2)u_{yy} = 0$.
(ii) The above equation is the compatibility condition for the existence of a function $\Phi(x,y)$ such that $\Phi_{xx} = (1 + u_x^2)(1 + u_x^2 + u_y^2)^{-1/2}$, $\Phi_{xy} = u_x u_y(1 + u_x^2 + u_y^2)^{-1/2}$, $\Phi_{yy} = (1 + u_y^2)(1 + u_x^2 + u_y^2)^{-1/2}$.
(iii) Each Φ as above satisfies the equation $\Phi_{xx}\Phi_{yy} - \Phi_{xy}^2 = 1$ for all $(x,y) \in \mathbb{R}^2$. Moreover, Φ is a quadratic polynomial.

Schrödinger operators and Schrödinger evolution equations

2.144 Prove that the Schrödinger operator $-\Delta + V$ with $V \in L_{\text{loc}}^2(\mathbb{R}^n)$ and $V \ge 0$ is essentially selfadjoint on $C_0^\infty(\mathbb{R}^n)$.

2.145 Consider the operator $H_0 = -\Delta$ in $L^2(\mathbb{R}^n)$ and $(Vf)(x) = \bullet_U(x)f(x)$ where U is an open set in $\mathbb{R}^n$. Define the selfadjoint operator H_∞ on $L^2(U)$ by $\|H_\infty^{1/2}f\|^2 = \|H_0^{1/2}f\|^2$. Identify H_∞ with the Laplacian with the Dirichlet boundary conditions on ∂U.

2.146 It is well known that the Schrödinger operator $H = -\Delta + V$ with a potential V satisfying the condition $\lim_{|x| \to \infty} |x| V(x) = 0$ has no strictly positive eigenvalues (the Kato theorem). Show that even for $V \in C_0^\infty(\mathbb{R}^n)$ H may have 0 as an eigenvalue.

2.147 The Wigner–von Neumann example. Verify that $u(r) = (r^{-1} \sin r) \times (1 + g(r)^2)^{-1}$ is an eigenfunction with eigenvalue 1 of the Schrödinger operator $-\Delta + V$ with the radial potential $V(r) = (1 + g(r)^2)^{-2}(-32 \sin r)(g(r)^3 \cos r - 3g(r)^2 \sin^3 r + g(r) \cos r + \sin^3 r)$, where $g(r) = 2r - \sin 2r = 4 \int_0^r \sin^2 x \, dx$. This example shows that there may exist positive eigenvalues even if the potential is bounded and vanishes at infinity.

2.148 Consider the Schrödinger operator $-\Delta + \lambda V$ on $\mathbb{R}^n$, $n = 1,2$, where $\lambda > 0$, $V \le 0$ but $V \ne 0$ and V belongs to $C_0^\infty(\mathbb{R}^n)$. Show that for every $\lambda > 0$ this operator has a negative eigenvalue.

2.149 Give an example of a potential $V \in C^\infty(\mathbb{R}^n)$ such that a solution u of the Schrödiner equation $-\Delta u + Vu = 0$ is in $L^2(\mathbb{R}^n)$ but $u(x)$ does not tend to zero as x goes to infinity.

2.150 Using the stationary phase method, determine the asymptotics for $t \to \infty$ of solutions of the Schrödinger equation $i\psi_t = -\psi_{xx} + V\psi$, where V is a continuous potential with compact support contained in $[-a,a]$.

2.151 Consider the unitary group $\exp(it\Delta)$ on $L^2(\mathbb{R}^n)$ generated by the Laplacian. Show that for every $f \in L^2(\mathbb{R}^n)$

$$\lim_{t \to \infty} \|\exp(it\Delta)f(x) - (4\pi it)^{-n/2} \exp\left(\frac{ix^2}{4t}\right) \hat{f}\left(\frac{x}{2t}\right)\|_2 = 0$$

Interpret this result as asymptotically classical behavior of a quantum free particle.

2.152 Prove the following estimates for the Schrödinger equation $iu_t = \Delta u$ in $\mathbb{R}^n$, $u(x,0) = u_0(x)$: $|u(x,t)| \le (4\pi t)^{-n/2} \|u_0\|_1$, $\|u(t)\|_q \le ct^{n/2 - n/p} \|u_0\|_p$ for $p^{-1} + q^{-1} = 1$, $1 \le p \le 2$, and the space-time estimates $\|u\|_{2 + 4/n, 2 + 4/n} \le c\|u_0\|_2$, $\|u\|_{r,p+1} \le c\|u_0\|_2$ for $1 \le p \le 1 + 4/(n - 2)$, $r = 4(p + 1)/n(p - 1)$.

2.153 Show that if in the formal asymptotic solution as $h \to 0$

$$\psi(x;h) = \varphi(x;h) \exp\left(\frac{iS(x)}{h}\right), \quad \text{Im } S = 0, \quad \varphi(x;h) = \varphi_0(x) + \sum_{j=1}^{n} h^j \varphi_j(x)$$

of the Schrödinger equation $h^2\psi''(x) + (E - V(x))\psi(x) = 0$, the first term $\varphi_0(x) \neq 0$, then S satisfies the Hamilton-Jacobi equation $(S')^2 + V - E = 0$, hence $\varphi_0(x) = \text{const}(E - V(x))^{-1/4}$.

2.154 The first term of the quasiclassical asymptotics of the solution of the Cauchy problem for the Schrödinger equation

$$ih\frac{\partial\psi}{\partial t} = -\left(\frac{h^2}{2m}\right)\Delta\psi + V(q,t)\psi, \quad \psi(q,0) = \exp\left(\frac{i}{hS_0(q)}\right)\varphi_0(q)$$

equals

$$\exp\left(\frac{i}{hS(q,t)}\right)(J(q,t))^{-1/2}\varphi_0(q_0(q,t))$$

Show that the function $\rho(q,t) = J^{-1}(q,t)$ satisfies the continuity equation

$$\frac{\partial\rho}{\partial t} + \text{div } \rho \ \nabla S(q,t) = 0$$

2.155 Show that if the initial condition $u_0 \in L^1(\mathbb{R}^n)$ of the Cauchy problem for the Schrödinger equation $iu_t + \Delta u = 0$, $u(x,0) = u_0(x)$, is of compact support, then $u(t)$ is in $C^\infty(\mathbb{R}^n)$ for $t > 0$.

2.156 Prove that for the Schrödinger equation $iu_t = \Delta u$ in $\mathbb{R}^n$

$$\left(x_j - it\frac{\partial}{\partial x_j}\right)u, \quad j = 1, \ldots, n$$

is a solution whenever u is. Draw as a corollary that if $\exp(\,|\,x\,|\,)u_0 \in L^2$, then $u(t) \in C^\infty$ for $t > 0$.

2.157 Show that for the nonlinear Schrödinger equation $iu_t - \Delta u + f(u) = 0$ with $uf(u) \leq (2 + 4/n)F(u)$ for all $u \in \mathbb{R}$ (here $f(u) = g(|\,u\,|^2)u$, $G' = g$, $G(0) = 0$, $F(u) = \frac{1}{2}G(|\,u\,|^2))$ and with the Cauchy data $u(x,0) = \varphi(x)$ satisfying $E(\varphi) : \int(\frac{1}{2}|\,\nabla\varphi\,|^2 + F(\varphi))\,dx < 0$, no smooth solution can exist for all time.

2.158 Discuss conformal invariance for the nonlinear Schrödinger equation in $\mathbb{R}^n$, $iu_t - \Delta u + f(u) = 0$, where $f(u) = g(|\,u\,|^2)u$.

2.159 Let $A_1, \ldots, A_m$ be $n \times n$ symmetric matrices. Prove that the Fourier multiplier B with the symbol $\exp(i \sum_{j=1}^m \xi_j A_j)$ satisfies $\| Bu \|_p \le C \| u \|_p$ for $p \ne 2$, $1 \le p \le \infty$, if and only if the matrices $A_1, \ldots, A_m$ commute. Interpret this result as a negative result for symmetric hyperbolic systems. Remember the Hörmander theorem on the Schrödinger equation: $Bu = \mathscr{F}^{-1}(\exp(ix \mid \xi \mid^2)\mathscr{F}u)$ is bounded in L^p norm if and only if $p = 2$. Write the Maxwell system $E_t = c \operatorname{rot} H$, $H_t = -c \operatorname{rot} E$ ($\operatorname{div} E = \operatorname{div} H = 0$), in a similar form. Conclude on the well-posedness of the Cauchy problem for this system.

2.160 Pointwise behavior of solutions of the Cauchy problem for the Schrödinger equation. Show that for sufficiently smooth (make it precise!) initial data f the solution u of the Schrödinger equation $iu_t = \Delta u$ in $\mathbb{R}^n \times \mathbb{R}$ converges a.e. to f as $t \to 0$.

Parabolic equations

2.161 A fundamental solution of the iterated heat equation $(\partial/\partial t - \Delta)^k u = \delta$ may be obtained multiplying by $t^{k-1}/(k - 1)!$ a fundamental solution for $k = 1$. Prove it.

2.162 Show that if u is a solution of the heat equation $u_t = u_{xx}$, then for $t > 0$, so is its Appell transform

$$v(x,t) = (4\pi t)^{-1/2} \exp\left(\frac{-x^2}{4t}\right) u\left(\frac{x}{t}, \frac{-1}{t}\right)$$

2.163 Show that the formula $u(x,t) = \lambda \exp(-t\lambda^{-2}) \sin(x/\lambda)$ defines a solution of the heat equation $u_t = u_{xx}$ that approaches zero for $t \ge 0$, but not for $t < 0$, in the limit as $\lambda \to 0$.

2.164 Prove that if a function $u = u(x,t)$ is continuous in a domain D together with its space derivative u_x and satisfies the relation $\int_\Gamma (uv_2 - u_x v_1) = 0$ for every simple closed rectifiable curve Γ in D, where (v_1, v_2) is the exterior unit normal vector, then u satisfies the heat equation $u_t = u_{xx}$ in D.

2.165 Let $B = B(x_0, t_0, c)$ denote the level set of the fundamental solution of the heat equation shifted to (x_0, t_0) : $\{(4\pi(t - t_0))^{-1/2} \exp(-(x - x_0)^2/4(t - t_0)) = c\}$, $t > t_0$, and $Q(x,t) = cx^2(4x^2t^2 + (2t - x^2)^2)^{-1/2}$. Show that every solution of the heat equation satisfies the parabolic mean value property $u(x_0, t_0) = \int_B Q(x_0 - x, t_0 - t)u(x,t) \, ds$ for all $c \ge c_0$ (c_0 is such that $B(x_0, t_0, c)$ is contained in the domain of definition of u). Conversely, each continuous function satisfying the above mean value property is a solution of the heat equation.

2.166 It is well known that the heat operator $P = \partial/\partial t - \partial^2/\partial x^2$ in $\mathbb{R}^2$ is hypoelliptic. Show that there is a nonlinear change of variables in $\mathbb{R}^2$ such that P is transformed into an operator which is not formally hypoelliptic.

2.167 Give an example of a function $u : \mathbb{R} \times [0,T] \to \mathbb{R}$ not vanishing identically which solves the heat equation $u_t = u_{xx}$ with the zero initial condition $u(x,0) = 0$.

2.168 A function $u = u(x,t)$ is called a Holmgren function if it satisfies the inequalities $| \partial^n u/dx^n | \le KC^n n!$ and $| \partial^n u/\partial t^n | \le KC^n(2n)!$ for some $K, C > 0$. Is a Holmgren function necessarily quasianalytic?

2.169 Show that solutions of the Cauchy problem for the equation $\partial^q u/\partial t^q = a^{-1} \partial^p u/\partial x^p$, $p > q$, $a > 0$, are not unique.

2.170 Prove that the solution of the Cauchy problem for the heat equation in $\mathbb{R}^n \times [0,T]$ (or, more generally, for a uniformly parabolic equation with Hölder continuous coefficients) $u_t = \Delta u$, $u(x,0) = 0$, is unique in the class of functions u satisfying the condition $\int_0^T \int_{\mathbb{R}^n} | u(x,t) | \exp(-k | x |^2)\, dx\, dt < \infty$ for some positive k.

2.171 Let L be a parabolic operator

$$\frac{\partial}{\partial t} - \left(\sum_{i,j=1}^n a_{ij} \frac{\partial^2}{\partial x_i \partial x_j} + \sum_{j=1}^n b_j \frac{\partial}{\partial x_j} + c \right)$$

with continuous coefficients in $\mathbb{R}^n \times (0,T]$ satisfying the conditions $| a_{ij} | \le M$, $| b_j(x,t) | \le M(1 + | x |)$, $c(x,t) \le M(1 + | x |^2)$ for some constant M. Prove that there exists at most one solution of the Cauchy problem $Lu = f(x,t)$, $u(x,0) = \varphi(x)$, which satisfies $| u(x,t) | \le B \exp(\beta | x |^2)$ for some positive constants B, β.

2.172 Prove that given a positive solution u of the heat equation in $\mathbb{R}^n$ the limit $\lim_{(y,t)\to(x,0)} u(y,t)$ is equal a.e. to the Radon-Nikodym derivative of the representing measure μ with respect to the Lebesgue measure:

$$u(x,t) = c_n t^{-n/2} \int_{\mathbb{R}^n} \exp\left(\frac{- | x - y |^2}{4t} \right) du(y)$$

Here (y,t) is in the parabolic nontangential approach region $\{(y,t) : 0 < t < T, | y - x | < \alpha t^{1/2}\}$, $\alpha > 0$.

2.173 The Widder theorem: Let $u = u(x,t)$ be defined and continuous for $x \in \mathbb{R}$, $0 \le t < T$, and let u_t, u_x, u_{xx} exist and be continuous for $x \in \mathbb{R}$, $0 < t < T$. Assume that $u_t = u_{xx}$ for $x \in \mathbb{R}$, $0 < t < T$, $u(x,0) = f(x)$, $u(x,t) \ge 0$. Then

$u(x,t)$ is determined uniquely for $x \in \mathbb{R}$, $0 < t < T$, is real analytic in x, and is represented by the convolution of f with the Gauss-Weierstrass kernel.

2.174 It is well known from the (Aronson-)Widder theorem that each positive solution of the heat equation $u_t = \Delta u$ in $\mathbb{R}^n \times (0,T)$ is the convolution of the Gauss-Weierstrass kernel with a positive Borel measure μ such that $\int_{\mathbb{R}^n} \exp(- \mid x \mid^2 / 4T) \, d\mu(x) < \infty$. Give examples of solutions of the heat equation which exist only for $t < T$, that is, positive solution u defined in $\mathbb{R}^n \times (0,T)$ with no continuation beyond T to any solution of the heat equation.

2.175 Show that if $u_0 \in L^\infty(\mathbb{R}^n)$ satisfies the condition $\lim_{R \to \infty} (\omega_n R^n)^{-1}$ $\int_{\{\mid x \mid \leq R\}} u_0(x) \, dx = A$, then $\lim_{t \to \infty} u(x,t) = A$ uniformly on compacts of $\mathbb{R}^n$, where u is the bounded solution of the heat equation $u_t = u_{xx}$ with the initial condition $u(x,0) = u_0(x)$. Give an example of $u_0 \in L^\infty(\mathbb{R}^n)$ such that the solution u does not stabilize when t tends to ∞.

2.176 Prove that if $u = u(x,t)$, $x \in \mathbb{R}^n$, $t \geq 0$, is the solution of the Cauchy problem for the heat equation $u_t = \Delta u$ with $u_0 = u(x,0)$ satisfying $\int \mid u_0 \mid < \infty$, $\int u_0 = 0$, then $\lim_{t \to \infty} \int u(x,t) \, dx = 0$. Try to derive this result for more general parabolic equations in $\mathbb{R}^n$.

2.177 Consider positive solutions of the heat equation $u_t = u_{xx}$ in a rectangle $\{\mid x \mid < b, -c < t < T\}$ for some $b, c > 0$. Prove that for any $0 < a < b$ there exists a constant K such that the (analogue of the Harnack) inequality $u(x,t) \leq Ku(y,s)$ holds for all positive solutions u, $-a < x,y < a$, $0 < s \leq t < T$.

2.178 Let u be a solution of the one-dimensional heat equation $u_t = u_{xx}$ in an open subset. Show that for any point in this set there exist constants C, M such that $\mid \partial^k u / \partial t^k \mid \leq CM^k (2k)!$ for all nonnegative integers k.

2.179 Let $L_p^s(\mathbb{R}^n)$ denote the space $(I - \Delta)^{-s/2} L^p(\mathbb{R}^n)$ and $\partial/\partial t - \Delta_x$ the heat operator in $\mathbb{R}^n \ni (t,x)$, $x \in \mathbb{R}^{n-1}$. Show that $(\partial/\partial t - \Delta_x)u \in L_p^s$, $1 < p < \infty$, implies
(i) $u \in L_p^{s+1}$;
(ii) $\Delta_x u \in L_p^s$;
(iii) $\nabla_x u \in L_p^{s+1/2}$.

2.180 Consider a smooth solution u of the heat equation on $\mathbb{T} \times [0,1]$: $u_t - u_{xx} = f(x,t)$ with a smooth function f. Show that $[u_{xx}]_\vartheta + [u_t]_\vartheta \leq C[f]_\vartheta$ where the constant C does not depend on f and $[.]_\vartheta$, $0 < \vartheta < 1$, denotes the norm

$$[v]_\vartheta = \max_{\mathbb{T} \times [0,1]} \mid v \mid + \sup_{x \neq y, 0 \leq t \leq 1} \frac{\mid v(x,t) - v(y,t) \mid}{\mid x - y \mid^\vartheta}.$$

2.181 L is a parabolic operator in $\mathbb{R}^n \times (0,T]$ with continuous coefficients, and $|a_{ij}(x,t)| \leq M(1 + |x|^2)$, $|b_j(x,t)| \leq M(1 + |x|)$, $c(x,t) \leq M$ for some constant M. Prove: If $Lu \leq 0$ in $\mathbb{R}^n \times (0,T]$, $u(x,t) \geq -N(1 + |x|^q)$ in $\mathbb{R}^n \times [0,T]$ for some positive constants N and q, then $u(x,0) \geq 0$ implies $u(x,t) \geq 0$ everywhere.

2.182 Let G_m be the set $\{(x,t) \in \mathbb{R}^n \times \mathbb{R} : |x|^2 + |t| \leq m^2, t < 0\}$, $m > 0$. Show that any solution u of the heat equation in $C^{2,1}(\overline{G_{R+r}})$ satisfies the estimate (of Bernstein type) $|\nabla u|^2 \leq cr^{-2} \max_{\overline{G_{R+r}}} |u|^2$ on G_R with a constant $c = c(n)$.

2.183 Let u be a solution of the heat equation in $\mathbb{R}^n \times (-\infty,T)$ satisfying the estimate $|u(x,t)| \leq C(1 + |x|^2 + |t|)^{q/2}$ with some $q \geq 0$. Show that u is a polynomial of degree $\leq [q]$.

2.184 Show that solutions of the problem $u_t = u_{xx}$ for $0 < x$ and all t, $u(0,t) = f(t)$, $u_x(0,t) = g(t)$, do not depend continuously on their data f and g.

2.185 Show that solutions of the problem $u_t = u_{xx}$ for $0 < x < \infty$, $0 < t$, $u_x(0,t) = 0$, $u(x,0) = 0$, are not unique.

2.186 Show that the initial value problem $u_t = u_{xx}$, $u(x,0) = 0$, $u(0,t) = 0$, $u(1,t) = t^{-3/2}e^{-1/4t}$, has many C^∞ solutions for $(x,t) \in (0,1) \times (0,\infty)$.

2.187 Show that solutions of the problem $u_t = u_{xx}$, $0 < x < 1$, $-\infty < t < \infty$, $u(0,t) = g(t)$, $u(1,t) = h(t)$, are not unique in general, unless a growth condition such as $g, h = \mathbb{O}(\exp(c|t|))$ for $-\infty < t \leq 0$ and some constant $c < \pi^2$ is satisfied.

2.188 Consider the problem $u_t = u_{xx}$ for $0 < x < 1$ and $t > 0$, $u(0,t) = g(t)$, $u(1,t) = h(t)$ for $t > 0$, $u(a,t) = \varphi(t)$ for $0 < t_0 < t$, $0 < a < 1$, and $u(x,t)$ uniformly bounded if $0 \leq x \leq 1$ and $t \geq 0$. Show that for a a rational number the above problem has infinitely many solutions, while for a an irrational number it can have at most one solution.

2.189 Consider a solution u of the heat equation $u_t = \Delta u$ in $\Omega \times (0,\infty)$, where Ω is a bounded domain in $\mathbb{R}^n$, vanishing on $\partial\Omega \times [0,\infty)$. Prove that $\lim_{t\to\infty} u(x,t) = 0$ uniformly in $x \in \Omega$.

2.190 Discuss the asymptotic behavior for $t \to \infty$ of the solutions of the mixed boundary value problems for the heat equation $u_t = u_{xx}$ in $(0,1) \times (0,\infty)$, $u(x,0) > 0$ given, $u_x(0,t) = 0$, $u_x(1,t) + cu(1,t) = 0$ for $c > 0$ (the Newton law of cooling) and $c < 0$ respectively.

2.191 Let

$$L = \sum a_{ij}(x,t)\frac{\partial^2}{\partial x_i\, \partial x_j} + \sum b_j(x,t)\frac{\partial}{\partial x_j} + c(x,t) - \frac{\partial}{\partial t}$$

be a parabolic operator in the strip $0 \leq t \leq T$. Prove: if $Lu = 0$ in the comple-
ment of $D \times [0,T]$ in $\mathbb{R}^n \times [0,T]$, if $u(x,0) = 0$ in the complement of D in $\mathbb{R}^n$,
if $\partial u/\partial N := \Sigma_{i,j=1}^n a_{ij}(x_0,t) \cos(\nu x_0, x_j) \partial u/\partial x_i = u$ on $\partial D \times (0,T]$, and if
$\sup_{0 \leq t \leq T} |u(x,t)| \to 0$ as $|x| \to \infty$, then $u \equiv 0$ in the complement of
$D \times [0,T]$.

2.192 Let G be a closed bounded subset of $\mathbb{R}^n$ with smooth boundary. Give a
necessary and sufficient condition for the existence of a smooth (C^∞) solution
$u : \mathbb{R}^+ \times G \to \mathbb{R}$ of the heat equation $u_t - \Delta u = f$, where f is a given C^∞
function on $\mathbb{R}^+ \times G$, with the conditions $u(x,t) = 0$ if $t = 0$ or $x \in \partial G$. Solve
the same problem for the wave equation $u_{tt} - \Delta u = f$, $u(x,0) = u_t(x,0) = 0$
for all $x \in G$, $u(x,t) = 0$, for all $x \in \partial G$.

2.193 Prove the parabolic version of the Hopf theorem: Suppose that u is a
solution of the heat equation (or, more generally, a parabolic equation with
regular coefficients) on a domain $D_T = \{(x,t) : s_1(t) < x < s_2(t), 0 < t \leq T\}$,
where s_1 and s_2 are Lipschitz continuous, u is continuous up to the boundary of
D_T and is not identically constant in each D_t, $0 < t \leq T$. If u assumes its
maximum value at $(s_2(t_0),t_0)$, $t_0 > 0$, then

$$\liminf_{x \to s_2(t_0)^-} \frac{u(x,t_0) - u(s_2(t_0),t_0)}{x - s_2(t_0)} > 0$$

Formulate similar assertions in the case of a minimum of u and extrema attained
on the left part of the boundary $\{s_1(t): 0 < t \leq T\}$.

2.194 As in the case of elliptic equations, the existence of a barrier (a super-
parabolic positive function vanishing at a given point of the parabolic boundary)
implies that solutions of the heat equation are continuous at this point of the
parabolic boundary. Show that for the domain D bounded by the curve
$\{x^2 = -4(1 + \varepsilon)t \log |\log |t||\}$ and the line $\{t = t_0 < 0\}$, there cannot exist
a barrier at the origin.

2.195 Show the existence of a solution of the problem $u_t = u_{xx}$, $0 < x < 1$,
$t > 0$, $u_x(0,t) = v(t)$, $u_x(1,t) + u(1,t) = w(t)$, $v' = \alpha v + \beta w$, $w' = \gamma v +
\delta w + \int_0^1 u(x,t) \, dx$ with a given $u(\cdot,0) \in L^2(0,1)$ and real $v(0)$, $w(0)$, Moreover,
$u(x,t) \geq 0$, $v(t) \geq 0$, $w(t) \geq 0$, if the initial data are positive.

2.196 Show that the parabolic degenerate equation (describing gene drift) $u_t =
(x(1 - x)u)_{xx}$, $0 < x < 1$, $t > 0$, generates an analytic semigroup in the
space $\{v : (0,1) \to \mathbb{R}, \int_0^1 x(1 - x) |v(x)|^2 \, dx < \infty \}$. Moreover, show the
$\lim_{t \to \infty} \|u(.,t)\| = 0$.

2.197 Construct a fundamental solution of the Kolmogorov equation
$u_t = u_{xx} + xu_y$, $t > 0$, $u(.,t) \to \delta(x_0,y_0)$ as $t \to 0^+$.

2.198 Determine a positive function a and a function u such that they satisfy the equation $u_t = a(t)u_{xx}$ for $0 < x < 1$ and $t > 0$, $u(0,t) = u(1,t) = 0$, $u(x,0) = \sin \pi x$, $0 < x < 1$ and $u(1/2,t) = h(t)$, $0 < t$, where the last condition (an interior temperature measurement) is a necessary overspecification of data.

2.199 Consider the heat equation with sources $u_t - \Delta u + qu = 0$, where $q = q(x)$ is a bounded function, $x \in \mathbb{R}^n$, $t > 0$. Show that if $u(x,0)$ is positive and continuous, $\lim_{|x| \to \infty} u(x,0) = 0$, then $u(x,t)$ is also positive. Prove that $u(x,t)$ is C^∞ for all x and $t > 0$.

2.200 Let $P : \mathbb{R}^+ \times \mathbb{R} \times \mathcal{B}orel(\mathbb{R}) \to [0.1]$ satisfy
(i) $P(t,x,\Gamma)$ is a Borel function in x for fixed t, Γ;
(ii) $P(t,x,\Gamma)$ is a probability measure in Γ for fixed, t, x;
(iii) $P(t + s, x, \Gamma) = \int_{-\infty}^{\infty} P(t,x,dy)P(s,y,\Gamma)$ for all t, s, x, Γ;
(iv) $P(0,x,\Gamma) = 1$ or 0 according as $x \in \Gamma$ or $x \notin \Gamma$;
(v) $P(t,x,\Gamma)$ is absolutely continuous in Γ for fixed t, x.
Define $p : \mathbb{R}^+ \times \mathbb{R} \times \mathbb{R} \to [0,\infty)$ by $P(t,x,\Gamma) = \int_\Gamma p(t, x,y) \, dy$. Assume further that
(vi) $\partial p/\partial x$, $\partial^2 p/\partial y^2$ exist and are continuous in $\mathbb{R}^+ \times \mathbb{R} \times \mathbb{R}$;
(vii) $\lim_{t \to \infty} t^{-1} \int_{|x-y| \geq \delta} p(t,x,y) \, dy = 0$ for each $\delta > 0$, uniformly in $x \in \mathbb{R}$;
(viii) $\lim_{t \to \infty} t^{-1} \int_{|x-y| < \delta} (y - x)p(t,x,y) \, dy = m(x)$ and $\lim_{t \to 0} t^{-1} \int_{|x-y| < \delta} (y - x)^2 p(t,x,y) \, dy = \sigma^2 (x)$ exist uniformly in $x \in \mathbb{R}$, for each $\delta > 0$, and are continuous.
The $p(t,x,y)$ is the fundamental solution of the equation $u_t = \frac{1}{2}\sigma^2(x)u_{xx} + m(x)u_x$.

2.201 Consider the free boundary problem $u_{1t} = a_1^2 u_{1xx}$, $0 < x < \xi(t)$, $u_{2t} = a_2^2 u_{2xx}$, $\xi(t) < x < \infty$, $0 < t < \infty$, $u_1(\xi(t),t) = u_2(\xi(t),t)$, $u_1(0,t) = U_1$, $u_2(\infty,t) = U_2$, $(k_1 u_{1x} - k_2 u_{2x}) |_{x = \xi(t)} = \rho \, d\xi/dt$, $0 < t < \infty$, $u_2(x,0) = U_2$, $\rho > 0$, describing the evolution of the temperatures in two-phase medium. Initially the half-space $x > 0$ contains a liquid of temperature $U_2 = $ const and the plane $x = 0$ is kept at the temperature $U_1 < U_2$, U_1 below the solidification point of the liquid. Determine the solidification front $\xi(t)$ and u_1, u_2.

2.202 Some comparison theorems can be proved for weakly coupled parabolic systems. Show that the maximum principle fails if the coupling occurs in the first derivative terms.

2.203 Show that the Burgers equation $u_t + uu_x = \mu u_{xx}$, $\mu > 0$, transforms into $v_t = \mu v_{xx} + c(t)v$ via $v = \exp(-(2\mu)^{-1} \int u \, dx)$ and further, via an appropriate transform, into the heat equation.

2.204 Prove the uniqueness of solutions of the first initial-boundary value problem for a parabolic quasilinear equation $Lu = f(x,t,u,\nabla_x u)$ with the Dirichlet boundary data in the case when f is nondecreasing in u. The operator L is assumed to have continuous coefficients.

2.205 Give an example of nonuniqueness of solutions of the boundary value problem for a nonlinear parabolic equation with nonlinear terms satisfying the Hölder condition.

2.206 Prove that any nonconstant stationary solution of the equation $u_t = \Delta u + u^2$ must be unstable.

2.207 Consider the parabolic problem (arising in optimal control of nuclear reactors) $u_t = ku_{xx} + au - bu^2$, $0 < x < 1$, $t > 0$, $u(0,t) = u(1,t) = 0$, $u(x,0) = u_0(x) \geq 0$, where $u_0 \in H_0^1(0,1)$, a, b, $k \geq 0$. Show the global existence of the solution. Prove that this equation generates a dynamical system in $\{v \in H_0^1(0,\pi) : v(x) \geq 0\}$.

2.208 Consider the parabolic problem $u_t = \Delta u + \lambda u^3$, $\partial u/\partial v + au = 0$ on the boundary of a bounded open subset Ω of $\mathbb{R}^n$. Show that if $\lambda \leq 0$, there exists a global solution for all $t > 0$. This solution tends to 0 in L^∞ and $W^{1,2}$ as $t \to \infty$.

2.209 Prove that the initial-boundary value problem $u_t = u_{xx} + au - bu^3$, $0 < x < \pi$, $t > 0$, $u(0,t) = u(\pi,t) = 0$, $u(x,0) = u_0(x)$, is well posed in the Sobolev space $H_0^1(0,\pi)$. Moreover, if $t \to \infty$, the solution $u(t)$ converges in H_0^1 to a solution of the stationary problem $\varphi_{xx} + a\varphi - b\varphi^3 = 0$, $\varphi(0) = \varphi(\pi) = 0$. There are $2n + 1$ such steady-state solutions if $n^2 < a \leq (n + 1)^2$. If $0 < a \leq 1$, the trivial solution is stable. If $a > 1$, then there is a pair φ, $-\varphi$ of stable solutions which are of constant sign for $x \in (0,\pi)$, the others being unstable.

2.210 Show that the equation $u_t = u_{xx} + u - bu^3$, $b > 0$, $x \in \mathbb{R}$, $t > 0$, has a stationary solution $u = \varphi(x)$ such that $\lim_{x \to \pm\infty} \varphi(x) = \pm b^{1/2}$, $\varphi'(x) > 0$. Prove that if the norm $\| u(.,0) - \varphi \|$ in H^1 is sufficiently small, then $\| u(.,t) - \varphi(\cdot + c) \| \leq K \exp(-\beta t) \| u(.,0) - \varphi \|$ for some c and K, $\beta > 0$.

2.211 Show that solutions of the parabolic problem $u_t = u_{xx} + u^3$, $0 < x < \pi$, $t > 0$, $u(0,t) = u(\pi,t) = 0$, $u(x,0) = u_0(x)$, may blow up in a finite time.

2.212 More on blowup for a nonlinear heat equation. Consider the equation $u_t = \Delta u + |u|^{p-1} u$ in $D \times (0,T)$, $u = 0$ on $\partial D \times (0,T)$, where D is an n-dimensional domain and $1 < p < (n + 2)/(n - 2)$ or $n \leq 2$. Prove that $\lim_{t \to T} (T - t)^{-1/(p-1)} u(a + y(T - t)^{1/2}, t) = 0$ or $\pm k$, uniformly for $|y| \leq C$, with $k = (p - 1)^{-1/(p-1)}$ according to whether the solution u has the existence time $> T$ or the solution blows up at $t = T$.

2.213 Show that the problem $u_t = \Delta u + u^\alpha$, $\alpha \in (0,1)$, $x \in \mathbb{R}^n$, $u(x,0) = 0$, has infinitely many solutions.

2.214 Consider the equation $u_t = \Delta u + f(u)$ on $\mathbb{R}^n$, where $f : \mathbb{R} \to \mathbb{R}$ is a C^1 function such that $f(0) = 0$, $f'(0) < 0$. Show that the trivial solution is asymptotically stable in the space of bounded uniformly continuous functions in $\mathbb{R}^n$. Moreover, if $f > 0$ on $(\alpha,0)$ and $f < 0$ on $(0,\beta)$ for some $\alpha < 0 < \beta$, inf $u(x,0)$ $= w_-(0) > \alpha$, sup $u(x,0) = w_+(0) < \beta$, $w_\pm(t)$ are the solutions of $w' = f(w)$, then for all x and $t \geq 0$, $w_-(t) \leq u(x,t) \leq w_+(t)$, $\lim_{t\to\infty} w_\pm(t) = 0$.

2.215 Let $u \neq 0$ be a classical solution of the problem $u_t = \Delta u + Q(u)$, $u(x,0)$ $= u_0(x) \geq 0$, $x \in \Omega \subset \mathbb{R}^n$, $u_0 \in C(\Omega)$, sup $u_0 < \infty$, $u(x,t) = u_1(x,t) \geq 0$, $t \in$ $(0,T)$, $x \in \partial\Omega$, $u_1 \in C(\partial\Omega \times [0,T))$, sup $u_1 < \infty$, $Q(0) = 0$, Q is C^1 and $Q(r) > 0$ for $r > 0$. Show that u is bounded if and only if $\int_1^\infty Q(r)^{-1} \, dr = \infty$.

2.216 Consider the following boundary value problem: $u_t = \Delta u + Q(u)$, $t >$ 0, $x \in \Omega$, Ω is a bounded domain in $\mathbb{R}^n$ with smooth boundary, $u(x,0) = u_0(x)$ > 0, $u_0 \in C(\overline{\Omega})$, $\partial u/\partial \nu = 0$ on $\partial\Omega$, and $Q''(r) > 0$ for $r > 0$, $\int_1^\infty Q(r)^{-1} \, dr <$ ∞. Show that each solution $u(x,t)$ exists on a bounded time interval $[0,T)$ and $\lim_{t\to T} u(x,t) = \infty$.

2.217 Consider the following problem: $u_t = \Delta u + Q(u)$, $t > 0$, $x \in \Omega \subset \mathbb{R}^n$, $u(x,0) = u_0(x) \geq 0$, $u_0 \in C(\overline{\Omega})$, $u_0 \neq 0$, $u(x,t) = 0$ for $x \in \partial\Omega$, $t > 0$, where Ω is a bounded domain with smooth boundary. Let λ_1 be the first eigenvalue of the problem $\Delta\psi + \lambda\psi = 0$, $\psi = 0$ on $\partial\Omega$, and let $\psi_1 > 0$ be the first eigenfunction normalized so that $\int_\Omega \psi_1(x) \, dx = 1$. Assume that $Q(r) - \lambda_1 r >$ 0 for $r \geq \delta > 0$ and some constant δ, $\int_\delta^\infty (Q(r) - \lambda_1 r)^{-1} \, dr < \infty$, $Q \in$ $C^2(\mathbb{R}^+)$ and $Q''(r) \geq 0$ for $r > 0$. Show that if $E_0 = \int_\Omega u_0(x)\psi_1(x) \, dx \geq \delta$, then the solution $u(x,t)$ of this problem exists on the interval $[0,T)$, $T \leq$ $\int_{E_0}^\infty (Q(r) - \lambda_1 r)^{-1} \, dr < \infty$ and $\lim_{t\to T} u(x,t) = \infty$.

2.218 Verify that solutions of the parabolic problem $u_t = u_{xx} + f(x,u)$, $0 < x$ $< \pi$, $u = 0$ for $x = 0$ and $x = \pi$, where $f(x,u) = \sin x \cdot g(u_1,u_2) + \sin 2x$ $\times h(u_1,u_2)$, $u_j = (2/\pi) \int_0^\pi u(x) \sin jx \, dx$, $j = 1, 2$, may be as complicated as solutions of an arbitrary dynamical system in the plane for suitably chosen g and h.

2.219 Consider the equation $u_t = \Delta u + \sum_{j=1}^n a_j (x,t) \partial u/\partial x_j + b(x,t)u$ in a bounded smooth domain $\Omega \subset \mathbb{R}^n$, where a_j and b are continuous functions such that $|a_j(x,t)| \leq A$, $b(x,t) \leq B$ on $\Omega \times \mathbb{R}^+$. Suppose that $u \, \partial u/\partial \nu \leq 0$ on $\partial\Omega$ $\times \mathbb{R}^+$ and $u(\cdot,0) \in L^p(\Omega) \cap L^\infty(\Omega)$ for some p, $1 \leq p < \infty$. Prove that if $\sup_{t\geq 0} \|u(\cdot,t)\|_p$ is finite, then $\sup_{x\in\Omega,t\geq 0} |u(x,t)|$ is also finite.

2.220 Study the equation $u_t = u_{xx} + \lambda^2 f(u)$, $0 < x < 1$, $u(0,t) = u(1,t) = 0$, where $f(u) = u(u - b)(1 - u)$ and $\lambda > 0$ is the bifurcation parameter.

2.221 Prove that for the (simplified) Fitz-Hugh–Naguymo equation $u_t = u_{xx} + f(u)$, $x \in \mathbb{R}$, $t > 0$, $f(u) = u(1 - u)(u - a)$, where $0 < a < 1/2$, there exists a traveling wave solution of the form $u(x,t) = \varphi(x + Vt)$ for some $V > 0$.

2.222 Determine traveling wave solutions $u(x,t) = \varphi(x + ct)$ of the Burgers equation $u_t + uu_x = \nu u_{xx}$, $x \in \mathbb{R}$, $t > 0$, $\nu > 0$.

2.223 Show that for the reaction-diffusion equation $u_t - d \Delta u + g(u) = 0$ on a bounded open subset Ω of $\mathbb{R}^n$, with the homogeneous Dirichlet boundary condition, $F(u) = \frac{1}{2}d\|\nabla u\|_2^2 + \int_\Omega G(u)$, where $G' = g$, $G(0) = 0$, is a Liapunov function.

2.224 Prove a local existence and uniqueness result for the problem from modeling of liquid crystals: $u_t - \Delta u + (u,h(t))^2 u - (u,h(t))h(t) = 0$ in a bounded domain $\Omega \subset \mathbb{R}^n$, $n = 1,2,3$, $(u,v) = 0$ and $(\partial u/\partial v) \times v = 0$ on the boundary Γ of Ω (which is supposed to be smooth), where $u = (u_1, \ldots, u_n)$ and $h \in C^1(\mathbb{R}^+;\mathbb{R}^n)$.

2.225 One says that a solution of some evolution equation quenches in time T if some norm of the solution itself remains bounded while some norm of its derivative becomes unbounded on the interval $(0,T)$. For example, the solution of $u_t = 1/(1 - u)$, $u(0) = 0$, given by $u(t) = 1 - (1 - 2t)^{1/2}$, quenches in time $T = 1/2$. Prove that a similar phenomenon can occur for the parabolic equation $u_t = u_{xx} + c/(1 - u)$, $0 < x < 1$, $t > 0$, $u(x,0) = u(0,t) = u(1,t) = 0$. More precisely, if the constant c is sufficiently large, one of the quantities $u_t(1/2,t)$, $u_{xx}(1/2,t)$ becomes unbounded in finite time.

2.226 Consider the evolution equation associated with the p-Laplacian $u_t = \nabla \cdot (\mid \nabla u \mid^{p-2} \nabla u)$ in an open subset of $\mathbb{R}^n$, $p > 2$, written in an abstract form as $u_t = A(u)$, $-\int A(u)u \, dx = \int \mid \nabla u \mid^p$. Let $\Lambda(t) = \int A(u)u/(\int u^2)^{p/2}$. Give a formal argument (that is, ignoring all existence of solutions questions, etc.) that
(i)   If $u(x,0) \neq 0$, then $u(x,t) \neq 0$.
(ii)   There exists $\lim_{t \to \infty} \Lambda(t) = \Lambda$.
(iii) $\|u(t)\|_2 \simeq C(u_0)t^{-\sigma}$, $\sigma = \sigma(p,n)$, as $t \to \infty$.
(iv) $u(t)/\|u(t)\|_2$ tends to a limit v as $t \to \infty$, $A(v) = \Lambda v$, $\|v\|_2 = 1$, $\int_0^\infty \|u(t)/\|u(t)\|_2 - v\|_2^2 \, dt < \infty$.

2.227 Discuss the existence of selfsimilar solutions of the boundary value problem for the porous medium equation $Au_t = (uu_x)_x$, $A > 0$, $u(x,0) = P_0$, $0 < x < \infty$, $u(0,t) = P_1 < P_0$, $0 \leq t < \infty$.

2.228 Verify that the porous medium equation $u_t = \Delta(u^m)$, $m > 1$, in $\mathbb{R}^n$ has a family of selfsimilar radial solutions of the form

$$U(x,t;M) = t^{-\alpha}(A - B \mid x \mid^2 t^{-2\beta})_+^{1/(m-1)},$$

$M \geq 0$ a parameter, the so-called Barenblatt solutions, with compact support in x for all t. Compute the limit of $U(x,t;1)$ as $m \to 1$.

2.229 Let u be a positive solution of the porous medium equation $u_t = \Delta(u^m)$ in $\mathbb{R}^n$ and let $v = m(m - 1)^{-1}u^{m-1}$ be the corresponding (to the density u) pressure. Prove the Aronson-Bénilan inequality $\Delta v \geq -C/t$ in $\mathcal{D}'(\mathbb{R}^n \times \mathbb{R}^+)$, where $C = (m - 1 + 2/n)^{-1}$. Moreover, show that $v_t \geq -(m - 1)Cv/t$.

Remark: This result is sharp, since the above inequalities become equalities for the Barenblatt solutions.

2.230 Prove the following estimate for the pressure $v = m(m - 1)^{-1}u^{m-1}$ in the porous medium equation $u_t = \Delta(u^m)$: $|v_x(x,t)|^2 \leq 2/((m + 1)t)\|v(\cdot,t)\|_\infty$.

2.231 A formal computation with respect to the Cauchy problem for the porous medium equation $u_t = (|u|^m \operatorname{sgn} u)_{xx}$, $(x,t) \in \mathbb{R} \times \mathbb{R}^+$, $u(x,0) = u_0(x)$ when $m > 0$ and $m \neq 1$, shows that, if $u(x,t)$ is the solution corresponding to the initial value u_0, then $v(x,t) = \lambda^{1/(m-1)}u(x,\lambda t)$, $\lambda > 0$, is the solution corresponding to the initial value $\lambda^{1/(m-1)}u_0$. This motivates the definition of the homogeneous nonlinear semigroup $T(t)$ on a Banach space (here on $L^1(\mathbb{R})$) which should satisfy the condition $T(t)(\lambda^{1/(m-1)}u_0) = \lambda^{1/(m-1)}T(\lambda t)u_0$ for $\lambda > 0$ and $t \geq 0$. Prove that if such $T(t)$ satisfies moreover the Lipschitz condition $\|T(t)f - T(t)g\| \leq L\|f - g\|$ and $T(t)(0) = 0$, then for all t, $h > 0$, $\|T(t + h)f - T(t)f\| \leq 2L\|f\| \, |1 - (1 + h/t)^{1/(m-1)}|$, hence $\lim\sup_{h\to 0}\|h^{-1}(T(t + h)f - T(t)f)\| \leq 2L\|f\|/(t|m - 1|)$.

2.232 It is known (see the references in Answers) that there is one-to-one correspondence between positive solutions of the porous medium equation $u_t = \Delta(u^m)$, $m > 1$, in $\mathbb{R}^n \times (0,T)$ and positive Borel measures satisfying the condition $\sup_{R\geq 1} R^{-(n+2/(m-1))} \mu\{|x| \leq R\} < \infty$. Exhibit some solutions of the porous medium equation which blow up in finite time, that is, $\lim_{t\to T} u(x,t) = \infty$ for all x.

Hyperbolic equations

2.233 Show that the thrice differentiable function $u(x,t)$ is a solution of the wave equation $u_{tt} = u_{xx}$ if and only if u satisfies the difference equation $u(x - k, t - h) + u(x + k, t + h) = u(x + h, t + k) + u(x - h, t - k)$ for all h, $k > 0$.

2.234 Show that the solvability condition for the equation $u_{xy} = 0$ with the boundary condition $u(\vartheta) = \varphi(\vartheta)$ on the unit circle is $\varphi(\vartheta + \pi) + \varphi(\vartheta) = \varphi(-\vartheta + \pi) + \varphi(-\vartheta)$. Then there is an infinite number of solutions and the difference of two such solutions is a constant multiple of $(x^2 + y^2 - 1)$.

2.235 Show by constructing an example that the boundary value problem for the wave equation $u_{xy} = 0$ in the unit square, $u|_{x=0} = \varphi_0(y)$, $u|_{x=1} = \varphi_1(y)$, $u|_{y=0} = \psi_0(x)$, $u|_{y=1} = \psi_1(x)$ is not well posed.

2.236 Suppose that u satisfies the wave equation in $\mathbb{R}^3$ and $u(x,y,0)$ is a smooth function, $0 \leq u(x,y,0) \leq 1$, $u_t(x,y,0) = 0$. Does it imply that $0 \leq u \leq 1$ for all $t \geq 0$?

2.237 Show that formally

$$u(x,t) = \sum_{k=0}^{\infty}\left(\frac{(\Delta^k f)(x)t^{2k}}{(2k)!} + \frac{(\Delta^k g)(x)t^{2k+1}}{(2k+1)!}\right)$$

solves the wave equation $u_{tt} = \Delta u$ with the conditions $u(x,0) = f(x)$, $u_t(x,0) = g(x)$. Determine a class of functions such that the above formula can be justified. Does this formula work for $f, g \in C_0^\infty(\mathbb{R}^n)$?

2.238 Show that the wave equation $u_{tt} = \Delta u$ in $\mathbb{R}^n \times \mathbb{R}$ is conformally invariant; that is, it is invariant under the group of conformal transformations preserving Lorentz angles. Give an explicit expression for the associated conserved quantities.

2.239 By performing Lorentz transformations on the identity

$$u(x,2t) = \frac{\partial(t(4\pi)^{-1}\int_{|y-x|=t} u(y,t)\, dS(y))}{\partial t}$$

valid for any solution of the wave equation $u_{tt} = \Delta u$ in the three-dimensional case, solve the characteristic initial value problem for that equation in closed form. Determine precisely what data are needed to calculate the solution and relate the answer to the Huygens principle. Derive the above identity directly from the Poisson formula for the solutions of the wave equation.

2.240 Prove that the fundamental solution of the wave equation vanishing for $t < 0$ must have its support in the forward light cone.

2.241 Suppose that the plane $z = 0$ in $\mathbb{R}^3$ is made of an elastic material (a "bell" in the homogeneous medium where the sound propagation velocity equals c). After a localized strike, a circular wave front in this elastic sheet propagates at the velocity a from the point of the excitation. Does this wave propagate in the whole space?

2.242 Assume that there exist smooth functions $\alpha(r) > 0$ and $\delta(r) > 0$, $\delta(0) = 0$, $\alpha(0) = 1$, such that for any polynomial f the function $\alpha(r)f(t - \delta(r))$ is a radially symmetric solution of the n-dimensional wave equation. Show that $n = 1$ or $n = 3$; moreover, for $n = 1$, $\alpha(r) \equiv 1$. This exercise shows that symmetric distortionless wave propagation is possible only in dimensions one and three. Moreover, in one dimension there is no attenuation.

2.243 Let us call the family of functions $u = g(x,t)S(\varphi(x,t))$ with some g and φ and arbitrary S "relatively undistorted progressing waves" ($\varphi(x,t)$ is the phase function, g is the attenuation factor). Prove that in the case $x \in \mathbb{R}$, $t \in \mathbb{R}$, the only partial differential equations which admit relatively undistorted progressing wave families in both space directions are $u_{xy} = 0$ and equations equivalent to it (here $t = y$).

2.244 Show that the solution of the Cauchy problem $u_{tt} = \Delta u$ in $\mathbb{R}^2 \times \mathbb{R}$, $u\mid_{t=0} = \varphi$, $u_t\mid_{t=0} = \psi$, with the initial data supported in the ball $\{\,|\,x\,| \le R\}$ is analytic in the cone $\{\,|\,x\,| < t - R,\, t > R\}$. Moreover, show that u satisfies the estimate $|\,u(x,t)\,| \le Ct^{-1/2}(1 + |\,t - |\,x\,|\,|^{1/2})^{-1}$. Observe that if $\varphi = 0$ and $\psi \ge 0$, $\psi \ne 0$, then $u(x,t) \ge C_0 t^{-1/2}(1 + |\,t - |\,x\,|\,|^{1/2})^{-1}$ for some constant $C_0 > 0$. Show that the solution of the Cauchy problem for the wave equation in $\mathbb{R}^3 \times \mathbb{R}$ with compactly supported initial data satisfies the estimate $|\,u(x,t)\,| \le C/t$.

2.245 Prove that for the solutions of the Cauchy problem for the wave equation $u_{tt} = \Delta u$ in $\mathbb{R}^n \times \mathbb{R}$ with the initial data $u(x,0)$, $u_t(x,0)$ in $\mathscr{S}(\mathbb{R}^n)$ the following estimates hold:
(i)   for every N and $\varepsilon > 0$ there exists a constant $C_{N,\varepsilon}$ such that $|\,u(x,t)\,| \le C_{N,\varepsilon}(1 + |\,x\,| + |\,t\,|)^{-N}$ if $|\,x\,| \ge (1 + \varepsilon)\,|\,t\,|$;
(ii)   for every $\varepsilon > 0$ there exists C_ε such that $|\,u(x,t)\,| \le C_\varepsilon(1 + |\,t\,|)^{-(n-1)}$ if $|\,x\,| \le (1 - \varepsilon)\,|\,t\,|$;
(iii) $|\,u(x,t)\,| \le C(1 + |\,t\,|)^{-(n-1)/2}$ for all x, t and some C.

2.246 Prove the estimate $|\,u(x,t)\,| \le ct^{-(n-1)/2}\,||\,u_1\,||_{W^{[n/2],1}}$ for the wave equation $u_{tt} = \Delta u$ in $\mathbb{R}^n$, $u(x,0) = 0$, $u_t(x,0) = u_1(x)$.

2.247 Prove that if $C = \int_{-\infty}^{\infty} \lambda\, dE_\lambda$ is a selfadjoint operator in a Hilbert space $(H, |\cdot|)$, then the abstract wave equation $u_{tt} + C^2 u = 0$ admits equipartition of energy, that is, $\lim_{t \to \pm\infty} |\,u_t(t)\,| / |\,Cu(t)\,| = 1$, if and only if for all $h \in H$, $\lim_{t \to \pm\infty} \int_{-\infty}^{\infty} \exp(it\lambda)\, d\,|\,E_\lambda h\,|^2 = 0$.
Remark: The condition in this theorem is satisfied if the spectral measure E_λ of C is absolutely continuous. The classical example is the wave equation $u_{tt} = \Delta u$ in $\mathbb{R}^n \times \mathbb{R}$, $H = L^2(\mathbb{R}^n)$, where $|\,u_t\,|^2$ is the kinetic energy, $|\,\nabla u\,|^2$ is the potential energy, and the equipartition of energy gives $\lim_{t \to \pm\infty} |\,u_t\,|^2 = \lim_{t \to \pm\infty} |\,\nabla u\,|^2 = \frac{1}{2}(|\,u_t(0)\,|^2 + |\,\nabla u(0)\,|^2)$.

2.248 Prove the averaged version of equipartition of energy for the generalized wave equation: If C is a selfadjoint operator in a Hilbert space $(H, |\cdot|)$ and u satisfies the equation $u_{tt} + C^2 u = 0$, then

$$\lim_{t \to \pm\infty} T^{-1} \int_0^T |\,u_t(s)\,|^2\, ds = \lim_{t \to \pm\infty} T^{-1} \int_0^T |\,Cu(s)\,|^2\, ds$$

$$= \frac{1}{2}(|\,u_t(0)\,|^2 + |\,Cu(0)\,|^2)$$

2.249 Prove equipartition of energy for the damped generalized wave equation $u_{tt} + \varepsilon u_t + C^2 u = 0$, where $\varepsilon > 0$; that is, show that $\lim_{t \to \infty} |\,u(t)\,| / |\,Cu(t)\,|$

$= 1$. Here $|u(t)|^2 + |Cu(t)|^2 = c \exp(-\varepsilon t)$, $t \to \infty$, which is very different from the conservative case treated in the preceding problems.

2.250 The Taylor example. There exist a strictly convex bounded open set $\Omega \subset \mathbb{R}^n$ with smooth boundary and a solution of the wave equation $u_{tt} = \Delta u$ in $\Omega \times \mathbb{R}$ such that $u \in C^\infty(\Omega \times \mathbb{R}) \backslash C^\infty(\overline{\Omega} \times \mathbb{R})$.

2.251 Give an example of a strictly hyperbolic operator (with bad coefficients' behavior at infinity) such that not every point $(x,t) \in \mathbb{R}^{n+1}$ has a finite dependence domain.

2.252 Study the propagation of singularities for solutions of the system of three-dimensional elasticity $(u = (u_1,u_2,u_3))$ $u_{tt} - (\lambda + u) \nabla(\text{div } u) - \mu \Delta u = 0$ where λ, $\mu > 0$ are the Lamé constants.

2.253 Solve the wave equation $u_{tt} = u_{xx}$ with the conditions $u = h$, $\partial u/\partial v = g$ given on the line $t = kx$, $k > 0$. What happens for $k = 1$ (a version of the Goursat problem)?

2.254 Solve the Goursat problem $u_{xy} = 0$, $y > ax$, $x > 0$, $a < 0$; $u \mid_{y=ax} = 0$, $u \mid_{x=0} = 0$.

2.255 Prove that the mixed problem $u_{xy} = 0$, $y > ax$, $x > 0$, $a < 0$; $u \mid_{y=ax} = \varphi_1(x)$, $u_y \mid_{y=ax} = \varphi_2(x)$, $u \mid_{x=0} = \varphi_3(y)$, where φ_1, $\varphi_3 \in C^2[0,\infty)$, $\varphi_2 \in C^1[0,\infty)$, has at most one solution. Give a necessary and sufficient condition that this solution is twice continuously differentiable in $\{y \geq ax, x \geq 0\}$.

2.256 Find the general solution of the equation $u_{xy} = u$.

2.257 Show that a solution of the problem $\Delta_{st}U - (s^2 + t^2)U = 0$, $0 < s < (2R)^{1/2}$, $\lim U(s,t) = F(t)$, where $F \in L^2(\mathbb{R})$ and convergence is in the L^2 metric, is of the form $\sum_{n=1}^\infty c_n \exp((t^2 - s^2)/2)H_n(s)H_n(it)$. $H_n(z) = (-1)^n \exp(z^2) d^n(\exp(-z^2))/dz^n$ is the nth Hermite polynomial.

2.258 Find the general solution of the equation $u_{tt} + t^\alpha \Delta u = 0$, $\alpha > 0$, $u = u(x,t)$, $x \in \mathbb{R}^n$, which is analytic off the hyperplane $t = 0$.

2.259 Show that small oscillations of a vertically hanged thread rotating at the angular velocity ω are described by the equation $u_{tt} = g(xu_x)_x + \omega^2 u$. Solve this equation for $\omega = 0$.

2.260 Prove the Ásgeirsson theorem: If u is a continuous solution of the equation $\Delta_x u = \Delta_y u$ in a neighborhood of the set $K = \{(x,y) : x,y \in \mathbb{R}^n, |x| + |y| \leq R\}$, then $\int_{|x|=R} u(x,0) \, dS(x) = \int_{|y|=R} u(0,y) \, dS(y)$. Moreover, if $n > 1$ is odd, this suffices to assume that u is a solution in a neighborhood of ∂K.

2.261 Prove that the series $\sum_{k=0}^\infty \lambda^k Z_{2k-2}$, where

$$Z_a(x,t) = 2^{1-a}\left(\Gamma\left(\frac{a}{2}\right)\Gamma\left(\frac{a+1-n}{2}\right)\right)^{-1}(t^2 - |x|^2)^{(a-n-1)/2}$$

when $t \geq |x|$ and 0 otherwise, $x \in \mathbb{R}^n$, defines an entire function of the complex variable λ, with their values in $\mathcal{D}'(\mathbb{R}^{n+1})$, which is a fundamental solution of the operator $\partial^2/\partial t^2 - \Delta - \lambda$.

2.262 Study the Dirichlet problem for the damped wave equation $u_{tt} = u_{xx} + cu$ in the rectangle $[0,a] \times [0,b]$.

2.263 Consider the Cauchy problem for the linear Klein-Gordon equation $u_{tt} = (\Delta - m^2)u$, $m > 0$, in $\mathbb{R}^n \times \mathbb{R}$ with the initial data $u(x,0) = u_0(x)$, $u_t(x,0) = u_1(x)$ such that $\hat{u}_0, \hat{u}_1 \in C_0^\infty$. Show that the solution (which is a so-called regular wave packet) satisfies the following estimates;
(i) for some $\rho < 1$ and every N there exists a constant C_N such that $|u(x,t)| \leq C_N (1 + |x| + |t|)^{-N}$ if $|x| > \rho t$;
(ii) there exists a constant C such that $|u(x,t)| \leq C(1 + |t|)^{-n/2}$ for all x, t;
(iii) $\int_{\mathbb{R}^n} |u(x,t)| \, dx \leq C(1 + |t|)^{n/2}$.

2.64 Discuss the solutions of the telegraph equation (for the potential or the current in a telegraphic line) $u_{xx} = au_{tt} + 2bu_t + cu$.

2.265 Consider the damped wave equation $u_{tt} + 2\alpha u_t - \Delta u = 0$ in a bounded open set $\Omega \subset \mathbb{R}^n$ with the Dirichlet boundary conditions $u = 0$ on $\partial\Omega$. Write this equation as a system for the vector $<u,u_t>$ and show that this system generates a strongly continuous group in a suitable function space.

2.266 Determine the decay rates as $t \to \infty$ for the norms of solutions of the abstract damped second-order equation $u_{tt} + \delta u_t + Au = 0$ in a Hilbert space H, where $u : [0,\infty) \to H$, A is a selfadjoint operator in H, $A \geq \alpha I$ for some $\alpha > 0$.

2.267 Determine the decay rates as $t \to \infty$ for the norms of solutions of the abstract strongly damped second-order equation $u_{tt} + Bu_t + Au = 0$ in a Hilbert space H, where $u : [0,\infty) \to H$, A and B are selfadjoint commuting operators in H, $A \geq \alpha I$, $B \geq \beta I$ for some α, $\beta \geq 0$.

2.268 Solve the equation $x^2 u_{xx} = y^2 u_{yy}$.

2.269 Determine the solution of the equation $u_{xx} - u_{tt} + \lambda x^{-1} u_x = 0$ with the conditions $u(\ell_1,t) = u(\ell_2,t) = 0$, $0 < \ell_1 < \ell_2$, $u(x,0) = g(x)$, $u_t(x,0) = h(x)$, $\ell_1 \leq x \leq \ell_2$.

2.270 Solve the equation $u_{tt} + (2n - 1)t^{-1}u_t = \Delta u$ with $x \in \mathbb{R}^n$ and the initial conditions $u(x,0) = v(x)$, $u_t(x,0) = 0$.

2.271 Let A generate a C_0 group $T(t)$ on a Banach space X and let $f \in D(A^2)$. Then u defined by

$$u(t) = \Gamma\left(\frac{\rho+1}{2}\right)\left(\Gamma\left(\frac{\rho}{2}\right)\Gamma\left(\frac{1}{2}\right)\right)^{-1}\int_{-1}^{1}(1-s^2)^{\rho/2-1}\,T(st)\,f\,ds$$

satisfies the abstract Euler-Poisson-Darboux equation $u_{tt} + \rho t^{-1}u_t - A^2u = 0$, $t > 0$, $\rho > 0$ and $u(0) = f$, $u_t(0) = 0$.

2.272 Verify that the problem $t^2u_{xx} - u_{tt} + (4n+1)u_x = 0$, $u(x,0) = f(x)$, $u_t(x,0) = 0$, $0 \le x \le 1$, has the unique solution

$$u(x,t) = \sum_{k=0}^{n} \pi^{1/2}\,t^{2k}\left(k!(n-k)!\Gamma\left(n+\frac{1}{2}\right)\right)^{-1}\frac{\partial^k f(x+t^2/2)}{\partial x^k}$$

2.273 Show that the general solution of the equation $2(2n+1)^{-1}xu_{xx} - u_{yy} + u_x = 0$ is

$$u(x,y) = \frac{\partial^{n-1}\,(x^{-1/2}\,(F_1((2(2n+1)x)^{1/2}+y) + F_2((2(2n+1)x)^{1/2}-y)))}{\partial x^{n-1}}$$

where F_1 and F_2 are arbitrary functions.

2.274 Show that the general solution of the equation $a^{-2}u_{tt} = u_{xx} + 2x^{-1}u_x - n(n+1)x^{-2}u$ is

$$u(x,t) = x^n\left(x^{-1}\frac{\partial}{\partial x}\right)^n (x^{-1}(F_1(x-at) + F_2(x+at)))$$

where F_1 and F_2 are arbitrary functions.

2.275 Find the general solution of the Euler-Darboux equation $u_{xy} - \alpha(x-y)^{-1}u_x + \beta(x-y)^{-1}u_y = 0$ if α and β are nonnegative integers or $0 < \alpha, \beta < 1$, $\alpha + \beta \ne 1$.

2.276 Verify that the operators

$$L_1 = \frac{\partial}{\partial x_1} + \frac{\partial}{\partial x_2}, \qquad L_2 = x_1\frac{\partial}{\partial x_1} + x_2\frac{\partial}{\partial x_2}$$

$$L_3 = x_1^2\frac{\partial}{\partial x_1} + x_2^2\frac{\partial}{\partial x_2}, \qquad L_4 = u\frac{\partial}{\partial u}, \qquad L_5 = \varphi(x_1,x_2)\frac{\partial}{\partial u}$$

(where φ is an arbitrary solution of the equation below) transform solutions into solutions of the Euler-Poisson equation

$$\frac{\partial^2 u}{\partial x_1 \, \partial x_2} + a(x_1 - x_2)^{-2} u = 0, \qquad a = \text{const.}$$

2.277 Show that the (characteristic) problem for the hyperbolic system $\partial^2 u_1/\partial x^2 + \partial^2 u_1/\partial y^2 + 2\,\partial^2 u_2/\partial x \, \partial y = 0, 2\,\partial^2 u_1/\partial x \, \partial y + \partial^2 u_2/\partial x^2 + \partial^2 u_2/\partial y^2 = 0$ in the domain bounded by the lines $x - y = 0$, $x + y = 0$, $x > 0$, with the conditions $u_j(x,x) = f_1^j(x)$, $u_j(x, -x) = f_2^j(x)$, $x \ge 0$, $f_1^j(0) = f_2^j(0)$, $j = 1,2$, is not well posed.

2.278 Consider a quasilinear hyperbolic equation $au_{xx} + 2bu_{xy} + cu_{yy} + d = 0$ $(a,b,c,d = a,b,c,d(x,y,u,u_x,u_y), b^2 > ac)$ and its characteristics described by the equations $y_\xi - \lambda_+ x_\xi = 0$, $y_\eta - \lambda_- x_\eta = 0$, where $\lambda_\pm = a^{-1}(b \pm (b^2 - ac)^{1/2})$. Establish the identity

$$\det \begin{bmatrix} x_{\xi\eta} & y_{\xi\eta} & u_{\xi\eta} \\ x_\xi & y_\xi & u_\xi \\ x_\eta & y_\eta & u_\eta \end{bmatrix} = \frac{d(x_\xi y_\eta - x_\eta y_\xi)^2}{2(b^2 - ac)^{1/2}}$$

2.279 Show that the solution of the nonlinear Cauchy problem $u_{xy} = u^2$, $u(x,-x) = 6$, $u_x(x,-x) = u_y(x,-x) = 12$, becomes infinite on the line $x + y = 1$ and hence does not exist in the large.

2.280 Show that the equation $u_{tt} - u_{xx} + uu_x = 0$ has discontinuous solutions.

2.281 Verify that the following functions solve the Liouville equation $u_{tt} - u_{xx} = g \exp u$, $g > 0$:
(i) $\log(\alpha^2(1 - a^2)/ 2g \cosh^2(\alpha/2(x - x_0 - at)))$, $0 \le a \le 1$;
(ii) $\log(8\varphi'(x + t)\psi'(x - t) /g(\varphi(x + t) - \psi(x - t))^2)$ for arbitrary $\varphi, \psi \in C^3$ such that $\varphi'\psi' > 0$;
(iii) $\frac{1}{2}(u_0(x + t) + u_0(x - t)) - \log(\cos^2((g/8)^{1/2} \int_{x-t}^{x+t} \exp(u_0(\xi)/2) \, d\xi))$ if the absolute value of the argument of $\cos^2$ is less than $\pi/2$.

2.282 Find all simple wave solutions of the equation $u_{tt} = (1 + u_x)^2 u_{xx}$ with $u(x,0) = h(x)$.

2.283 Show that for the nonlinear wave equation $u_{tt} - \Delta u + u^3 = 0$ in $\mathbb{R}^3 \times \mathbb{R}$, $E(t) = \frac{1}{2}\int(|\nabla u|^2 + u_t^2 + \frac{1}{2}u^4) \, dx$ and $Q(t) = \int((t^2 + |x|^2)\frac{1}{2}(u_t^2 + |\nabla u|^2 + \frac{1}{2}u^4) + 2tu_t \, x \cdot \nabla u + 2tu_t u - u^2) \, dx$ are constant in time.

2.284 Prove that the nonlinear wave equation $u_{tt} - \Delta u + cu^p = 0$ in $\mathbb{R}^n \times \mathbb{R}$ is conformally invariant if and only if $c = 0$ or $p = 1 + 4/(n - 1)$.

2.285 Assume that $uf(u) \le (2 + \varepsilon)F(u)$ for all $u \in \mathbb{R}$ and some $\varepsilon > 0$ for the nonlinearity in the wave equation $u_{tt} - \Delta u + f(u) = 0$ ($F' = f$, $F(0) = 0$). Prove that if the initial data satisfy $E(u) = \int(\frac{1}{2}u_t^2 + \frac{1}{2}| \nabla u |^2 + F(u))\, dx \mid_{t=0}$ < 0 and $\int uu_t\, dx \mid_{t=0} > 0$, then the solution of the nonlinear wave equation with those data does not exist for all time.

2.286 Prove that the nonlinear Klein-Gordon equation $u_{tt} - u_{xx} + m^2u + 2\varepsilon u^3 = 0$ has for $\varepsilon = -1$ nonglobal solutions (they explode in finite time). **Remark:** If $\varepsilon = 1$ then the energy integral $E(t) = \frac{1}{2}\int_{\mathbb{R}}(| u_x |^2 + | u_t |^2 + m^2 | u |^2 + \varepsilon | u |^4)\, dx$ in constant and this provides a necessary a priori bound for global existence of solutions of the Cauchy problem.

2.287 Suppose that φ is twice differentiable on $[0,T)$, h is locally integrable, $\varphi'' \ge h(\varphi)$, $\varphi(0) > 0$, $\varphi'(0) > 0$, and for a positive function H, $H' = h$, $(1 + H)^{-1/2}$ is integrable at infinity. Show that T must be finite.
  Use the above lemma to prove the Glassey theorem: Consider a nonlinear wave equation $u_{tt} - \Delta u = f(u)$ on a bounded domain $\Omega \subset \mathbb{R}^n$ with $u = 0$ on $\partial\Omega$, f a convex function, $f(s) - \mu s > 0$ for $s > 0$ (where μ is the first eigenvalue of $-\Delta$ on Ω), $(F(s) - \mu s)^{-1/2}$ integrable at infinity, $F' = f$. If $\int u(x,0)\psi(x)\, dx > 0$, $\int u_t(x,0)\psi(x)\, dx > 0$, where $\psi \ge 0$ is the first eigenfunction of $-\Delta$ on Ω, then the norm of $u(t)$ in $L^2(\Omega)$ must explode in finite time.

2.288 Prove that the Cauchy problem for the nonlinear wave equation $u_{tt} - u_{xx} = u^k$, $k \in \mathbb{N}$, does not have the global solution for certain $u_0(x) = u(x,0)$, $u_1(x) = u_t(x,0)$.

2.289 Consider the nonlinear wave equation $u_{tt} - u_{xx} + u + \varepsilon u^3 = 0$, $0 < \varepsilon \ll 1$, which has for $\varepsilon = 0$ the periodic wave solution $u = \sin \vartheta$, $\vartheta = kx - (k^2 + 1)^{1/2}t$. Show, after Stokes, that for $0 < \varepsilon \ll 1$ these periodic solutions survive and can be approximated by $u(x,t,\varepsilon) = \sin \theta + \varepsilon u_1 (\theta) + \varepsilon^2 u_2(\theta) + \cdots$, where $\theta = kx - (k^2 + 1)^{1/2}(1 + \varepsilon\omega_1 + \varepsilon^2\omega_2 + \cdots)t$. Calculate ω_1 and u_1. Deduce the initial conditions $u(x,0,\varepsilon)$, $u_t(x,0,\varepsilon)$ which would produce this solution to $\mathcal{O}(\varepsilon)$. Derive an exact expression for this periodic solution.

2.290 Show that for the damped sine-Gordon equation $u_{tt} + \alpha u_t - \Delta u + \beta \sin u = f$ in a bounded open subset Ω of $\mathbb{R}^n$, with the homogeneous Dirichlet boundary condition, $F(u,u_t) = \frac{1}{2}\|\nabla u\|_2^2 + \frac{1}{2}\|u_t\|_2^2 - \beta \int_\Omega \cos u - \int_\Omega fu$ is a Liapunov function.

2.291 Consider the bifurcation problem for the damped wave equation $u_{tt} + 2\alpha u_t - u_{xx} = \lambda f(u)$, $0 < x < \pi$, $t \ge 0$, $u = 0$ at $x = 0$ and $x = \pi$, where $\alpha > 0$, $\lambda > 0$ is a parameter, f is twice continuously differentiable, $f(0) = 0$, $f'(0) = 1$, $\lim \sup_{|x| \to \infty} f(x)/x \le 0$, $\operatorname{sgn} f''(x) = -\operatorname{sgn} x$.

Semigroups and singular perturbation problems

2.292 Give an example of a uniformly bounded strongly continuous semigroup which is not a semigroup of contractions.

2.293 Give an example of a dissipative operator with no proper dissipative extension but not m-dissipative.

2.294 Consider the space $C_{0,\infty}(\mathbb{R})$ of continuous functions vanishing at 0 and ∞ and the differentiation operator $D = d/dx$. Show that D and $-D$ are both accretive on the natural domain of definition, but only one of them generates a contraction semigroup.

2.295 If V is an increasing continuous function on $\mathbb{R}$, the formula $(T(t)f)(x) = f(x - t)\exp(-V(x) + V(x - t))$ defines a one-parameter semigroup of the space $C_0(\mathbb{R})$ of all continuous functions on $\mathbb{R}$ such that $\lim_{x\to\pm\infty} f(x) = 0$. Describe the infinitesimal generator A of this semigroup.

2.296 Give an example of a C_0 semigroup of contractions $T(t)$ on a Banach space X such that the domain of its infinitesimal generator $\{x \in X : \lim_{t\to 0} t^{-1}(T(t)x - x) \text{ exists}\}$ is strictly smaller than its Favard class $\{x \in X : \sup_{t>0} \|t^{-1}(T(t)x - x)\| < \infty\}$.

2.297 Let $A_n = n^2 \sin nx \, d/dx$ on the space $C_0(\mathbb{R})$. Show that A_n generates a C_0 semigroup $T_n(t)$ on $C_0(\mathbb{R})$. Show also that for each $f \in C_0(\mathbb{R})\backslash\{0\}$, $\lim_{n\to\infty}\|T_n(t)f - f\| = 0$ for each $t \geq 0$ while $\lim_{n\to\infty}\|A_n f\| = \infty$.

2.298 Define the bounded operators $T(t)$ on $C[0,1]$ with the sup norm by $(T(t)f)(x) = x^t f(x) - x^t f(0)\log x$, $T(0) = I$. Does this semigroup satisfy the C_0-condition (joint continuity with respect to $(t,f) \in [0,\infty) \times C[0,1]$)?

2.299 Let $\{P_t\} = \{e^{t(P-I)}\}$ be a semigroup of Markov operators on $L^1(\mathbb{R}^n)$ generated by the linear Boltzmann equation $du/dt = (P - I)u$, where P is a Markov operator on L^1; that is, if $f \geq 0$ then $Pf \geq 0$ and $\|f\| = \|Pf\|$. Show that if the trajectory $\{P_t f\}$ of an $f \in L^1$ is weakly precompact, then $P_t f$ strongly converges as $t \to \infty$.

2.300 Assume that P is a Markov operator on L^1 and $f_0 > 0$ is a fixed point of P. Show that for all $f \in L^1$ the strong limit of $e^{t(P-I)}f$ exists as $t \to \infty$.

2.301 Consider the semigroup $T(t)$ on $C_0(\mathbb{R})$ defined by $(T(t)f)(x) = \exp(ix^2 t)(1 + x^2)^{-t}f(x)$. Study the differentiability of $T(t)$.

2.302 Show that the operator C defined by $(Cf)(x) = (-\log(1 + |x|) + ix)f(x)$ generates a strongly continuous semigroup $\exp(tC)$ on $L^2(\mathbb{R})$ and $t \mapsto \exp(tC)$ has exactly k continuous derivatives when $k < t \leq k + 1$, $k = 0,1,2,\ldots$.

2.303 Let $S(t)$ be a strongly continuous semigroup which is differentiable for $t > 0$ and let A be the infinitesimal generator of $S(t)$. Prove that if $\lim \sup_{t \to 0} t \| AS(t) \| < 1/e$, then A is a bounded operator and $S(t)$ can be extended analytically to the whole complex plane. Show that the constant $1/e$ is the best possible.

2.304 Prove that if for a strongly continuous semigroup $S(t)$, $\lim \sup_{t \to 0} \| I - S(t) \| < 2$, then $S(t)$ is analytic in a sector around the nonnegative real axis.

2.305 Give an example of a C_0 semigroup of bounded operators with no analytic vector.

2.306 Consider the spaces $L_s^2(\mathbb{R}) = \{ f \in L^2(\mathbb{R}) : | x |^s f(x) \in L^2(\mathbb{R}) \}$ and the operators $A_s : L_s^2 \to L^2$ defined by $(A_s f)(x) = - | x |^s f(x)$ and $(B_s f)(x) = (- | x |^s + ix) f(x)$, $B_s : L_{\min(s,1)}^2 \to L^2$. Show that A_s generate analytic semigroups while B_s for $0 < s < 1$ do not. Nevertheless, $\exp(tB_s) f(x) = \exp((- | x |^s + ix)t) f(x)$, $0 \le t < \infty$, is a strongly continuous semigroup and even $t \mapsto \exp(tB_s)$ is a C^∞ operator-valued function, satisfying the estimate $\| B_s \exp(tB_s) \| \le Ct^{-1/s}$ for $0 < s < 1$ and some $C > 0$.

2.307 Give an example of a strongly continuous semigroup $S(t)$ with the infinitesimal generator A such that $\sup\{ \mathrm{Re}\, \lambda : \lambda \in \sigma(A) \} \le -1$ while $\| S(t) \|$ does not decay exponentially.

2.308 Let ρ be a positive, nondecreasing, and differentiable function for $r \ge 0$ and $\int_0^\infty \rho(r)^{-1} dr < \infty$. Define $\rho(r) = \rho(0)$ for $r < 0$. Show that the operator $L = \rho(x_2) \, \partial/\partial x_1 + \rho(x_1) \, \partial/\partial x_2$ has an m-dissipative realization which is not the closure of L restricted to $\mathscr{D}(\mathbb{R}^2)$.

2.309 Prove that if $T(t)$ is a one-parameter group of isometries generated by the operator A and $S(t)f = \int_{-\infty}^\infty (2\pi t)^{-1/2} \exp(-s^2/2t) T(s) f \, ds$, then $S(t)$ is a one-parameter contraction semigroup with the generator $\frac{1}{2} A^2$.

2.310 Consider a 2-periodic smooth function ρ on $\mathbb{R}$ such that $\rho((\epsilon, 1 - \epsilon)) = \{1\}$ for some small $\epsilon > 0$, $\rho((1,2)) = \{0\}$, and ρ is monotone on $(0, \epsilon)$ and $(1 - \epsilon, 1)$. Define the following operators on the space $C_0(\mathbb{R})$ of the continuous functions on $\mathbb{R}$ which vanish at $\pm\infty$: $A_1 u = \rho x^3 u'$ on $D(A_1) = \{ u \in C_0(\mathbb{R}) : \rho x^3 u' \in C_0(\mathbb{R}) \}$, $A_2 u = (1 - \rho) x^3 u'$ on $D(A_2) = \{ u \in C_0(\mathbb{R}) : (1 - \rho) x^3 u' \in C_0(\mathbb{R}) \}$, $A_3 u = x^3 u'$ on $D(A_3) = \{ u \in C_0(\mathbb{R}) : x^3 u' \in C_0(\mathbb{R}) \}$; the derivatives are taken in the sense of distributions. Prove that $-A_1$ and $-A_2$ generate continuous semigroups of contractions $S_1(t)$ and $S_2(t)$, but the expression $(S_1(t/n) S_2(t/n))^n u$ which appears in the Lie-Trotter formula does not always converge in $C_0(\mathbb{R})$. (Obviously, $-A_3$ does not generate any semigroup, even in the nonlinear sense.)

2.311 Show that if an operator B generates a C_0 group $T(t)$ in a Banach space X, then $aI + B^2$ generates a cosine function C_a; that is, $C_a(s + t) + C_a(t - s) =$

$2 C_a(s)C_a(t)$ for all s, $t \in \mathbb{R}$, $C_a(0) = I$, and $C_a(t)x$ is continuous for all $x \in X$. Give explicit formulas for C_a.

Remark: Cosine functions generate solutions of well-posed Cauchy problems for second-order evolution equations such as $u_{tt} = Au$.

2.312 Prove that if C is a cosine function (see the above problem) then its generator A generates a C_0 semigroup $T(t)x = (\pi t)^{-1/2} \int_0^\infty \exp(-y^2/4t)C(y) \, x \, dy$. T is an analytic semigroup (when the considered space is complex) but not all operators generating analytic semigroups generate cosine functions.

2.313 Give an example of an evolution equation $u_t + Bu = f(u)$ in a Hilbert space, where the spectrum of B is contained in $\{Re \; \lambda \geq \delta > 0\}$, f has compact range and f is continuously differentiable, and there exists $u_0 \in X$ such that the solution $u(t)$ is uniformly bounded and is not differentiable in t.

2.314 Let A be a symmetric operator in a Hilbert space, u a solution of the equation $u'(t) = Au(t)$. Show that $\log \|u(t)\|$ is convex.

2.315 Consider a function $\varphi : [0, \infty) \to H$ with its values in a Hilbert space such that $\|d\varphi/dt + A\varphi\| \leq n(t)(A\varphi, \varphi)$ for a selfadjoint, strictly positive operator A with compact inverse and $n \in L^1(0, \infty) \cap L^2(0, \infty)$. Show that $(A\varphi, \varphi)/\|\varphi\|^2$ tends to an eigenvalue Λ of A as $t \to \infty$. Moreover, $\exp(\Lambda t)\varphi(t)$ tends to a nonzero eigenvector of A. Draw as corollaries results on the asymptotic behavior of solutions of parabolic equations (and inequalities).

2.316 Let $(H, \|.\|)$ be a Hilbert space and A be a positive selfadjoint unbounded operator in H with domain $D(A) \subset H$. Consider a function $w \in L^\infty(0,T;D(A^{1/2}))$ $\cap L^2(0,T;D(A))$ that satisfies the relation $dw(t)/dt + Aw(t) = h(t,w(t))$, $0 < t < T$, where $\|h(t,w(t))\| \leq k(t)\|A^{1/2}w(t)\|$ for a.e. t, $k \in L^2(0,T)$. Prove the following backward uniqueness result: If $w(T) = 0$, then $w(t) \equiv 0$ for all $0 \leq t \leq T$.

2.317 Consider the singular Cauchy problem $c(t)x' + \lambda x = 0$, $t > 0$, $x(0^+) = x_0$, $c(0) = 0$, $c(t) > 0$ for $t > 0$. Describe the behavior of solutions. Generalize to the case of the evolution equation $c(t)u_t + Au = f$ with a linear operator A which is the infinitesimal generator of an analytic semigroup.

2.318 Consider the singular perturbation problem $\varepsilon \Delta u = au_x + bu_y$, where a and b are constants, $\varepsilon > 0$, on a bounded domain in $\mathbb{R}^2$, with the Dirichlet boundary conditions. Prove that $\lim_{\varepsilon \to 0} u(x,y,\varepsilon) = v(x,y)$ exists and v satisfies the limit equation $av_x + bv_y = 0$.

2.319 Consider the elliptic singular perturbation problem $\varepsilon \Delta u = xu_x + yu_y$ on the unit disk in the plane, with the Dirichlet boundary condition $f = f(\vartheta)$, $0 \leq \vartheta < 2\pi$. Prove that the boundary layer approximation for $\varepsilon \to 0$ is $u \simeq \alpha + (f(\vartheta) - \alpha) \exp(-(1 - r)/\varepsilon)$, $r \to 1$, for some α.

2.320 Determine the solution to $\mathcal{O}(1)$ of the singular perturbation problem $\varepsilon(u_{xx} + u_{yy}) = u_y$, $0 < \varepsilon \ll 1$, on the unit square with boundary conditions $u_y(x,0) = 0$, $u(0,y) = 1$, $u(1,y) = -1$, $u(x,1) = 0$.

2.321 Consider the hyperbolic singular perturbation problem $\varepsilon(u_{xx} - u_{tt}) = au_x + bu_t$ with the initial conditions $u(x,0) = F(x)$, $u_t(x,0) = G(x)$, $x \in \mathbb{R}$. Show that for $b/\mid a \mid > 1$ there is an asymptotic expansion of $u(x,y,\varepsilon)$, $\varepsilon \to 0$, while the remaining cases lead to instability phenomena.

2.322 Work out the first term of the asymptotic expansion for the Riemann function

$$A(\xi,\tau;x,t) = \exp\left(\frac{\tau - t}{2\varepsilon^2}\right) J_0\left(\frac{(\varepsilon^2(x - \xi)^2 - (t - \tau)^2)^{1/2}}{2\varepsilon^2}\right)$$

of the hyperbolic equation $u_{xx} - \varepsilon^2 u_{tt} - u_t = 0$ in the limit as $\varepsilon \to 0$. Derive from it the expression

$$\lim_{\varepsilon \to 0} \frac{A(\xi,\tau;x,t)}{2\varepsilon} = \frac{1}{2}(\pi(t - \tau))^{-1/2} \exp\left(\frac{-(x - \xi)^2}{4(t - \tau)}\right)$$

for the fundamental solution of the heat equation $u_t = u_{xx}$.

2.323 Study the singular perturbation problem for the Burgers equation $\varepsilon u_{xx} = u_t + uu_x$ as $\varepsilon \to 0$.

2.324 Show that the solutions $u = u_\varepsilon$, $\varepsilon > 0$, of the Cauchy problem for the Burgers equation $u_t + uu_x = \varepsilon u_{xx}$, $u(x,0) = \operatorname{sgn} x$, which are continuous for $t \geq 0$, $\mid x \mid + t \neq 0$, and continuously differentiable for $t > 0$, converge to a (unique) solution of $u_t + uu_x = 0$, which is continuous for $t \geq 0$, $\mid x \mid + t \neq 0$, and continuously differentiable for $t \neq \mid x \mid$.

2.325 Suppose that A is an operator generating a strongly continuous semigroup and 0 is in the resolvent of A. Show that the solutions of the initial value problem $\varepsilon u'_\varepsilon(t) = Au_\varepsilon(t) + f(t)$, where $\varepsilon > 0$, f is a continuous function, approximate $-A^{-1}f(t)$ uniformly on compact subsets of $(0,\infty)$ as ε tends to zero.

2.326 Consider the (control) problem of finding u in $L^2(\mathbb{R}^n)$ such that given $T > 0$, v in $L^2(\mathbb{R}^n)$, and $\eta > 0$, $\|S(T)u - v\| \leq \eta$, where $S(t)$ is the semigroup solving the heat equation $u_t = \Delta u$ in $\mathbb{R}^n$. Obviously, one cannot take $u = S(-T)v$ since this is, in general, meaningless. However, if $\|w - v\| \leq \eta$ and $w \in \mathcal{F}^{-1}(\mathcal{D}(\mathbb{R}^n))$, then $u = S(-T)w$ is a solution. Unfortunately, this solution is far from being satisfactory from the computational point of view: u is very sensitive to small changes of w. Justify the following approach (a quasireversibility method): Let $u_t = \Delta u + \varepsilon \Delta^2 u$, where $\varepsilon > 0$, and denote by $S_\varepsilon(t)v$ the solution of the above equation for $t \le 0$. It can be shown that $v = \lim_{\varepsilon \to 0} S(T)S_\varepsilon(-T)v$ and that the formula $u = S_\varepsilon(-T)v$ will pose no computational problems.

2.327 Prove the regular perturbation theorem for second-order evolution equations: Suppose that A and B are commuting nonnegative selfadjoint operators on a Hilbert space H, B is injective, $\sigma_p(A^2 B^{-1})$ is bounded from above, and $u_\varepsilon \in C^2(\mathbb{R}^+;H)$, $\varepsilon > 0$, is the unique solution of the well-posed Cauchy problem $(u_\varepsilon)_{tt} + \varepsilon A(u_\varepsilon)_t + Bu_\varepsilon = 0, t \ge 0, u_\varepsilon(0) = f, (u_\varepsilon)_t(0) = g \in D(A^2) \cap D(B^2)$. Then $u_\varepsilon(t) = v_0(t) + \mathcal{O}(\varepsilon)$ where $v_0 \in C^1(\mathbb{R}^+; H)$ satisfies $(v_0)_{tt} + Bv_0 = 0$, $t \ge 0$, $v_0(0) = f$, $(v_0)_t(0) = g$, and the term $\mathcal{O}(\varepsilon)$ is uniform for t in compact intervals.

2.328 Prove the singular perturbation theorem for second-order evolution equations: Suppose that A and B satisfy the conditions as in the regular perturbation theorem above and $0 \notin \sigma_p(A)$, $\sigma_p(A^2 B^{-1})$ is bounded away from zero. For $\varepsilon > 0$ let $v_\varepsilon \in C^2(\mathbb{R}^+;H)$ be the unique solution of the well-posed Cauchy problem $\varepsilon(v_\varepsilon)_{tt} + A(v_\varepsilon)_t + Bv_\varepsilon = 0, t \ge 0, v_\varepsilon(0) = f, (v_\varepsilon)_t(0) = g \in D(A^2) \cap D(B^2)$. Then $v_\varepsilon(t) = v_0(t) + \mathcal{O}(\varepsilon)$, where $v_0 \in C^1(\mathbb{R}^+ H)$ satisfies $A(v_0)_t + Bv_0 = 0$, $t \ge 0$, $v_0(0) = f$, and the $\mathcal{O}(\varepsilon)$ term is uniform for t in compact intervals.

2.329 Consider the Klein-Gordon equation $\hbar^2 v_{tt} = \hbar^2 c^2 \Delta v - m^2 c^4 v$, where $\hbar$ is the Planck constant divided by 2π, c is the velocity of light. Let $v_\varepsilon = \exp(imc^2 t/\hbar)v(x,t)$. Then v_ε is a solution of $\hbar^2(2mc^2)^{-1}(v_\varepsilon)_{tt} + i\hbar(v_\varepsilon)_t - \hbar^2(2m)^{-1} \Delta v_\varepsilon = 0, t \in \mathbb{R}$. Taking $\varepsilon = 1/c^2$ prove that there exists v_0 such that $(v_0)_t = i\hbar(2m)^{-1} \Delta v_0$, $v_0(0) \in H^4(\mathbb{R}^n)$ and $\int |v_\varepsilon(x,t) - v_0(x,t)|^2 dx \le C(T)\varepsilon$ for all $T > 0$ and $|t| \le T$. This means that as the velocity of light becomes infinite, the Klein-Gordon equation has the correct nonrelativistic limit: the Schrödinger equation.

Miscellaneous equations and systems

2.330 Solve the Pfaffian equation $a^2 y^2 z^2 \, dx + b^2 z^2 x^2 \, dy + c^2 x^2 y^2 \, dz = 0$.

2.331 Verify that the equation $yz(y + z) \, dz + xz(x + z) \, dy + xy(x + y) \, dz = 0$ is integrable and find its solution.

2.332 Let P, N be the projection on the plane xy of a point M on a surface and the intersection of the normal to this surface at M with the plane xy respectively. Determine all the surfaces $z = z(x,y)$ such that the angle NOP is constant.

2.333 Find the general solution of the system $(x - a)(px + qy - 2z) = (z - c)(p - x), (y - b)(px + qy - 2z) = (z - c)(q - y)$, where $p = z_x, q = z_y$.

2.334 Solve the equation $p^2/x^2 + q^2/y^2 - 2/z^2 = 0$, where $z = z(x,y)$, $p = \partial z/\partial x$, $q = \partial z/\partial y$. Show that each integral surface contains a characteristic lying on a sphere centered at the origin.

2.335 Let $\vartheta = \vartheta(x,y,z)$ be a solution of the equation $(\partial \vartheta/\partial x)^2 + (\partial \vartheta/\partial y)^2 + (\partial \vartheta/\partial z)^2 = 1$. Prove that $\vartheta(x,y,z) = C$ describes a family of parallel surfaces.

2.336 Prove the Lie theorem: Each integral curve of the first-order partial differential equation $F(x,y,z,z_x,z_y) = 0$ has second-order tangency with any integral surface of the same equation at their tangent points.

2.337 Describe the orbits of the differential system $\partial x_1/\partial t_1 = 1$, $\partial x_1/\partial t_2 = 0$, $\partial x_2/\partial t_1 = 0$, $\partial x_2/\partial t_2 = 1$, $\partial x_3/\partial t_1 = -x_2/(x_1^2 + x_2^2)$, $\partial x_3/\partial t_2 = x_1/(x_1^2 + x_2^2)$ in $\mathbb{R}^3 \backslash \{x_1 = x_2 = 0\}$.

2.338 Show that the system $\partial v/\partial x = F(x,y,\partial u/\partial x,\partial u/\partial y)$, $\partial v/\partial y = G(x,y,\partial u/\partial x,\partial u/\partial y)$, where F and G are holomorphic near the origin and $\partial G/\partial p \neq 0$ ($p = \partial u/\partial x$, $q = \partial u/\partial y$) is equivalent to one second-order equation for u if and only if the functions F_q/G_p, $(G_q - F_p)/G_p$, $(pG_u - qF_u + G_x - F_y + FG_v - GF_v)/G_p$ are independent of v.

2.339 Discuss the relation of the problem of integrating the differential equation $X\,dx + Y\,dy + Z\,dz = 0$ to the theory of irrotational fields. In particular, give the geometric significance of the usual condition for integrability $X(\partial Z/\partial y - \partial Y/\partial z) + Y(\partial X/\partial z - \partial Z/\partial x) + Z(\partial Y/\partial x - \partial X/\partial y) = 0$.

2.340 Show that the solution of the nonlinear equation $u_x + u_y = u^2$ passing through the initial curve $x = t$, $y = -t$, $u = t$, becomes infinite along the hyperbola $x^2 - y^2 = 4$.

2.341 Solve the partial differential equation $(u_x y - u_y x)^2 = c^2 (1 + u_x^2 + u_y^2)$, where c is a given constant.

2.342 For the equation $u_y = u_x^3$ find a solution with $u(x,0) = 2x^{3/2}$. Prove that every solution regular for all x and y is linear.

2.343 Solve the equation $uu_x + u_y = 1$ with the condition $u = 0$ when $x = y$. What happens if $u = 0$ is replaced by $u = 1$?

2.344 Consider the inviscid Burgers equation $u_t + uu_x = 0$ with the initial condition $u(x,0) = -x \exp(1 - x^2)$. Show that there is no continuous solution on $(0,T) \times \mathbb{R}$ for $T > 1$.

2.345 Verify that the inviscid Burgers equation $u_t + uu_x = 0$ with the initial conditions
(i) $u(x,0) = 0$, $x < 0$ and $u(x,0) = 1$, $x > 0$;
(ii) $u(x,0) = -1$, $x < 0$ and $u(x,0) = 1$, $x > 0$
is solved (in the weak sense) by

(i) $u_1(x,t) = 0$ for $x < t/2$, $u_1(x,t) = 1$, $x > t/2$, and $u_2(x,t) = 0$ for $x < 0$,
$u_2(x,t) = x/t$, $0 < x < t$, $u_2(x,t) = 1$, $x > t$;
(ii) $u_\alpha(x,t) = -1$ for $2x < (1 - \alpha)t$, $u_\alpha(x,t) = -\alpha$, $(1 - \alpha)t < 2x < 0$,
$u_\alpha(x,t) = \alpha$, $0 < 2x < (\alpha - 1)t$, $u_\alpha(x,t) = 1$, $(\alpha - 1)t < 2x$.
Find in both cases a unique solution which satisfies the entropy condition
$a^{-1}(u(x + a, t) - u(x,t)) \le E/t$ for all $a > 0$, $t > 0$ and a constant E independent of x, t, and a.

2.346 Consider the piecewise continuous functions $u = u(x,t)$ which solve in
the distribution sense the equation $\partial u/\partial t + \partial F(u)/\partial x = 0$, where $F \in C^1(\mathbb{R})$.
Show that on the curve $x = x(t)$ of discontinuities of u the Rankine-Hugoniot
condition is satisfied, $dx/dt = (F(u^+) - F(u^-))/(u^+ - u^-)$, where u^+, u^- are
the right (respectively left) limits of u at the point of discontinuity.

2.347 Prove that the hyperbolic conservation law $u_t + f(u)_x = 0$ generates a
semigroup of contractions in $L^1(\mathbb{R})$.

2.348 Prove that if f is a convex function, then the semigroup generated by the
equation $u_t + f(u)_x = 0$ is weakly continuous.

2.349 Prove the Murat-Tartar compensated compactness theorem: Let Ω be an
open subset of $\mathbb{R}^n$ and sequences of vector functions (u_k) and (v_k) satisfy the
following conditions: $u_k, v_k : \Omega \to \mathbb{R}^n$, $u_k \rightharpoonup u$ weakly in $L^2(\Omega)^n$, $v_k \rightharpoonup v$
weakly in $L^2(\Omega)^n$, $\|u_k\|_2 + \|\operatorname{div} u_k\|_2$ and $\|v_k\|_2 + \|\operatorname{rot} v_k\|_2$ are uniformly
bounded. Then $(u_k,v_k) \to (u,v)$ in the sense of distributions (or in the sense of
weak convergence of measures), $(\cdot,\cdot)$ is the inner product in $\mathbb{R}^n$.
Remark: In general, one cannot expect convergence of (u_k,v_k) without supplementary assumptions when u_k, v_k are only weakly convergent.

2.350 Consider the differential equation $u_t + f(u)_x = 0$, where f is a sufficiently smooth function. Let $\Omega \subset \mathbb{R}^2$ and a sequence (u^ε) of solutions converge
to u weak* in $L^\infty(\Omega)$. Does u satisfy the same equation?

2.351 Show that the differential equation $u_t = \lambda u - xu_x$ generates a dynamical
system in the space $\{v : \mathbb{R}^+ \to \mathbb{R}, v(0) = 0\}$, $S(t)v(\cdot,0) = u(\cdot,t)$. Moreover, if
$\lambda < 1$, then $\lim_{t\to\infty} |S(t)v| = 0$. If $\lambda \ge 2$, then there exists $v = v(.,0)$ such
that the ω-limit set of v under $S(t)$ is a nonempty compact set which does not
contain periodic points.

2.352 Show that the solution of the linear hyperbolic Cauchy problem $u_t - x^2u_x$
$+ v + tx = 0$, $v_t + v_x + (1 - tx)u + t = 0$, $u(x,0) = x$, $v(x,0) = 0$, becomes singular on the hyperbola $tx = 1$. Explain why this happens.

2.353 Show that the study of the asymptotic behavior of solutions to the system
$u_t + u_x = 0$, $v_t - v_x = 0$, $0 < x < 1$, $t > 0$, $u(0,t) = v(0,t)$, $v(1,t) = f(u(1,t))$,
$t \ge 0$, reduces to that of the difference equation $z(t + 1) = f(z(t))$, $t \ge 0$.

2.354 Show that the solution of the nonlinear initial value problem $u_t + u_x - u + w = 1$, $v_t + 2u_x - w_x - 2u + w - v^2 + 2vw - w^2 = 0$, $w_t + 2u_x - w_x - 2u + w = 0$, $u(x,0) = 1$, $v(x,0) = 3$, $w(x,0) = 2$, has a singularity on the line $t = 1$.

2.355 Prove that the solutions of the Carleman system (a discrete velocity model related to the Boltzmann equation) $u_t = -\alpha u_x + v^2 - u^2$, $v_t = \alpha v_x + u^2 - v^2$ approximate the solutions of $u_t = \frac{1}{4}(d^2/dx^2) \log u$. More precisely, if $A(u,v) = (-u_x, v_x)$ and $B(u,v) = (v^2 - u^2, u^2 - v^2)$, then the closure of the operator $A + \alpha B$ generates a semigroup $T_\alpha(t)$ on $L^1(\mathbb{R}) \times L^1(\mathbb{R})$. For $u \geq 0$ and $u \in L^1(\mathbb{R})$ $\lim_{\alpha \to \infty} T_\alpha(\alpha t)(u,u) = (T(t)u, T(t)u)$, where $T(t)$ is the semigroup on $\{u \in L^1(\mathbb{R}) : u \geq 0\}$ generated by the closure of $\frac{1}{4}d(u_x/u)/dx = \frac{1}{4}(d^2/dx^2) \log u$.

2.356 Consider the initial value problem for the symmetric hyperbolic system $\partial u/\partial t = \sum_{j=1}^n a_j(x) \, \partial u/\partial x_j + b(x)u$, $u(x,0) = u_0(x)$, $x \in \mathbb{R}^n$, $u = (u_1, \ldots, u_N)$, where $a_j(x)$, $b(x)$ are $N \times N$ matrices, $a_j(x)$ being Hermitian, a_j are bounded together with their first derivatives, b_j are continuous and bounded. Prove that this system has a unique solution defined for all t.

2.357 (i) Letting $u = (\partial \varphi/\partial x_1, \ldots, \partial \varphi/\partial x_n, \varphi)$ show that the Klein-Gordon equation $\partial^2 \varphi/\partial t^2 = (\Delta - m^2)\varphi$ may be written as a symmetric hyperbolic system (in the sense of Friedrichs) such as in the preceding problem.
(ii) If there is a uniformly positive definite symmetric matrix $c(x)$ such that $ca_j = 0$ and $cb + b^*c = -(b + b^* + 2\sum_{j=1}^n \partial a_j/\partial x_j)$, then the energy $H(u) = \frac{1}{2}((u,u) + (u,cu))$ is conserved.
(iii) Write the Maxwell equations as a Hamiltonian system.

2.358 Consider the $m \times m$ vector differential operator $L = \sum_{j=1}^n A_j(x) \, \partial/\partial x_j + B(x)$ on the n-dimensional torus, where A_j are Hermitean matrices and $K(x) \geq c_0 I$ for some $c_0 > 0$ and $K(x) = \frac{1}{2}(B(x) + B^*(x)) - \frac{1}{2}\sum_{j=1}^n (\partial/\partial x_j)A_j(x)$. Show that the equation $Lu = f$ has a unique solution $u \in L^2$ for every $f \in L^2$. Prove that $u \in H^s$ if $f \in H^s$ and if the following condition (P_s) is satisfied:

$\forall \xi \in S^{n-1}$,

$$K(x) + s \sum_{1 \leq j,k \leq n} \left(\frac{\partial}{\partial x_k}\right) A_j(x)\xi_k\xi_k \geq c_s I$$

for some $c_s > 0$.
Study the solution $u \in L^2$ of $(-\sin x(d/dx) + m)u = (\sin x)^m$ on $\mathbb{T}^1$ with $m > 1/2$. Discuss the optimal character of these regularity results.

2.359 Prove the Weinstein mean value theorem for (regular) solutions of the Tricomi equation $yu_{xx} + u_{yy} = 0$, $\int_0^\pi u(x,y)(\sin \vartheta)^{1/3} \, d\vartheta = u(0,0) \int_0^\pi (\sin \vartheta)^{1/3} \, d\vartheta$, where the integral is taken along the curve $\rho^2 = x^2 + \frac{4}{9}y^3 = a^2$ with a real a and $\vartheta = \arccos(x/\rho)$.

2.360 Show that the general solution of the Reid-Burt equation $xyu_{xy} - y^2u_{yy} - 2xu_x + 2yu_y + c(x,y,u) - \gamma u^{-1}(xyu_xu_y - y^2(u_y)^2) = 0$, where $c = -2u$ $\log |u|$ if $\gamma = 1$ and $c = -2(1 - \gamma)^{-1}u$ if $\gamma = $ const $\neq 1$, can be written as $u = \exp(y^2F_1(x) + yF_2(xy))$ if $\gamma = 1$ and $u = (y^2F_1(x) + yF_2(xy))^{1/(1-\gamma)}$ if $\gamma \neq 1$.

2.361 Prove the following analogue of the Green identity:

$$U(x,y,z) = (4\pi)^{-1} \int_G (\varphi(\text{rot rot } U - k^2U) + \text{grad } \varphi \text{ div } U) \, dm$$

$$- (4\pi)^{-1} \int_\Sigma ((\nu \times \text{rot } U)\varphi$$

$$+ (\nu \times U) \times \text{grad } \varphi + (\nu U) \text{ grad } \varphi) \, d\sigma$$

where $\varphi = e^{ikr}/r$, $r^2 = (x - \xi)^2 + (y - \eta)^2 + (z - \zeta)^2$. Using this formula find a representation of the solutions E, H of the Maxwell equations rot $H = -ik_0\varepsilon E + 4\pi c^{-1}j$, $k_0 = \omega c^{-1}$, $k = k_0(\varepsilon u)^{1/2}$, rot $E = ik_0\mu H_0$, div $H = 0$, div $E = 4\pi\rho/\varepsilon$, in the interior of the region G bounded by the surface Σ.

2.362 Show that the Du Bois–Reymond lemma from the calculus of variations does not hold for double integrals.

2.363 The Zermelo navigation problem. Consider a ship sailing from a harbor A to a harbor B. Her velocity is constant and equal to 1, and the sea streams are described by the functions u, v depending on x, y, t. Determine and discuss the Euler equations for the problem of minimization of the time of the voyage.

2.364 Consider an isotropic, homogeneous, conducting body, ohmically heated by the passage of direct current between perfectly conducting electrodes on its surface, the rest of the surface being electrically and thermally insulated. Show that the maximum temperature in the body depends only on the potential difference between the electrodes, the electrode temperature, and the electrical and thermal conductivities, in general functions of the temperature, and is independent of the size and shape of the body and of the electrode configuration.

2.365 Show that the von Kármán system describing oscillations of a thin plate $\Delta^2w = \lambda[F,w] + [w,f]$, $\Delta^2f = -[w,w]$ in $\Omega \subset \mathbb{R}^2$, $w = w_x = w_y = 0, f = f_x = f_y = 0$ on $\partial\Omega$, where $[f,g] = f_{xx}g_{yy} + f_{yy}g_{xx} - 2f_{xy}g_{xy}$ and F is a given function, can be put in the variational form. Study bifurcation of solutions as the real parameter λ varies.

2.366 Consider the system of von Kármán equations for the equilibrium state of a thin elastic shell S with some initial curvature described by the functions k_1, k_2 (that measure the Gaussian curvature of S) $\Delta^2f = -\frac{1}{2}[w,w] - (k_1w_x)_x - (k_2w_y)_y$, $\Delta^2w = [f,w] + (k_1f_x)_x + (k_2f_y)_y + Z$. Here $[f,g]$ is defined as in the preceding problem and $Z = \lambda\Psi_0$ is chosen so that, together with the conditions $w = w_x = w_y = 0, f_{\nu\tau} = \lambda\Psi_1, f_{\tau\tau} = \lambda\Psi_2$ on the boundary of the projection of S

onto $\mathbb{R}^2$, the system has a solution $(w,f) = (0,\lambda F_0)$ for all λ, τ and v are the tangential and the normal directions, respectively; Ψ_1 and Ψ_2 represent edge stresses. Observe that $\Psi_0 = (k_2 F_{0y})_y - (k_1 F_{0x})_x$ where λF_0 solves $\Delta^2 F = 0$ with the boundary conditions as above for f. Writing a tentative solution for the full system as $w = w, f = F + \lambda F_0$, one finds the system $\Delta^2 F = -\frac{1}{2}[w,w] - (k_1 w_x)_x - (k_2 w_y)_y$, $\Delta^2 w = [F,w] + \lambda[F_0,w] + (k_1 F)_x + (k_2 F)_y$, with the boundary conditions $w = w_x = w_y = 0, F_x = F_y = 0$. Prove the existence of a solution of this latter problem using variational methods.

2.367 In the one-dimensional unsteady flow of a compressible fluid the velocity u and the density ρ satisfy the equations $u_t + uu_x + \rho^{-1} p_x = 0$, $\rho_t + u\rho_x + \rho u_x = 0$. If the law connecting the pressure with the density ρ is $p = k\rho^2$, show that $u_t + uu_x + 2cc_x = 0$, $2c_t + 2uc_x + cu_x = 0$, where $c^2 = dp/d\rho$. Prove that on the characteristics given by $dx = (u + c)\, dt\ u + 2c$ is constant.

2.368 Verify that for the system of gas dynamics equations with polytropic equation of state

$$\frac{\partial u}{\partial t} + u\frac{\partial}{\partial x} + \rho^{-1}\frac{\partial p}{\partial x} = 0$$

$$\frac{\partial \rho}{\partial t} + u\frac{\partial \rho}{\partial x} + \rho\frac{\partial u}{\partial x} = 0$$

$$\frac{\partial p}{\partial t} + u\frac{\partial p}{\partial x} + \gamma p\frac{\partial u}{\partial x} = 0$$

the following operators transform solutions into solutions:

$$L_1 = \frac{\partial}{\partial t}, \qquad L_2 = \frac{\partial}{\partial x}, \qquad L_3 = t\frac{\partial}{\partial x} + \frac{\partial}{\partial u}$$

$$L_4 = t\frac{\partial}{\partial t} + x\frac{\partial}{\partial x}, \qquad L_5 = -t\frac{\partial}{\partial t} + u\frac{\partial}{\partial u} - 2\rho\frac{\partial}{\partial \rho},$$

$$L_6 = \rho\frac{\partial}{\partial \rho} + p\frac{\partial}{\partial p}$$

and moreover for $\gamma = 3$

$$L_7 = t^2\frac{\partial}{\partial t} + tx\frac{\partial}{\partial x} + (x - tu)\frac{\partial}{\partial u} - t\rho\frac{\partial}{\partial \rho} - 3tp\frac{\partial}{\partial p}$$

2.369 Introduce polar coordinates r and ϑ to describe the viscous flow in a wedge $\{|\vartheta| < \alpha\}$ by seeking solutions of the Navier-Stokes equations $\rho u_t + \rho u u_x + \rho v u_y + p_x = \mu u_{xx} + \mu u_{yy}$, $\rho v_t + \rho u v_x + \rho v v_y + p_y = \mu v_{xx} + \mu v_{yy}$, $u_x + v_y = 0$, with an exclusively radial velocity q given by the formula $q = f(\vartheta)/r$, where f satisfies the ordinary differential equation $\mu \rho^{-1}(f^{(3)} + 4f') + 2ff' = 0$. Analyze the loss of boundary conditions at the walls of the wedge for inviscid flow by representing f in terms of an elliptic integral of the form $\vartheta = (3\mu/2\rho)^{1/2} \int ((f_1 - f)(f_2 - f)(f_3 - f))^{-1/2} \, df$ and by examining boundary layers where $f(\vartheta)$ does not converge uniformly in the limit as $\mu \to 0$.

2.370 Shock waves are discontinuity surfaces across which are the jumps in the flow quantities u, v, p, and ρ described by the Navier-Stokes equations such that mass, momentum and energy are conserved. Let the subscripts 1 and 2 indicate values of u, v, p, and ρ on the two different sides of a shock wave. In the case of one-dimensional flow show that the conservation laws, when put in divergence form, yield the Rankine-Hugoniot shock conditions $\rho_1 q_1 = \rho_2 q_2$, $\rho_1 q_1^2 + p_1 = \rho_2 q_2^2 + p_2$, $\frac{1}{2}q_1^2 + \gamma(\gamma - 1)^{-1} p_1/\rho_1 = \frac{1}{2}q_2^2 + \gamma(\gamma - 1)^{-1} p_2/\rho_2$, where $q_j = u_j - U$, $j = 1,2$, $p = $ const ρ^γ, and U stands for the velocity at which the shock wave itself advances.

2.371 Study 2π-periodic solutions of the Hopf turbulence model $u_t = -v * v - w * w - u * 1 + \mu u_{xx}$, $v_t = v * u + v * a + w * b + \mu v_{xx}$, $w_t = w * u - v * b + w * a + \mu w_{xx}$, where $(f * g)(x) = (2\pi)^{-1} \int_0^{2\pi} f(x - y)g(y) \, dy$, a and b are given 2π-periodic even functions, as the positive parameter μ tends to zero.

2.372 Consider the Minea system (a model or rather a caricature of hydrodynamic equations) $u_1' + u_1 + \delta(u_2^2 + u_3^2) = 1$, $u_2' + u_2 - \delta u_1 u_2 = 0$, $u_3' + u_3 - \delta u_1 u_3 = 0$, where $\delta > 0$ plays the role of the Reynolds number in fluid mechanics. Prove that $\lim \sup_{t \to \infty} |u(t)| \leq 1$, $u = (u_1, u_2, u_3)$. Study stationary solutions and their stability.

2.373 Prove (giving only a formal argument, that is, ignoring existence and regularity questions) that for two-dimensional Navier-Stokes equations in a bounded domain $\Omega \subset \mathbb{R}^2$ $\rho(u_t + (u \cdot \nabla)u) - v \Delta u + \nabla p = f$, div $u = 0$, $u = 0$ on $\partial\Omega$, the relation $\lim \sup_{t \to \infty} \|u(t)\|_{L^2} \leq C$ is satisfied for a constant C independent on the initial data.

2.374 Study bifurcations of space homogeneous solutions of the Brusselator system $X_t = D_1 \Delta X - (B + 1) X + X^2 Y + A$, $Y_t = D_2 \Delta Y + BX - X^2 Y$.

2.375 Consider the Lotka-Volterra system with dissipation $u_1' = v_1 \Delta u_1 + u_1(a - u_2)$, $u_2' = v_2 \Delta u_2 + u_2(u_1 - b)$, in a bounded domain $\Omega \subset \mathbb{R}^n$, $\partial u_1/\partial v = \partial u_2/\partial v = 0$ on $\partial\Omega$. Show that the space of positive functions in

$W^{1,p}(\Omega;\mathbb{R}^2)$ is invariant under the flow generated by this system. The space homogeneous functions form an invariant manifold on which this system reduces to the well-known Lotka-Volterra system $u_1' = u_1(a - u_2), u_2' = u_2(u_1 - b)$. Prove that each solution of this partial differential equation (PDE) system converges to a solution of the ordinary differential equation (ODE) system as t tends to infinity.

2.376 Show that (formally, that is, modulo existence, uniqueness, and smoothness questions) the Hopf bifurcation theorem applies to the following system of diffusion equations: $u_t = u_{xx} - q^2 u$, $v_t = v_{xx} - q^2 v$, $u_x(1,t) = v_x(0,t) = 0$, $u_x(0,t) = -pqf(v(0,t))$, $v_x(1,t) = pq(1 - f(u(1,t)))$, where p and q are positive parameters, $f(u) = u^2/(1 + u^2)$.

2.377 Give an example of a function f such that the ODE system $u' = f(u,v)$, $v' = 0$, has a global solution for arbitrary (u_0, v_0), but for certain initial data the corresponding reaction-diffusion system $u_t - u_{xx} = f(u,v)$, $v_t - v_{xx} = 0$, has solutions that blow up.

2.378 Consider the system $n_t = D \Delta n - \varepsilon n f(T)$, $T_t = \Delta T + qn f(T)$, arising in combustion theory, in a bounded smooth domain Ω in $\mathbb{R}^3$, $\partial n/\partial v = 0$ and $T = 1$ on $\partial \Omega$, $f(T) = \exp(-H/T)$, H, D, q, $\varepsilon > 0$. Show that if $n(x,0) \geq 0$, $T(x,0) \geq 0$, then n, T remain positive for all $t \geq 0$. Moreover, $\lim_{t\to\infty}(n,T)(t) = (0,1)$ in $H^1(\Omega;\mathbb{R}^2)$.

2.379 Prove that for the system (from combustion theory) $u_{1t} - d_1 \Delta u_1 - (u_2)^P h(u_1) = 0$, $u_{2t} - d_2 \Delta u_2 - (u_2)^P h(u_1) = 0$, where $h(s) = |s|^\gamma \exp(-\alpha/s)$, and p, α, γ are positive constants, $\gamma < 1$, the region $\{(u_1, u_2) : u_1 \geq 0, 0 \leq u_2 \leq 1\}$ is an invariant region.

2.380 Prove that the Varma-Amundson system describing a tubular chemical reactor $y_t = y_{xx} - Py_x + \beta zr(y) + \gamma(y_\alpha - y)$, $Lz_t = z_{xx} - Pz_x - zr(y)$, where $r(y) = \alpha \exp(-\delta/y)$, α, β, γ, δ, y_a, P, $L \geq 0$, with the boundary conditions $y_x = P(y - 1)$, $z_x = P(z - 1)$ for $x = 0$, $y_x = 0$, $z_x = 0$ for $x = 1$, generates a dynamical system in $\{(y,z) \in H_0^1((0,1);\mathbb{R}^2) : y(x) \geq 0, z(x) \geq 0 \text{ for } 0 \leq x \leq 1\}$.

2.381 Find solitary wave solutions of the Benjamin-Bona-Mahony equation $u_t + u_x + uu_x - u_{xxt} = 0$.

2.382 Consider the system of reaction-diffusion equations $u_t = D \Delta u + \sum_{j=1}^n A_j(x,u) \partial u/\partial x_j + f(u)$, where $u = (u_1, \ldots, u_m)$ is defined on $\Omega \times \mathbb{R}^+$, $\Omega \subset \mathbb{R}^n$ being a bounded smooth domain, D is a constant positive definite matrix, and the A_j are continuous matrices. Suppose that under the Neumann condition $\partial u/\partial v = 0$ on $\partial \Omega$ the system admits a compact invariant region $\Sigma \subset \mathbb{R}^m$, that is, $0 \in \Sigma$ and if $u_0(x)$ is in the interior of Σ for all $x \in \Omega$, then the solution $u(x,t)$ is in Σ for all $x \in \Omega$ and $t > 0$. Let $\sigma = d\lambda - M - a(m\lambda)^{1/2}$ be positive, where λ is the smallest positive eigenvalue of $-\Delta$ on Ω with the homogeneous Neumann condition, $a = \sup\{\, |\, A_j(x,u)\, |: x \in \Omega, \, u \in \Sigma, \, j = 1, \ldots, n\}$, d denotes the smallest eigenvalue of the matix D, and $M = \max\{\, |\, \nabla f(u)\, |: u \in \Sigma\}$. Prove that if $u_0 \in \Sigma$, then $\|\nabla u(.,t)\|_2 \le c_1 e^{-\sigma t}$, $\|u(.,t) - \bar{u}(t)\|_2 \le c_2 e^{-\sigma t}$ for some c_1, c_2, where $\bar{u}$, the average of u over Ω, satisfies the system of ordinary differential equations $d\bar{u}/dt = f(\bar{u}) + g(t)$, $\bar{u}(0) = |\Omega|^{-1} \int_\Omega u_0(x)\, dx$, where $|\, g(t)\, | \le c_3 e^{-\sigma t}$.

2.383 Verify that

$$u_1(x,t) = -\frac{2}{(\cosh(x - 4t))^2}$$

and

$$u_2(x,t) = -12\frac{3 + 4\cosh(2x - 8t) + \cosh(4x - 64t)}{(\cosh(3x - 36t) + 3\cosh(x - 28t))^2}$$

are solutions of the Korteweg–de Vries equation $u_t - 6uu_x + u_{xxx} = 0$ which correspond to one and two solitons. Find the asymptotics of u_2 as either $\xi = x - 16t$ or $\zeta = x - 4t$ is kept fixed and $t \to \pm\infty$.

2.384 Prove that the Kuramoto-Sivashinsky equation $u_t + v u_{xxxx} + u_{xx} + \frac{1}{2}(u_x)^2 = 0$ in $(0,1)$, $v > 0$, $u_x = u_{xxx} = 0$ at $x = 0$ and $x = 1$, generates a flow with a global attractor.

HINTS, ANSWERS, REFERENCES

Ordinary Differential Equations

1.2 *Answer:* No. [He] Ch. 3, Sec. 1.

1.3 [T-P] 3. 3.

1.4 B. Ricceri, *C. R. Acad. Sci. Paris* 295, 1982, 245–248.

1.5 *Answer:* No, since formally $x(t) = \sum_{k=1}^{\infty} (k - 1)! \, t^k$, which is a divergent series. The solution is $x(t) = \exp(-1/t) \int_t^0 u^{-1} \exp(1/u) \, du$ for $t < 0$, $x(t) = \exp(-1/t) \int_t^1 u^{-1} \exp(1/u) \, du$ for $t > 0$, $x(0) = 0$. A. D. Bruno, *Local Methods in Nonlinear Differential Equations*, Nauka, Moskva, 1979; Springer, Berlin, 1989, Ch. II, Sec. 1.4, Example 3.

1.6 *Answer:* For instance, $t^2 x' + x = 0$. [T] Ex. 11. 8.

1.7 *Answer:* Clearly $x'(0) = 1$ and for all $t, x(t) > 0$. Taking the logarithm of both sides and letting $x'(t) = \exp(u(t))$, one obtains $u = -t \exp(2u)$. If $|t| < 1/(2e)$, $C = \{u : |u| = 1/2\}$, then for every point on C $|t \exp(2u)| \leq |u|$. Thus, by the Rouché theorem, one sees that the equation above, viewed as an equation in u, has a unique root in the interior of C. Then, by the Lagrange theorem (E. T. Whittaker, G. N. Watson, *A Course in Modern Analysis*, Cambridge 1927, Sec. 7. 32), we obtain the power series expansion $x'(t) = 1 + \sum_{n=1}^{\infty}(-1)^n((2n + 1)^{n-1}/n!)t^n$. Integrating term by term and using the initial condition $x(0) = 0$, one obtains $x(t) = \sum_{n=0}^{\infty}(-1)^n((2n + 1)^{n-1}/(n + 1)!)t^{n+1}$. The final series converges for $|t| < 1/(2e)$, corresponding to the fact that $t = -1/(2e)$ is a branch point of the equation. A generalization: The equation $x' \exp(t(x')^p) = 1$

is solved by $x(t) = \sum_{n=0}^{\infty} (-1)^n (pn + 1)^{n-1} t^{n+1}/(n + 1)!$. O. G. Ruehr; C. C. Rousseau, *SIAM Review* 15, 1973, 384–385, Problem 72-8.

1.8 *Hint*:   Pass to an integral equation. See for a solution in a subset of the space $C[-h,h]$ whose elements satisfy conditions $| y(t) - x_0 | \leq$ const $\cdot | t - s |$. Recall the Ascoli-Arzelà theorem and use the Schauder fixed point theorem.

1.9 [PSV] Ch. III, Sec. 3. 1. Compare with 1.616.

1.10 M. Hirsch, C. Pugh, *Proc. Symp. Pure Math.*, vol. XIV, AMS, 1970, 133–164.

1.11 J. Sotomayor, *Boletim Soc. Brasil. Mat.* 4, 1973, 55–59.

1.12 [Di] 10. 5. 3.

1.13 *Hint*:   Let $f(t,x) = 0$ for $t = 0$, $x \in \mathbb{R}$; $f(t,x) = 2t$ for $0 < t \leq 1$, $x < 0$; $f(t,x) = 2t - 4x/t$ for $0 < t \leq 1$, $0 \leq x \leq t^2$; $f(t,x) = -2t$ for $0 < t \leq 1$, $t^2 < x < \infty$. [C-L] Ch. 2, Sec. 3; [So] Ch. I, Ex. 28.

1.14 and **1.15** E. R. van Kampen, *Amer. J. Math.* 63, 1941, 371–376.

1.16 [C-L] Ch. 2, Ex. 3.

1.17 *Hint*:   For a suitable $M > 0$, consider the successive approximations $u_0(t) = M(t - a)$, $u_{n+1} = \int_a^t \omega(s, u_n(s))\, ds$ and prove that $u_n \to 0$ and $\sup_{n,\, m \geq k} | x_n(t) - x_m(t) | \leq u_k(t)$ for $k \geq 1$, $t \in [a, a + \varepsilon]$. [PSV] Ch. I, Sec. Ex. 3.

1.18 [So] Ch. I, Ex. 27.

Remark: Other information on successive approximations for ordinary differential equations can be found in Z. Athanassov, *Math. Japonica* 35, 1990, 351–367.

1.19 *Hint*: $x' = | x |^{-3/4} x + t \sin(\pi/t)$. [C-L] Ch. 1, Ex. 12.

1.20 [PSV] Ch. III, Sec. 2. 2, Ex. 2.

1.21 [PSV] Ch. III, Sec. 2. 2, Ex. 1.

1.22 [Br] Ch. 1, Sec. 1. 13, Ex. 8.

1.23 *Answer*: $x(t) = t^2/2$ for $t \leq 2$ and $x(t) = 2 \exp(t - 2)$ for $t > 2$. [Bu] Ch. I, Ex. 7.

1.25 [H-S] Ch. 8, Sec. 7, Ex. 5.

1.26 S. Wallach, *Amer. J. Math.* 70, 1948, 345–350; [H] Ch. 3, Sec. 6, Ex. 6. 4.

1.27 *Hint*: $f(t, (x,y)) = (t \cos t + \sin t, t^2 \cos t + 2 t \sin t)$, $x(t) = t \sin t$, $y(t) = t^2 \sin t$, $t_1 = \pi$, $t_2 = 2\pi$. [So] Ch. 1, Ex. 14 and 16.

1.28 [So] Ch. I, Ex. 25; [Di] 10. 5. 1.

1.29 [So] Ch. I, Ex. 17.

1.30 [Di] 10. 5. 1.

1.31 [So] Ch. I, Ex. 6.

1.32 *Hint:* Let $x_1(t)$, $x_2(t)$ be the solutions of this problem. Show that $(\mid x_1(t) - x_2(t) \mid)' \leq 0$. [H] Ch. III, Sec. 6, Th. 6. 2.

1.33 *Hint:* Use the implicit function theorem.

1.34 *Hint:* $f(t,x,y) = y^2 - 2y + 4x - 4t + 1$, $t_0 = 0$, $x_0 = 0$, $y_0 = 1$.

1.35 [B1] Ch. 3. 4, Ex. 1.

1.36 *Answer:* The unique solution is given in parametric form by $x(t) = -2z'(t)/z(t)$, $x(t) = z(t)^{-2} \int_t^\infty z(s)^2 \, ds$, where $z(t) = \mathrm{Ai}(\mu(t - \alpha))$ is the Airy function, $\mu = 2^{-1/3}$, $-\mu\alpha$ is the first negative zero of $\mathrm{Ai}'(s)$. A. Tissier, R. J. Driscoll, P. J. Bushell, *Amer. Math. Monthly* 96, 1989, 657–659, 6551.

1.37 *Answer:* All the solutions have the same limit as $t \to 0$. [Fi] 179, 180.

1.38 E. K. Haviland, *Amer. J. Math.* 54, 1932, 632–634; E. R. van Kampen, *Amer. J. Math.* 63, 1941, 371–376.

1.39 [H] Ch. 3, Sec. 6, Th. 6. 1, Ex. 6.1.

1.40 **Remark:** For $\omega(t,x) = x/t$ one has the uniqueness result of Nagumo and Perron. For further extensions of this theorem see F. Brauer, *Canad. J. Math.*, 11, 1959, 527–533; [C-L] Ch. 2, Sec. 2.

1.41 *Hint:* Use the Kamke criterion or the result of problem 1.110. Compare with 1.14. [H] Ch. 3, Sec. 6, Corollary 6. 1.

1.42 O. Perron, *Math. Zeit.* 28, 1928, 216–219; [H] Ch. 3, Sec. 6, Ex. 6.2.

1.43 *Hint:* Consider $g(t,x) = (1 - t)\mid x \mid^{1/2}$ and $f(t,x) = (1 - t) \times (\mid x \mid^{1/2} + 1)$.

1.44 *Hint:* Let $A_\lambda = \{x_\lambda(\varepsilon) :$ where $x_\lambda(t)$ is a solution of the problem$\}$, $\varepsilon > 0$. Show that if $\alpha \neq \beta$, then $A_\alpha \neq A_\beta$. For a generalization see H. Nakano, *Proc. Phys. Math. Soc. Japan* 14, 1932, 41–43.

1.45 and **1.46** E. R. van Kampen, *Amer. J. Math.* 59, 1937, 144–152.

1.48 *Hint:* For each $x_0 \in \mathbb{R}$ define the set $A(x_0) = \{x(1) : x(t)$ is a solution of the problem $x' = f(t,x)$, $x(t_0) = x_0\}$. For $x_0 \neq y_0$ $A(x_0) \cap A(y_0) = \varnothing$, hence the set $\{A(x_0) : x_0 \in \mathbb{R}\}$ is countable.

1.49 [So] Ch. I, Ex. 37.

1.50 [H] Ch. II, Sec. 4, Ex. 4. 3.

1.51 [H] Ch. II, Sec. 4, Th. 4. 1 and Ex. 4. 2.

1.52 *Hint:* Study the system in the plane $x' = x - y \mid y \mid^{1/2}/(x^2 + y^2)^{1/2}$, $x(0) = 1$, $y' = y - x \mid y \mid^{1/2}/(x^2 + y^2)^{1/2}$, $y(0) = 0$, in polar coordinates, leading to the so-called Peano brush. [PSV] Ch. III, Sec. 2. 3, Ex. 2.

1.53 [PSV] Ch. III, Sec. 2. 3, Theorem on the Peano Phenomenon.

1.54 C. Pugh, *Bull. AMS* 70, 1964, 580–583. Cf M. F. Bokshtein, *Sci. Reports Moscow State Univ.* 15, 1939, 3–72.

1.55 *Hint*: Consider the function $z(t) = \arctan x(t)$ satisfying the equation $z' \cos^{-2} z = \tan^2 z + t$ and $0 < z(t) < \pi/2$. The inequality $z' = \sin^2 z + t \cos^2 z \geq 1$ holds for $t \geq 1$, hence $z(t) \geq t - 1$ and $z(2.9) > \pi/2$.

1.56 [PSV] Ch. I, Sec. 5. 8, Ex. 4.

1.57 [PSV] Ch. I, Sec. 5. 8, Ex. 1; cf S. D. Chatterji, C. El-Hayek, *Amer. Math. Monthly* 98, 1991, 67–70, 6596.

1.58 [Fi] 237.

1.59 [Fi] 238.

1.60 [Fi] 239.

1.61 [Fi] 240.

1.62 [PSV] Ch. III, Sec. 2. 4, Ex. 4.

1.63 [PSV] Ch. I, Sec. 3. 2, Ex. 6. Compare with 1.136.

1.64 [C-L] Ch. 2, Ex. 4. 5; A. Wintner, *Amer. J. Math.* 47, 1945, 277–284.

1.65 *Hint*: $\left| x(t)x'(t) \right| \leq \varphi(t) \left| x \right|^2$. [H] Ch. X, Sec. 1, Th. 1. 1.

1.66 [H] Ch. III, Sec. 5, Corollary 5. 2.

1.67 [So] Ch. I, Ex. 30.

1.68 [Ga] Ch. 2, Sec. 2. 6, Ex. 15.

1.69 *Hint*: $\left| x(t) \right| \leq \left| x_0 \right| + \int_0^t \varphi_T(\left| x(\tau) \right|) \, d\tau$. [Ga] Ch. 2, Sec. 2. 6, Ex. 10.

1.70 [H-S] Ch. 9, Sec. 4, Ex. 3. Compare with 1.294.

1.71 [B1] Ch. 4. 6, Ex. 5.

1.72 [N-S] Th. 3. 22. For a generalization on noncompact manifolds see [Ir] 3. 44; for a generalization on Banach manifolds see P. L. Renz, *Indiana Univ. Math. J.* 20, 1971, 695–698.

1.73 [A-M] Sec. 2. 2. H.

1.74 [A-M] Sec. 2. 2. H.

1.75 *Hint*: Differentiate $x(t,x,y(s))$ with respect to s; next integrate the result over $[t_0,t]$. [PSV] Ch. I, Sec. 3. 2, Ex. 5.

1.76 *Hint*: Denote by $x_k \in [k\pi, (k + 2)\pi]$, $k \in \mathbb{Z}$, any point where f vanishes. If $x(0) \in [x_k,x_{k + 1})$ for some $k \in \mathbb{Z}$, then $x(t)$ belongs to the same interval for all t.

1.77 [R-M1] Ch. V, Ex. 9. 2.

1.78 [B-W] 5. 37. For example: C. R. MacCluer, D. A. Moran, *Amer. Math. Monthly* 70, 1963, 893, E 1549.

1.79 [B2] Part 1, Ex. 29. 1

1.81 *Answer:* For $m = 0$ $x = 0$ is the only singular integral ($a \neq 0$). [Ju] 41.

1.82 *Answer:* In the parametric form $t = -1/2 - p + C/(p - 1)^2$, $x = -p^2/2 + Cp^2/(p - 1)^2$, where $p = x'$ and $x = 0$, $x = t + 1$. Sketch these integral curves. [Ju] 42.

1.83 *Hint:* If $f(t + x)(P\, dt + Q\, dx) = dU$, $g(t - x)(P\, dt + Q\, dx) = dV$, then one gets as the integrability condition $\partial(g(t - x)U_t/f(t + x))/\partial x = \partial(g(t - x)U_x/f(t + x))/\partial t$.
Answer: $P = ((fg' - gf')/f^3)F(g/f)$, $Q = ((-fg' + gf')/f^3)F(g/f)$ with an arbitrary function F. The general solution is (implicitly) given by $f(t + x) = $ const $\cdot g(t - x)$. [Ju] 10.

1.84 [I] Ch. II, Ex. 2.

1.85 [B1] Ch. 5. Th. 3.

1.86 *Answer:* $x = \pm t$ are the only solutions independent of a, b, c. Now $z = (x - t)/(x + t)$ transforms this Riccati equation into $dz/z = 2\, dt/(at^2 + bt + c)$. [Ju] 8.

1.87 [Z] Ex. 33.

1.88 [B3] Ch. 1, Ex. 22; Z. Koshiba, S. Uchiyama, *Proc. Japan Acad.* 42, 1966, 696–701.

1.89 I. Rubinstein; W. W. Meyer, *SIAM Review* 20, 1978, 858–859, Problem 77-16.

1.90 *Hint:* If one knows a solution x_1 of the Riccati equation, then one can integrate it reducing to a Bernoulli equation (by the substitution $x = x_1 + z$) and then to a linear equation (setting $z = 1/Z$). The general solution is a homographic function of the integration constant. Homographic functions conserve the anharmonic ratio. For instance: [V] Sec. 67.

1.91 *Answer:* $y(z) = ((y_1(z) - y_3(z))y_2(z) - C(y_2(z) - y_3(z))y_1(z)/(y_1(z) - y_3(z) - C(y_2(z) - y_3(z)))$.
Hint: The cross ratio of four solutions $((z_1 - z_3)/(z_1 - z_4))/((z_2 - z_3)/(z_2 - z_4))$ is a constant. Another method: Consider the associated linear second-order equation $w'' - (f_2'/f_2 + f_1)w' + f_1 f_2 w = 0$, $y = -f_2^{-1}w'/w$. [Hi] (4. 2. 11).

1.92 *Hint:* If $x_1 \neq x_2$ are two solutions, then $(x_2' - x_1')/(x_2 - x_1) = x_1 + x_2 + q(t)$. If the third periodic solution exists, then $(x_2' - x_1')/(x_2 - x_1) - (x_3' - x_1')/(x_3 - x_1) = x_2 - x_3$. The left-hand side would be the derivative of a periodic function while the right-hand side is of constant sign.

1.93 *Answer:* $p^{-1} \int dx/(mx^2 + nx + p) = C - \int c(t)\, dt$, since $x' = -c(t)(mx^2/p + nx/p + 1)$. [KKM2] 236.

1.94 *Answer:* $u = xy = x^p/((p - 1)/b) + x^p/((2p - 1)/a) + x^p/((3p - 1)/b) + \cdots$ where $p = n + 2$. [Da] Ch. 3, Sec. 7.

1.95 *Hint*: The coefficients may be obtained as the solution to a system of linear equations with determinant of the Vandermonde type.

1.96 *Answer*: $(u' - 2bu)/au^2 = \pm(c + 1/c)$. Taking $z = (x - x_1)/(x - x_2)$ one gets the (easily integrable) equation $z'/z = \pm au(1/c - c)$. [Ju] 7.

1.97 [Ra] Ch. I, Sec. 10.

1.98 *Answer*: $dx_k/dt = -kx_k + kx_{k+1}$, $k = 1,2, \ldots$, which is a system of linear equations. [B2] Part 2, Sec. 8, Ex. 8. 1.

1.99 V. I. Mironenko, *Linear Dependence of Functions along the Solutions of Differential Equations*, Izd. BGU im. V. I. Lenina, Minsk, 1981, Ch. III, Sec. 3, Th. 1.

1.100 [Ra] Ch. I, Ex. 12.

1.101 H. Dulac, *Bull. Soc. Math. France* 51, 1923, Sec. 9, Lemma 2.

1.102 *Hint*: Compare v with the solutions of $x' = f(t,x) + 1/n$, $x(a) = x_0$. [PSV] Ch. III, Sec. 2. 4, Theorem on Differential Inequalities.

1.103 [PSV] Ch. III, Sec. 2. 4, Corollary.

1.104 *Hint*: Set $u^*(t) = \text{ess sup}_{0 \le s \le t} u(s)$ and show that $u(s) \le C_1 + C_2 \varepsilon^{-1/2} \int_0^t u^* + 2C_2 \varepsilon^{1/2} u^*$ for $\varepsilon > 0$ and $0 \le s < t$. Then take $\varepsilon^{1/2} = (4C_2)^{-1}$ and apply the usual Gronwall lemma. [Hr] Ch. A-II, Lemma 6.

1.105 [He] Lemma 7. 1. 1.

1.106 *Hint*: Multiplying the inequality by $\exp(-\int_t^s g)$ one obtains the relation $d(y(s) \exp(-\int_t^s g))/ds \le h(s) \exp(-\int_t^s g) \le h(s)$. Then integrate between s and $t + r$ and between t and $t + r$. [Te] Ch. III, Sec. 1. 1. 3, Lemma 1. 1.

1.107 **Remark:** This is a nonlinear version of the Gronwall-Bellman lemma, where $\Phi(u) \equiv u$. I. Bihari, *Acta Math. Acad. Sci. Hung.* 7, 1956, 81–94.

1.109 D. Willett, J. S. Wong, *Mh. Math.* 69, 1965, 362–367; [Ra] Ch. I, Th. 2. 2 (a generalization).

1.110 [Hi] Th. 1. 6. 2 and Ex. 1. 6. 2. Compare with 1.41.

1.111 *Hint*: Use the Cayley-Hamilton theorem or show the formula for diagonal matrices and next use the fact that diagonalizable matrices are dense in the set of all matrices.

1.112 [P-M] Ch. 2, Ex. 3.

1.113 [H-S] Ch. 6, Sec. 5, Ex. 5.

1.114 [H-S] Ch. 6, Sec. 5, Ex. 2.

1.115 [H-S] Ch. 13, Sec. 3.

1.116 [So] Ch. III, Ex. 26.

1.117 [H-S] Ch. 6, Sec. 5, Ex. 6.

1.118 [H-S] Ch. 9, Sec. 2, Ex. 3.

1.119 [H-S] Ch. 6, Sec. 5, Ex. 4.

1.120 [H-S] Ch. 6, Sec. 5, Ex. 3.

1.122 [P-M] Ch. 2, Ex. 1.

1.123 [So] Ch. III, Ex. 1.

1.124 [So] Ch. III, Ex. 3.

1.125 *Hint*: $d(\int_0^t A(s)\,ds)^m/dt = mA(t)(\int_0^t A(s)\,ds)^{m-1}$. [So] Ch. III, Ex. 16.

1.126 [KPPZ] Ch. 3, Sec. 14, Ex. 14. 3.

1.127 *Hint*: $x(t) = Y(t)x(0) + \int_0^t Y(t)Y^{-1}(s)f(s,x(s))\,ds$ and $|Y(t)| \leq M$, $|Y^{-1}(s)| \leq M$ for all $t \geq t_0$ and s. Estimate $u(t) = |x(t)| |Y(t)|^{-1}$. [Ce] 6. 2.

1.128 [So] Ch. III, Ex. 7.

1.129 A. Wintner, *Amer. J. Math.* 76, 1954, 183–190.

1.130 [De] Ch. 5, Ex. 4.

1.131 *Hint*: Consider the Wronskian of a fundamental matrix of the system. Recall the Liouville formula.
Answer: $x' = y(\sin(\log t) + \cos(\log t))$, $y' = x(\sin(\log t) + \cos(\log t))$, $1 \leq t < \infty$. The trace of the matrix of this system vanishes identically, while the sum of characteristic exponents is equal to 2.

1.132 *Hint*: Consider evolution of the scalar quantity $|x|^2$. [De] Ch. 3, Sec. 6; T. Ważewski, *Studia Math.* 10, 1948, 48–59.

1.133 [De] Ch. 2, Sec. 13.

1.134 [De] Ch. 3, Sec. 10. A particular case is in 1.139.

1.135 *Answer*: Consider the system $y_1' = -ay_1$, $y_2' = (\sin(\log t) + \cos(\log t) - 2a)y_2$ whose general solution is $y_1 = c_1\exp(-at)$, $y_2 = c_2\exp(t\sin(\log t) - 2t)$. If $a > 1/2$, every solution approaches zero as $t \to \infty$. For the (perturbing) matrix

$$B(t) = \begin{bmatrix} 0 & 0 \\ \exp(-at) & 0 \end{bmatrix}$$

the solution of the perturbed system is $x_1 = c_1\exp(-at)$, $x_2 = \exp(t\sin(\log t) - 2at)(c_2 + c_1 \int_0^t \exp(-\sin(\log s))\,ds)$. For $t = \exp((2n + 1/2)/\pi)$ the integral $\int_0^t \exp(-s\sin(\log s))\,ds$ is greater than $t(\exp(-2\pi/3) - \exp(-\pi))\exp(-e^{-\pi}t/2)$. Hence if $1 < 2a < 1 + \exp(-\pi/2)$ the solutions will be bounded only if $c_1 = 0$. A nonlinear version of this example is: $z_1' = -az_1$, $z_2' = (\sin(\log t) + \cos(\log t) - 2a)z_2 + z_1^2$, $z_1(0) = c_1$, $z_2(0) = c_2$, solved by $z_1 = c_1\exp(-at)$, $z_2 = \exp(t\sin(\log t) - 2at)(c_2 + c_1^2 \int_0^t \exp(-s\sin(\log s))\,ds)$. Observe that $z_2 \to 0$ as $t \to \infty$ only if $c_1 = 0$. [B1] Ch. 2. 5, Ch. 4. 9.

1.136 [B1] Ch. 2. 3, Th. 3. Compare with 1.63.

1.137 [Ce] 6. 7.

1.138 *Hint*: Show that $|x(t)| \le c |x(0)| \exp(-C\int_0^t g(s)\,ds)$ for some $C > 0$. [C-L] Ch. 13, Ex. 1.

1.139 [C-L] Ch. 3, Ex. 6.

1.140 and **1.141** Gengzhe Chang, J. A. Crow, N. Waller, *SIAM Review* 27, 1985, 456–453, Problem 84-15.

1.142 *Hint*: Consider the equation $x(t) = x_0 - \int_t^\infty A(s)x(s)\,ds$. [R-S], Ch. XI, App. 2, Sec. 8, Th. XI; J. Dollard, C. Friedman, *J. Functional Analysis* 28, 1978, 309–384.

1.143 *Answer*: No.

1.144 [T] Ex. 11. 9.

1.145 *Answer*: The problem $u'' - x^{-1}u' = 0$, $u(0) = u'(0) = 0$, has the two solutions $u \equiv 0$ and $u(x) = x^2$. [P-W] Ch. 1, Sec. 3, Th. 6 and Exercise 1.

1.146 [Br] Ch. 2, Sec. 28; [Bu] Sec. 17, Example.

1.147 *Answer*: $x(t) = A \exp((b - c)t)(b - \tanh t) + B \exp(-(b + c)t) \times (b + \tanh t)$, $b^2 = 1 + c^2$.
Remark: This equation arises as uncoupled portion of the variational equation along a homoclinic orbit for a dynamical system in $\mathbb{R}^4$. J. Gruendler; C. Gheorghiou, *SIAM Review* 26, 1984, 282–283, Problem 83-10.

1.148 *Answer*: (i) $q = p^2/4 + p'/2$, (ii) $q' + 2pq = 0$. [Ju] 21, 22.

1.149 *Answer*: $a = 6$; these solutions are of the form $(cx + d)/(x^2(x + 1))$. [Ju] 28.

1.150 *Answer*: $m = 1/4$ is the critical value of the parameter. See 1.146. [H] Ch. XI, Sec. 1, Ex. 1. 1c); [Ce] 5. 2.

1.151 [P-S], Sec. 7, Problem 62; [B2] Part 1, Ex. 8. 5.

1.152 [P-T] Ch. 5, Lemma 1; [I] Ch. V, Ex. 5; G. Darboux, *C. R. Acad. Sci. Paris* 94, 1882, 1456.

1.153 [C-L] Ch. 3, Ex. 15.

1.154 [I] Ch. VII, Ex. 15.

1.155 [I] 7. 5. A proof is in O. Perron, *Die Lehre von den Kettenbrüchen*, Teubner, Leipzig, 1913, Sec. 57.

1.156 [I] Ch. IX, Ex. 1.

1.157 [B1] Ch. 6, Sec. 8, Ex. 3, Th. 6.

1.158 [Sb] Ch. 4, Sec. 5, Example 13, 15; N. W. McLachlan, *Theory and Applications of Mathieu Functions*, Oxford University Press, New York, 1947; [Gb] Ch. II, Sec. 5. Compare references to 1.606.

1.159 *Hint*: Use the Green function.
Answer: $-m/2 \leq x \leq 0$, $-m/(3t) \leq x' \leq m/(3t)$. [Fi] 781.

1.160 *Hint*: Recall the Liouville formula.
Answer: $\int_0^\infty p(t)\, dt = \infty$. [Fi] 717.

1.161 *Hint*: N. P. Kupcov's example. Consider any $\delta \in (0,1)$ and the piece-wise constant function $p(t) = 4/T$ for $0 < t < c$, $p(t) = \delta/(T(T - c))$ for $c < t < T$. The integral in the Liapunov condition equals $4 + \delta$. For $c = T\delta/16$ the trace of the fundamental matrix of the equation (written as the system in $\mathbb{R}^2$) is $-2 - \delta/6 + o(\delta)$ for $\delta \to 0$, which implies instability. [De] Ch. 3, Sec. 19.

1.162 *Hint*: Integrate the equation $x''x' + a(t)xx' = 0$. [B1] Ch. 6. 6, Th. 4.

1.163 [B1] Ch. 6, Ex. 11.

1.164 [Ce] 3. 3; [B1] Ch. 6. 5, Th. 3. Compare with 1.135.

1.165 *Hint*: First consider the case $\varphi \equiv 0$ and (to establish boundedness of solutions) the expression $x^2 + (x')^2$. [B1] Ch. 6. 4, Th. 1. Compare with 1.63.

1.166 *Hint*: Compare this equation with its asymptotic equation $x'' + x = \cos t$. [Sb] Ch. 4, Sec. 5, Ex. 4.

1.167 [B1] Ch. 2, Sec. 2, Ex. 3.

1.168 *Hint*: Use 1.165. [B2] Ex. 5. 5.

1.169 [B3] Ex. 1. 14. 6.

1.170 *Hint*: Use the Liouville transformation $s = \int_0^t (1 + f(\tau))^{1/2}\, d\tau$ for the independent variable. [B3] Ex. 1. 21. 1.

1.171 *Hint*: Compare the first part of this problem with 1.164. [Ce] 3. 4.

1.172 *Answer*: The fundamental systems of solutions of these equations are: $\sin t - t \cos t$, $\cos t + t \sin t$ and $t^{-1} \sin t$, $t^{-1} \cos t$.
Remark: The proper generalization for nth order equations needs a supple-mentary condition; for example, the roots z_j of the algebraic equations $z^n + (c_1 + f_1(t))z^{n-1} + \cdots + (c_n + f_n(t)) = 0$ have negative real parts for all $t_0 \leq t < \infty$. [Ce] 3. 4, 3. 7.

1.173 *Hint*: Use the Liouville transformation (a change of the independent variable and then the elimination of the first derivatives by a change of the dependent variable).

Answer: (i) $x_1 = (t/\log t)^{1/2}(\cos(\frac{1}{2}\log^2 t - \frac{1}{8}\log\log t) + \mathcal{O}(\log^{-2} t))$,
(ii) $x_{1,2} = (1 \pm 3/32t^2 + 105/2048t^4 + \mathcal{O}(t^{-6}))\exp(\pm t^2)(2t)^{-1/2}$,
(iii) $x_1 = t^{1/4}(1 + 3/64t)\cos(2t^{1/2} + 3/16t^{-1/2}) + \mathcal{O}(t^{-5/4})$. [Fi] 748, 749, 750.

1.174 [B3] Ch. 1, Ex. 5.

1.175 *Hint*: Successive approximations for $x_1(t) = 1 + \int_t^\infty (t - s)f(s)x_1(s)\,ds$ and for $x_2(t) = t + \int_0^t sf(s)x_2(s)\,ds + t\int_t^\infty f(s)x_2(s)\,ds$, where a is chosen so that $\int_a^\infty t \mid f(t) \mid dt < 1/2$. [C-L] Ch. 3, Ex. 28; [B1] Ch. 6, Sec. 7, Th. 5.

1.176 *Hint*: Calculate the solution in the integral form or observe that $y(t) = a \sin t + b \cos t + \sum_{k=1}^\infty c_k/t^k$ is formally the general solution for appropriate c_k. [K-C] Ch. 1, Sec. 2, Problem 1.

1.177 [B1] Ch. 6, Ex. 20.

1.178 [KNK] Ch. 5, Sec. 5, Ex. 9.

1.179 *Hint*: Rewrite this equation as a system; diagonalize its matrix. The integrability condition on p' will appear to be a small (integrable) perturbation condition for the system in new variables. [De] Ch. 4, Sec. 9, Ch. 4, Ex. 2.

1.180 [R-M2] 5. 5. 2.

1.181 [KPPZ] Ch. 3, Sec. 14, Ex. 8.

1.182 [H] Ch. XI, Sec. 2.

1.183 *Hint*: Use the Prüfer transformation (see the preceding problem). [H] Ch. XI, Sec. 3, Th. 3. 1.

1.184 [C-L] Ch. 8, Ex. 2.

1.185 [H] Ch. XI, Sec. 3, Corollary 3. 1.

1.186 [So] Ch. 4, Ex. 6.

1.187 *Hint*: Apply the Sturm comparison theorem. For instance: [P1] Ch. V, Sec. 39, Ex. 2.
Remark: It can be proved (using the stationary phase method) that the exact asymptotics of one of solutions of this equation (the Airy function) is

$$\text{Ai}(-t) = \pi^{-1/2}t^{-1/4}\left(\cos\left(\frac{2t^{3/2}}{3} + \frac{\pi}{4}\right) + \mathcal{O}(t^{-3/2})\right) \quad \text{as } t \to \infty$$

and

$$\text{Ai}(-t) \approx (2\pi^{1/2})^{-1}(-t)^{-1/4}\exp\left(-\frac{2\mid t\mid^{3/2}}{3}\right) \quad \text{as } t \to -\infty$$

1.188 *Hint*: If not, then $\int_0^1 p(x')^2 x^{-2} + a\int_0^1 r = 0$. Show that x vanishes at two points t_1, t_2 and $\mid t_1 - t_2 \mid \leq 1$. [C-L] Ch. 8, Ex. 7.

1.189 *Hint*: Compare u with solutions of the equation $v'' + Mv = 0$. Extend the Sturm comparison theorem to the case of two different operators. [P-W] Ch. 1, Sec. 8, Th. 20 and Corollary.

1.190 *Hint*: $y = t^{1/2}x$ transforms the Bessel equation into $y'' + (1 - \alpha/t^2)y = 0$, where $\alpha = \mu^2 - 1/4$. Next use the preceding problem. [H] Ch. XI, Sec. 3, Ex. 3. 2; [Bu] Sec. 46, Th. 24.

1.191 M. Bôcher, *Bull. AMS* 1897, 207; [I] Ch. X, Ex.8.

1.192 [I] Ch. X, Ex. 8.

1.193 *Hint*: Integrate $xy'' - xy''$ between two consecutive zeros of x. [B1] Ch. 6. 10, Th. 8.

1.194 *Hint*: Represent y as the product uz. [Ra] Ch. III, Sec. 19, Th. 19. 1.

1.195 *Answer*: $ky^2y'' + k(y')^2(xy' - y) + (y - xy')^3 = 0$. [Da] Ch. 1, Sec. 4, Example 1.

1.196 [Da] Ch. 1, Sec. 4, Ex. 1.

1.197 [Br] Ch. 4, Sec. 4. 6, Ex. 12.

1.198 *Hint*: Write $y = x'$ and consider the two associated systems (i) $x' = y$, $y' = (x + y)^{1/2}$, (ii) $x' = y$, $y' = -(x + y)^{1/2}$. Show that (i) has exactly one solution with slope -1 at the origin, whereas (ii) has a one-parameter family of such solutions which fill up a region in the (x,y) plane. These are the only solutions corresponding to the desired asymptotic behavior. [B3] Ex. 4. 27. 1.

1.199 [Ga] Ch. 2, Sec. 2. 9, Ex. 8, 9.

1.200 [PSV] Ch. I, Sec. 5. 8, Ex. 2.

1.201 *Answer*: Assuming that $f' + 2x \neq 0$, one has $(f'' + 2)/(f' + 2x) = f'/f + x/f$. Integration yields $\log |f' + 2x| = \log |f| + \log |a| + \int x/f \, dx$ and hence $f'/f + 2x/f = c_1 \exp(\int x/f \, dx)$. Letting $x/f = u''/u'$ and integrating, one obtains $f(u')^2 = c_2 \exp(c_1 u)$. Substituting $f = xu'/u''$ and noting the identity $u''/(u')^3 = -d^2x/du^2$, one obtains the linear differential equation $d^2x/du^2 + x \exp(-c_1 u)/c_2 = 0$. Let $-2 \exp(-au/2)/(c_1 c_2^{1/2}) = t + c$, where c is a constant. Then $f = c_2 \exp(c_1 u)(dx/du)^2 = (dx/dt)^2$, and the equation for x becomes $d^2x/dt^2 + (t + c)^{-1} dx/dt + x = 0$. Hence, the general solution in parametric form is $x = AJ_0(t + c) + BY_0(t + c)$, $f = (AJ_1(t + c) + BY_1(t + c))^2$, where A and B are arbitrary constants and J, Y are Bessel functions. Note that the limit $c \to \infty$ corresponds to the solution $f(x) = c - x^2$. O. G. Ruehr, *SIAM Review* 16, 1974, 261–262, Problem 73-12.

1.202 *Hint*: Consider an integral form of this problem.

1.203 *Hint*: $F(u) = (u - 1)f(u)/f'(u)$. [Ju] 13.

1.204 M. A. Abdelkader, *SIAM Review* 8, 1986, 526–528, Problem 65-4.

1.205 *Hint*:   Given an arbitrary monotone increasing function φ of t, there exists an irrational number α such that the function $x(t) = (2 - \cos t - \cos \alpha t)$ satisfies a polynomial equation of the form $P(t,x,x',x'') = 0$ and $\lim \sup_{t\to\infty} x(t)/\varphi(t) \geq 1$. [B1] Ch. 5. 2, Th. 1; Ch. 5. 3, Th. 2.

1.206 [B2] Part 2, Ex. 2. 5; R. A. Smith, *J. London Math. Soc.* 36, 1961, 33–34.

1.207 *Hint*:   Let $z = xe^{-y}/y'$, $w = xy'$, and $x = e^t$. Then the equation takes the form $z' = z(3 - z - w)$, $w' = w(z - 1)$, which is similar to the predator-prey model, see 1.249.
Answer: For the above system $(0,0)$, $(3,0)$ are saddle points, $(1,2)$ is a stable focus. The axes $x = 0$, $y = 0$ are separatrices of $(0,0)$ and a separatrix of $(3,0)$ tends to $(1,2)$ as $t \to \infty$. Compare with 2.375.

1.208 *Hint*:   Introduce the new variables $\xi = x\psi(t)$, $\eta = y\psi(t)$, $d\tau = \sigma(\tau)dt$, where ψ and σ have to be determined.
Answer: $\mu(t) = \mu_0 (1 + 2\beta t + \gamma t^2)^{-1/2}$. [Du] Ch. IV, Sec. 2.

1.209 [R-M1] Ch. IV, Ex. 12. 17.

1.210 *Answer*:   If one sets $H = x^2 - y^2$, then $H(0) = 1$, $H'(0) = H''(0) = 0$. Continuing to compute, one obtains $H^{(3)} = 6x^2H'$. Thus $H^{(n)}(0) = 0$ for $n \geq 3$ also and consequently $H \equiv 1$. From $y'' = 2x^2y$, it follows that $y'' = 2y(1 + y^2)$, $((y')^2)' = ((y^2 + 1)^2)'$, $y' = y^2 + 1$, and so $y(t) = \tan t$, $x(t) = \sec t$ from $x^2 = y^2 + H$. L. Carlitz; C. Givens, *SIAM Review* 20, 1978, 859.

1.211 *Answer*: $2(p - q)^2(p + q) = (p' + q')^2$. [Ju] 34.

1.212 *Hint*:   Use the energy integral. Compare the limit circle–limit point theorem of Weyl in 1.501–1.505.

1.213 A. Kneser, *J. reine ang. Math.* 116, 1896, 178–212.

1.214 [S-C] Ch. VII, Sec. 4. 1; F. John, *Comm. Pure Appl. Math. 1*, 1948, 341–359.

1.215 *Hint*:   Suppose that $u(c) = M$ and consider an auxiliary function $u + \varepsilon(\exp(\alpha(x - c)) - 1)$ or $u + \varepsilon((x - a)^\alpha - (c - a)^\alpha)$.
Answer: The function $u = \cos x$ satisfies $u'' + g(x)u' = 0$ with $g(x) = -\cot x$, and u has a maximum at $x = 0$. [P-W] Ch. 1, Sec. 1, Th. 1, Ex. 1, 2.

1.216 *Hint*:   Consider $x'' = 3(1 + (x')^2)^{3/2}x^2$. P. Hartman, A. Wintner, *Amer. J. Math.* 73, 1951, 390–404.

1.217 *Hint*:   Use the Liapunov function $V(t,x,x') = \int_0^x f(s)\,ds + (x')^2/2q(t)$. [De] Ch. 4, Sec. 15.

1.218 J. E. Littlewood, *J. London Math. Soc.* 41, 1966, 491–496, 497–507.

1.219 *Hint:* Take the new variables $X = (a^2 + b^2)^{-1/2}(bx - ay)$, $Y = (a^2 + b^2)^{-1/2}(ax + by)$.
Answer: In these variables the system takes the form $X' = Y^2$, $Y' = -XY$, hence $X = c\tanh(c\tau)$, $Y = \pm c/\cosh(c\tau)$ where $c^2 = X^2 + Y^2$. [GDO] Ch. 1, Sec. 3. 1.

1.220 *Hint:* Show that $x(t) = -x(-t)$ and $y(t) = y(-t)$. [KNK] Ch. 8, Sec. 4, Ex. 12.

1.221 [H-S] Ch. 10, Sec. 4, Ex. 3; [H] Ch. 7, Th. 10. 2.

1.222 [H-S] Ch. 11, Sec. 5.

1.223 and **1.224** W. A. Coppel, *J. Differential Equations* 2, 1966, 293–304.

1.225 [Ma] Ch. I, Sec. 3, Ex. 3. 6.

1.226 *Hint:* Use the Poincaré-Bendixson theorem.

1.227 *Answer:* $X(x,y) = \partial/\partial x$, $Y(x,y) = \partial/\partial x + \varepsilon(y)\,\partial/\partial y$, where $\varepsilon(y) > 0$ for $|y| < 1$ and $\varepsilon(y) = 0$ for $|y| \geq 1$.

1.228 *Hint:* Assume that D is the domain enclosed by a periodic orbit. Observe that $\int_D \mathrm{div}(BP,BQ) = 0$. [ALGM1] Ch. VI, Sec. 12, Th. 31.

1.229 [ALGM1] Ch. VI, Sec. 12, Lemma 1.

1.230 [H] Ch. 7, Sec. 6, Ex. 6. 1.

1.231 [G-H] Ch. 1, Sec. 8, Corollary 1. 8. 5.

1.232 [RSC] Th. 2. 2. 2.
Hint: Use the Ważewski theorem (*ibidem* Th. 2. 2. 1).

1.233 L. Markus, H. Yamabe, *Osaka Math. J.* 12, 1960, 305–317.

1.234 [So] Ch. VII, Ex. 12. Compare with 1.284.

1.235 [So] Ch. VII, Ex. 13.

1.236 [H-S] Ch. 11, Sec. 5, Th. 2.

1.237 [ALGM1] Ch. IV, Sec. 8, Lemma 4.

1.239 [Ir] 2. 41.

1.240 [ALGM1] Supplementary chapter, Sec. 9.

1.241 [Re] Example 1. 3. 20.

1.242 [Re] Example 1. 3. 17.

1.243 *Hint:* $V(x,y) = (x^2 - 1)\exp(-y^2)$ is a Liapunov function for this system. [Re] Example 4. 1. 4.

1.244 *Hint:* Pass to polar coordinates. If $x' = A_m(x,y)$, $y' = B_m(x,y)$, then $dr/d\varphi = -rZ(\varphi)/N(\varphi)$, where $Z(\varphi) = A_m(\varphi)\sin\varphi + B_m(\varphi)\cos\varphi$, $N(\varphi) = B_m(\varphi)\sin\varphi - A_m(\varphi)\cos\varphi$, $r(\varphi) = r_0\exp(-\int_{\varphi_0}^{\varphi} Z(\varphi)/N(\varphi)\,d\varphi)$. [N-S] 4. 2.

1.245 *Hint*: In the new variables $u = x/t$, $v = y/t$, the system takes the form $dt/t = du/(f(u,v) - u) = dv/(g(u,v) - v)$.
Answer: $t^2 + x^2 + y^2 = $ const, so $t\,dt + x\,dx + y\,dy = 0$, or $1 + uf(u,v) + vg(u,v) = 0$. [Ju] 5.

1.246 *Answer*: $x = (C - 2t)^{-1/2} \cos(C_1 - (2(C - 2t))^{-1})$, $y = (C - 2t)^{-1/2} \sin(C_1 - (2(C - 2t))^{-1})$, where C, C_1 are arbitrary constants. [Er] Ch. V, Sec. 1, Example; N. P. Erugin, *Differencialnye Uravnenia* 15, 1979, 1894–1897.

1.247 *Answer*: With $x + iy = ze^{it}$ the system becomes $(ze^{it})' = -r(r^{\exp(it)} - z)$ Im z or $z' + iz + r(r - z)$ Im $z = 0$ (Riccati-like). Under the change of variable $z = -w'/r$ Im w it transforms to $(w'/r)' + iw'/r + r$ Im $w' = 0$, $r \in \mathbb{R}\backslash\{0\}$. If $w'/r = u - iv$ then $u = v'$, $v'' + (1 - v^2) v = 0$. Generally, for r defined throughout an interval (a,b), $x(t) + iy(t) = (v'(t) - iv(t))(c + \int_a^t r(s)v(s)ds)^{-1}e^{it}$, $a \le t \le b$, where v is an arbitrary real solution of the equation $v'' + (1 - v^2) v = 0$. This formula encompasses even the singular case $r = 0$ (if one allows v to take on a complex value whenever r vanishes). Another solution: Let $x = u_1 \sin t + u_2 \cos t$, $y = -u_1 \cos t + u_2 \sin t$. Then $u_1' = u_2 - ru_1^2$, $u_2' = (r^2 - ru_2 - 1)u_1$. If $r^2 = 1$, then $u_1 = (B + t)/T$, $u_2 = 1/T$ for $T = r(A + Bt + t^2/2)$. If $r^2 \ne 0$, then $u_1 = \mu A \cosh \tau/(\mu + rA \sinh \tau)$, $u_2 = \mu^2 A \sinh \tau/(\mu + rA \sinh t)$, $\mu = (r^2 - 1)^{1/2}$, $\tau = \mu(t + B)$. Note that if $A \to \infty$, $u_1 \to (\mu \cosh \tau)/r$, $u_2 \to \mu^2/r$, which is a solution. G. N. Lewis, O. G. Ruehr, D. F. Lockhardt, W. Weston Meyer, *SIAM Review* 23, 1984, 525–527.

1.248 Songling Shi, *Scientia Sinica* 23, 1980, 153–158.
Remark: This problem is connected with the XVIth Hilbert problem.

1.249 [HKW] Ch. 2, Sec. 3, Ex. 5.

1.250 S. Chandrasekhar, *An Introduction to the Study of Stellar Structure*, Chicago, University Press, 1939. For some recent results on the Emden-Fowler equation see: F. V. Atkinson, L. A. Peletier, *Nonlinear Analysis* 10, 1986, 755–776 and the references therein.

1.251 H. Weyl, *Ann. Math.* 43, 1942, 381–407.

1.252 *Hint*: Use the Dulac criterion with $B = b \exp(-\beta x)$; see 1.228.

1.253 *Hint*: $x(t) = \sin(t + c)$, $y(t) = \cos(t + c)$, c an arbitrary constant, is the unique periodic solution. To prove this show that if $x(t)$, $y(t)$ is a periodic solution, then $\int (1 - x^2 - y^2)y^2\,dt = 0$.

1.254 *Hint*: $dy/dx = -(x - y^2)/(xy)$ is the Bernoulli equation.
[ALGM1] Ch. IV, Sec. 7, Ex. 7.

1.255 *Hint*: $(x^2 + y^2)' \ge 0$.

1.256 *Hint*: Show that $x^2 + y^2 - x^6/3$ is a first integral of this system.
[A-P] Ch. 3, Ex. 6.

1.257 [A-P] Ch. 3, Ex. 29.

1.258 *Hint*: Take $D = \{x^2 + y^2 \le c\}$, where c is sufficiently large. [G-H] Ex. 1. 6. 2.

1.259 *Hint*: Use the Dulac criterion (problem 1.228) with $B(x,y) = x^{k-1}y^{h-1}$, where k and h are suitable constants.

1.260 For instance: [H-S] Ch. 7, Sec. 3; [Le] Ch. XI, Sec. 3.

1.261 [ALGM2] Ch. V, Sec. 14.

1.262 [G-H] Ex. 1. 5. 2.

1.263 *Answer*: $a < -1/2$. [Fi] 1048.

1.264 *Hint*: In polar coordinates this system takes the form $r' = r(r^2 - 1)\sin(r^2 - 1)^{-1}$ for $r \ne 1$, $r' = 0$ for $r = 1$ and $\varphi' = 1$.
Answer: $r_k = (1 + 1/k\pi)^{1/2}$ are periodic solutions. [N-S] 3.264.

1.265 [H-S] Ch. 13, Sec. 3.

1.266 [B-L] Ch. 4, Sec. 3, Example 5.

1.267 *Hint*: Put $yx^{-2} = u$. [B-L] Ch. 4, Sec. 3, Example 7.

1.268 *Hint*: Put $y/x = u$. [B-L] Ch. 4, Sec. 3, Example 6.

1.269 *Hint*: The curve $x^4 + y^2 = cx^2$ is a trajectory. [B-L] Ch 4, Sec. 3, Example 8.

1.270 [BNF] Ch. 3, Sec. 5, Example 1.

1.271 [S-C] Ch. V, Sec. 1. 5.

1.272 *Hint*: In polar coordinates one has $r' = -r^5$, $\varphi' = r^{-2}(\cos(1/r) + r\sin(1/r))$.
Answer: The curves $\varphi = r^{-1}\sin(1/r) + c$ are solutions of this system. [S-C] Ch. IV, Sec. 1. 7.

1.273 *Hint*: In polar coordinates this system becomes $r' = -3r\cos\varphi$, $\varphi' = \sin3\varphi$.
Answer: There are six rays and the curves $r\sin3\varphi = $ const, that is, the cubics $y(2x^3 - y^2) = c(x^2 + y^2)$. [S-C] Ch. 2, Sec. 2. 6; L. S. Liagina, *Uspekhi Mat. Nauk* 6:2, 1951, 171–183.

1.274 [G-H] Ex. 1. 8. 8.

1.276 [Bo] Ch. I, Sec. 1. V.

1.277 [R-M2] 4. 17.

1.278 [G-H] Ex. 1. 8. 1.

1.279 [AVK] Ch. III, Sec. 2.

1.280 [AVK] Ch. III, Sec. 2.

1.281 [Sb] Ch. 1, Sec. 2, Ex. 10.

1.282 For related topics see: [Ni] Introduction.

1.283 For instance: [Si] Ch. I, Sec. 6, Th. 2. 6.

1.284 [P-M] Ch. 1, Sec. 1.

1.285 [P-M] Ch. 1, Ex. 8.

1.287 [S1] Ch. VI, Th. VI. 7.

1.288 [P1] Ch. II, Sec. 1.

1.289 [P1] Ch. II, Sec. 1, Th. 1. 7.

1.290 *Answer*: The classical example is the solenoid of Vietoris and van Dantzig. [N-S] Ch. IV, Sec. 8. 15.

1.291 *Hint*: It suffices to construct an orientation-preserving C^1 diffeomorphism of the circle (the Denjoy example). A. Denjoy, *J. Math. pures appl.* 11, 1932, 333–375; [Tm] Th. 1. 6, Th. 1. 11 and Th. 1. 12.

1.292 *Answer*: The famous Schweitzer example gives a negative answer to the Seifert conjecture. Here it is worth noting the Seifert theorem that nonsingular vector fields on the three-dimensional sphere which are arbitrarily small perturbations of the Hopf vector field have periodic trajectories. [Tm] Th. 3. 4; P. Schweitzer, *Ann. Math.* 100, 1974, 386–400.

1.293 [Bn2] Ch. 2, Sec. 1, Th. 2. 2.

1.294 *Hint*: Let $X = \nabla U$, and let $x(t)$ be a trajectory of X. Observe that $dU(x(t))/dt \geq 0$. Compare with 1.70.

1.295 *Hint*: Consider a transversal section S at the point $p \in \gamma$. Show that $S \cap \gamma = P$ is a perfect set, hence P is finite or uncountable. Using the flow box theorem show that P is either finite or countable.

1.296 *Hint*: Let $A = \{c \in \mathbb{R} :$ there exists $k_n \to \infty, k_n \in \mathbb{N}, \varphi(k_n + c, p) \to p\}$. Prove that: A is closed and if $c_1, c_2 \in A$, then $c_1 + c_2 \in A$ and $c_1 + k \in A$ for each $k \in \mathbb{Z}$.

1.297 [P-M] Ch. 4, Ex. 28.

1.298 T. Nadzieja, *Proc. AMS* 86, 1982, 87–90.

1.299 [N] Sec. 4, Th. 1 and Th. 2.

1.300 [A-R] Ch. 7, Sec. 32, Perturbation theorem.

1.301 [P-M] Ch. 1, Ex. 2.

1.302 *Hint*: Construct a flow on the circle with exactly one equilibrium point. [G-H] Ex. 5. 2. 5.

1.303 [Ma] Ch. I, Sec. 3.

1.304 *Hint*: Prove that X preserves the volume form $\psi\omega$ if and only if $X\psi = -\psi \, \mathrm{div}_\omega X$. [Ma] Ch. I, Sec. 3, Ex. 3. 3.

1.305 *Hint*: Show that $(\rho x^2 + \sigma y^2 + \sigma(z - 2\rho)^2)' < 0$ for sufficiently large x, y, z; next use the fact that the divergence of the Lorenz flow is $-(\sigma + \beta + 1)$. B. A. Coomes, *J. Differential Equations* 82, 1989, 386–407; [Sp] Appendix C. See also: [Te] Ch. I, Sec. 2. 3; Ch. VI, Sec. 1, and problem 1.364.

1.306 *Hint*: $x^2 - 2z = 2A$, $y^2 + z^2 = B^2$ are integrals. [Sp] Appendix K.

1.307 [Re] Examples 4. 2. 5. 1, 2, 4.

1.308 [R-M1] Ch. IV, Ex. 12. 16.

1.309 *Hint*: $I = xy + yz + zx$ is a first integral, $x = y = z$ are singular points. $I = c$ is a cone for $c = 0$, a hyperboloid of revolution of one sheet for $c > 0$ and of two sheets for $c < 0$ around the line of singular points. [S-C] Ch. II, Complement 4; E. Kasner, *Trans. AMS* 27, 1925, 155–162.

1.313 [H-S] Ch. 9, Sec. 1, Ex. 3.

1.314 *Hint*: Consider $x' = -y + x(x^2 + y^2)$, $y' = x + y(x^2 + y^2)$ in polar coordinates.

1.315 *Hint*: Use polar coordinates. [ALGM2] Ch. VIII, Sec. 22, Example 8.

1.316 [S-C] Ch. II, Sec. 4. 6.

1.318 Remark: It can be shown that all Lotka-Volterra systems of the form $x' = x(a - by)$, $y' = y(cx - d)$ have monotone period functions. C. Chicone, F. Dumortier, *Proc. AMS* 102, 1988, 706–710. Compare with 1.569.

1.319 *Hint*: Observe that the trajectories enjoy some symmetry with respect to the coordinate axes. [N-S] 4. 6731.

1.320 [Fi] 998, 999.

1.321 *Answer*: For $\varepsilon > 0$ it is a nonhyperbolic (or weakly attracting) spiral sink, for $\varepsilon = 0$ a center, for $\varepsilon < 0$ a repelling source. [G-H] Ex. 1. 3. 1.

1.322 *Answer*: The unstable manifold is $\{y = x^2/3\}$; the stable one is $\{x = 0\}$. [G-H] Example 1. 3. 8.

1.323 *Hint*: The center manifold of class C^{2N} has the form $x = \sigma(y,z) = \sum_{m=1}^{N} 2^{m-1}(m - 1)! y^{2m}/\prod_{j=1}^{m} (1 - 2jz) + o(|y|^{2N})$ as $y \to 0$ and $z < 1/2N$. If $z < 0$, then $y \mapsto \sigma(y,z)$ is analytic at $y = 0$. If $z = 0$, then this function is smooth but not analytic. If $z > 1/2N$, it is not $2N$ times differentiable. [He] Remark to the Section 6. 2; S. van Strien, *Math. Z.* 166, 1979, 143–145.

1.324 *Hint*: Suppose that $\varphi(y,z) = \sum_{j+k>1} c_{jk} y^j z^k$. Compute $c_{jk} = (j + k)!(j!)^{-1} h_{j+k}$, where $h(y) = \sum_{j \geq 2} h_j y^j$.
Remark: A C^∞ center manifold is given by $\varphi(y,z) = \int_{-\infty}^{0} e^s h_1(y - sz)\, ds$, where h_1 is a bounded C^∞ function which coincides with h in a neighborhood of the origin. [M-MC] Ch. 2, Remark 2. 6. 2.

1.325 *Hint*: Consider the linearized system $x' = 0$, $y' = -y$, and center manifolds of these systems. [HKW] Ch. 1, Sec. 7, Ex. 6.

1.326 *Hint*: Such a center manifold would be defined by $y = \sum_{n=2}^{\infty}(n - 1)!x^n$. [HKW] Appendix A, Remark 3.

1.327 *Answer*: The x-axis is the only analytic center manifold. Piecing together any solution curve in the left half-plane with the positive half of the x-axis one obtains a smooth center manifold. [G-H] Example (3. 2. 1).

1.328 [N] Sec. 3.

1.329 *Hint*: Let $A(x,y) = (y,-x)$ and $\varphi : \mathbb{R} \to \mathbb{R}$ be a C^{∞} function whose Taylor series at the origin is 0 but which is not identically 0 in any neighborhood of the origin and let $X = \varphi(r)A$, where $r^2 = x^2 + y^2$. Consider the periods of the orbits of the vector fields X and A. [N] Sec. 3.

1.330 *Hint*: 0 is asymptotically stable, hence there is a Liapunov function V. $V^{-1}(\{1\})$ is diffeomorphic to S^1. First define h on $V^{-1}(\{1\})$; next extend h to the whole neighborhood of 0 in a natural way.
Remark: For $n > 2$ the problem is connected with the Poincaré conjecture.

1.331 *Hint*: Show that there exists a function $\lambda : \mathbb{R}^n \to \mathbb{R}$ such that $Df(x)X(x) = \lambda(x)Y(f(x))$. Next prove that for $v \neq 0$ there exists $\sigma(v) = \lim_{t \to 0} \lambda(tv)$ and $Df(0)DX(0)v = \sigma(v)DY(0)Df(0)v$. [P-M] Ch. 2, Ex. 13.

1.332 *Answer*: The Hartman example is $x_1' = 2x_1$, $x_2' = x_2 + x_1x_3$, $x_3' = -x_3$. This system is topologically but not differentiably equivalent to $y_1' = 2y_1$, $y_2' = y_2$, $y_3' = -y_3$.
For instance: [BVGN] Ch. X, Example 32. 2; [Sk] Remark 1. 8. 1; S. Sternberg, *Amer. J. Math.* 80, 1958, 623–631.
Remark: S. van Strien has shown (*Smooth linearization of hyperbolic fixed point without resonance condition*, preprint, TU Delft) that if f is C^2 then the conjugacy with the linearization is differentiable at the origin and satisfies the Hölder condition outside 0. In general, this conjugacy is not Lipschitz in a neighborhood of the origin.

1.334 *Answer*: There is no finite singular point. In homogeneous coordinates the singular points are: $(1,0,0)$, $(1,2,0)$, and $(1,-2,0)$. The first is a saddle of index -1; the second and the third are nodes of index 1. [Le] Ch. IX, Sec. 5. 15.

1.335 [KPPZ] Ch. 3, Sec. 13, Th. 13. 5.

1.336 [G-H] Ex. 1. 8. 9.

1.337 [KPPZ] Ch. 2, Sec. 7, Th. 7. 1.

1.338 [KPPZ] Ch. 2, Sec. 7, Th. 6. 3.

1.339 *Hint*: Use 1.338. [H] Ch. VII, Sec., Ex. 3. 3.

1.340 [KPPZ] Ch. 2, Sec. 2, Lemma 7. 2.

1.341 [KPPZ] Ch. 2, Sec. 7, Th. 7. 2.

1.342 [KPPZ] Ch. 2, Sec. 8, Th. 8. 3.

1.343 [KPPZ] Ch. 2, Sec. 8, Ex. 2.

1.344 [KPPZ] Ch. 2, Sec. 8, Th. 8. 4.

1.345 [KPPZ] Ch. 3, Sec. 14.

1.346 [G-H] Ex. 1. 9. 1.

1.347 [ALGM2] Ch. VIII, Sec. 22, Example 7.

1.348 [G-H] Ch. 3, Sec. 4, (3. 4. 1)–(3. 4. 4).

1.349 [A-P] Ch. 5, Example 5. 5. 3.

1.350 [M-MC] Ex. 3. 17, 3. 18.

1.351 *Hint*: This cannot be done directly as the radius of this limit cycle is about 2 as $\mu > 0$ is small. Transform the Van der Pol equation (or more generally $x'' + f(x)x' + g(x) = 0$) into a Liénard equation (here $X' = Y - F(X)$, $Y' = -G(x)$), so take $X = x$, $Y = y' + F(y)$, $F' = f$. [M-MC] Ch. 4B, Example 4B. 3.

1.352 [A-P] Ch. 5, Ex. 17.

1.353 *Hint*: Compute $(x^2 + y^2)'$. [B-L] Ch. 11, Sec. 4, Example 2.

1.354 *Hint*: $x^2 + y^2/\lambda = 1/\lambda$ is the periodic orbit. [B-L] Ch. 11, Sec. 4, Example 3.

1.355 Remark: This is so-called bifurcation from infinity phenomenon. [I-J] Ch. III, Sec. 3, Ex. 1. See also: S. Rosenblat, S. H. Davis, *SIAM J. Appl. Math.* 37, 1979, 1–19.

1.356 [I-J] Ex. V. 6.

1.357 *Answer*: The stationary solutions are $x = 0$ and $y = 0$, $x = 0$ and $y = \sigma - \mu$, $x = \mu$ and $y = 0$, $x = 2\sigma - 3\mu$ and $y = 2\mu - \sigma$. The points $x = 0$, $y = \sigma/3$ for $\mu = 2\sigma/3$ and $x = \sigma/2$, $y = 0$ for $\mu = \sigma/2$ are the secondary bifurcation points. [I-J] Ch. V, Sec. 9, Example V. 6.

1.358 *Hint*: Rewrite this system in polar coordinates: $r' = \mu r - r^{2k} \sin(1/r)$, $\theta' = 1$.
Answer: There exists a finite number of periodic solutions for each $\mu \neq 0$ and there are infinitely many periodic solutions for $\mu = 0$. [HKW] Ch. 1, Sec. 2, Remark 4.

1.359 *Answer*: There is no bifurcation. Multiplying the first equation by v and the second by u one obtains $\lambda \int_0^1 (u^2 + v^2)^2 \, dx = 0$.

Remark: This system is not of variational form. If it were, then for every eigenvalue λ of the linearized problem there would be a bifurcation. [A-K] M. Berger, Ch. VI, Sec. 3, Example 1.

1.360 *Hint*: $\lambda = a^2/4$ is a bifurcation value of the parameter. [AVK] Ch. II, Sec. 5. 4.

1.361 *Hint*:   Under arbitrarily small perturbations this limit cycle can be split into three (two stable and one unstable) limit cycles.

1.362 [HKW] Ch. 2, Sec. 3, Ex. 6.

1.363 [M-MC] Ch. 3A.

1.364 *Answer*:   For the "classical" choice of parameters, $\sigma = 10$, $b = 8/3$, $r = 470/19$, there is a subcritical bifurcation of (unstable) periodic orbits.
Remark: $r = 28$ above corresponds to the Lorenz attractor. [M-MC] Ch. 4B, Example 4B. 8, [HKW] Ch. 3, Sec. 6, Example 3; and references to 1.305.

1.365 *Answer*:   The following examples are due to C. L. Siegel. The system $x' = -y - x(x^2 + y^2)/2$, $y' = x - y(x^2 + y^2)/2$ does not have nontrivial periodic solutions.
Hint: In polar coordinates $rr' = -r^4/2$, hence $r^{-2} = t + $ const, which shows that there are no periodic solutions. The second example is defined by the Hamiltonian $H = (x_1^2 + y_1^2)/2 - (x_2^2 + y_2^2) + x_1 y_1 x_2 + (x_1^2 - y_1^2)y_2/2$. The matrix A has the eigenvalues $\pm i$ and $\pm 2i$. There exists a family of periodic solutions such that $\lim_{R \to 0} T_1(R) = \pi$, but there is no solution of minimal period close to 2π. The important feature of this example is the fact that the surfaces $H = $ const are not homeomorphic to spheres. [A-K] M. Berger, Ch. VI, Sec. 3, Example 2; [S-M].

1.366 [KNK] Ch. 7, Ex. 24.

1.367 [KNK] Ch. 7, Ex. 25, 26.

1.369 *Hint*:   Consider $x' = xg'(t)/g(t)$, where $g(t) = \sum_{n=1}^{\infty}(1 + n^4(t - n)^2)^{-1}$, $C = 2 + \sum_{n=1}^{\infty} n^{-4}$, $V(x,t) = x^2(C^2 + \int_t^{\infty} g^2(s)\, ds)g^{-2}(t)$. J. L. Massera, *Ann. Math.* 50, 1949, 705–721.

1.370 [BVGN] Ch. VI, Corollary 16. 2. 4.

1.371 *Hint*:   In polar coordinates $\varphi' = 0$, $r' = rg_t(t,\varphi)/g(t,\varphi)$, $g(t,\varphi) = \sin^4\varphi(\sin^4\varphi + (1 - t\sin^2\varphi)^2)^{-1} + (1 + \sin^4\varphi)^{-1}(1 + t^2)^{-1}$. J. L. Massera, *Ann. Math.* 50, 1949, 705–721.

1.372 *Hint*: $V = \sum_{j=1}^{n} \int_0^y f_j(s)\, ds$ in a suitable Liapunov function; recall the Sylvester criterion. [Bn1] Ch. I, Sec. 14, Example 7.

1.373 L. Markus, H. Yamabe, *Osaka Math. J.* 12, 1960, 305–317.

1.374 *Hint*:   Use the preceding problem. Show that $\lambda(x) < -\varepsilon$ for some $\varepsilon > 0$. L. Markus, H. Yamabe, *Osaka Math. J.* 12, 1960, 305–317.

1.375 L. Markus, H. Yamabe, *Osaka Math. J.* 12, 1960, 305–317.

1.376 *Answer*: The equation $x' = \sin^2 x$ has the general solution of the form $x(t) = \text{arccotan}(x_0 - t)$ for $x_0 \neq k\pi$ or $x = k\pi$ for $x_0 = k\pi$, $k \in \mathbb{Z}$. 0 is an unstable solution since $\lim_{t \to \infty} x(t) = \pi$ for every $0 < x_0 < \pi$. [De] Ch. 2, Sec. 7.

1.377 R. E. Vinograd, *Mat. Sb.* 41, 1957, 431–438; W. Hahn, *Stability of Motion*, Springer, Berlin, 1967.

1.378 *Hint*: $F(x,y) = (-y(x^2 + y^2)^{1/2}, x(x^2 + y^2)^{1/2})$.

1.379 *Answer*: The real part of these eigenvalues is equal to $(a - 2b)/2$. [RHL] V. 9. 1.

1.380 [RHL] IX. 3. 11.

1.381 [R-M2] 1. 9. 10.
Remark: Another example is $x' = -y \pm x(x^2 + y^2)$, $y' = x \pm y(x^2 + y^2)$.

1.382 *Hint*: $V(x,y) = x^2 + y^2$ is a Liapunov function. [A-P] Ch. 5, Example 5. 4. 3.

1.383 *Answer*: Equilibrium points are not structurally stable: there is a pair of imaginary eigenvalues.
Remark: Numerically one observes complicated nonperiodic solutions similar to those in the Lorenz system. A. S. Pikovskiĭ, M. I. Rabinovich, On strange attractors in physics, in *Nonlinear Waves*, ed. A. V. Gaponov, Nauka, Moskva, 1979.

1.384 [KNK] Ch. 7, Ex. 54.

1.385 [PSV] Ch. V, Sec. 2. 4. Theorem; C. Olech, *Contributions to Differential Equations*, v. I, 1963, 389–400, A. Gasull, J. Llibre, J. Sotomayor, *J. Differential Equations* 91, 1991, 327–335.

1.386 *Hint*: $V = (dx - by)^2 + 2 \int_0^{\check{x}}(df(s) - bg(s))\, ds$ is a good Liapunov function for this system. N. N. Krasovskiĭ, *Doklady AN SSSR* 88/3, 1953.
Remark: The third condition is important; N. N. Krasovskiĭ (*Applied Math. Mech.* 16/5, 1952) gave examples where this condition is not satisfied and the zero solution may be destabilized by arbitrarily small perturbations. [Bn1] Ch. I, Sec. 14, Example 1.

1.387 *Answer*: The origin is uniformly stable. [R-M2] 2. 2. 7.

1.388 [R-M2] 2. 10. 1.

1.389 *Hint*: $V(x,y) = (x^2 + y^2)$ is a Liapunov function for this system on the unit disk. [R-M2] 1. 4. 15.

1.390 [De] Ch. 3, Ex. 4.

1.391 *Hint*: Use the existence of a Liapunov function and the Grobman-Hartman theorem.

1.392 [De] Appendix, Ex. 14; L. L. Helms, C. R. Putnam, *J. Math. Mech.* 6, 1957, 901–903.

1.393 [R-M2] 1. 5. 6.
Hint: Use the Chetaev theorem (1. 5. 1 *ibidem*).

1.394 [R-M2] 1. 9. 1.

1.395 *Hint:* Taking the new variable $y = x' + \int_0^x g$ (the Liénard transformation) one gets the system $x' = y - \int_0^x g$, $y' = -f$. $V = y^2 + 2\int_0^x f$ is a suitable Liapunov function. [Bn1] I, Sec. 14, Example 3.

1.396 *Hint:* Use 1.374. L. Markus, H. Yamabe, *Osaka Math.* J. 12, 1960, 305–307.

1.397 [R-M2] 2. 10. 9.

1.398 [De] Ch. 4, Ex. 12.

1.399 *Hint:* The general solution is $x(t) = c \sin(ct + d)$ with some constants c and d. Consider two solutions with incommensurable c's.
Answer: The trivial solution $x(t)$ is stable, the others unstable. [Ce] Ex. 1. 4. 7.

1.400 *Answer:* For instance $V = (x')^2 + (x' + x)^2 + 4(1 - \cos x)$.
Caution: A natural Liapunov function $(x')^2/2 + (1 - \cos x)$ (the total energy) shows the stability only (not the asymptotic stability). [RHL] I. 6. 12.

1.401 *Hint:* $H(x,x') = (x')^2/2 + 2b \sin^2(x/2)$ is a Liapunov function. [C-L] Ch. 16, Ex. 3.

1.402 [R] Pb. 17.

1.403 *Answer:* This is an unstable solution. [KPPZ] Ch. 3, Sec. 13, Ex. 13. 6.

1.404 *Hint:* $x' = v$, $v' = -x + v(1 - 3x^2 - 2v^2)$; write this system in polar coordinates. [A-P] Ch. 3, Sec. 3. 9, Example 3. 9. 1.

1.405 [R-M2] 1. 9. 1.

1.406 *Hint:* Consider $U(x) = \exp(-|x|^{-1}) \sin(|x|^{-1})$.
Answer: No.

1.407 [H-S] Ch. 9, Sec. 3. Compare 1.406, 1.408.

1.408 *Hint:* Consider $U = x^5 \sin(x^{-1}) - y^2 - z^2$. [Ap] Ch. III, Sec. 207.

1.409 [Ga] Ch. 2, Sec. 2. 9, Proposition 14.

1.410 *Answer:* This example is due to Cherry. The origin is unstable. $p_1 = 2^{1/2} \sin(t - T)/(t - T)$, $p_2 = \sin2(t - T)/(t - T)$, $q_1 = -2^{1/2} \cos(t - T)/(t - T)$, $q_2 = \cos 2(t - T)/(t - T)$, is a solution whose initial value is close to the origin if T is sufficiently large. This solution blows up at time T. [Po] Ch. 3, Sec. 6.

1.411 [RHL] Ch. III. 5. 2.

1.412 [RHL] Ch. III. 4. 6.

1.413 [RHL] Ch. III. 2. 15.

1.414 *Answer:* The origin is stable in this example due to Wintner. [RHL] Ch. III. 2. 8.

1.415 *Answer:* Despite negativity of the potential energy for $q_1 = q_2 \neq 0$ the origin is stable. [RHL] Ch. III. 2. 13.

1.416 *Answer:* Unstable, even in the case of forces proportional to the kth power of the inverse of the distance. The total force is proportional to $F(x) = -\sum_{j=1}^{2^n} (x - a_j) \mid x - a_j \mid^{-k-1}$, where $a_1, \ldots, a_{2^n}$ are the vertices of the unit cube. The potential φ is given by $\varphi(x) = (k - 1)^{-1} \sum_{j=1}^{2^n} \mid x - a_j \mid^{-k+1}$. The origin is stable if and only if φ has a local maximum at $x = 0$. The Hessian of φ at 0 is $F_x(0) = -\sum_{j=1}^{2^n} \mid a_j \mid^{-k-3} (\mid a_j \mid^2 I - (k + 1) a_j a_j^T) = -n^{-(k+3)/2} 2^n (n - k - 1) I$. If $n > k + 1$, then the origin is stable; $n < k + 1$ implies the instability of the origin. If $n = k + 1$, the potential φ is a harmonic function which does not have maxima or minima. Hence 0 is unstable. O. Bottema, O. P. Lossers, *SIAM Review* 18, 1976, 118–119.

1.417 [RHL] Ch. V. 6. 1.–6. 4.

1.418 *Answer:* The position of median inertia axis is unstable, the others stable. [RHL] Ch. I. 4. 7, I. 5. 6; [A3] Ch. 6, Sec. 29. See 1.734.

1.419 [R-M1] 4. 9. 11.

1.420 *Hint:* Represent $f(r)$ as $r^k g(r) h(r)$ with $g(r) > 0$ for $r > 0$ and h which has positive roots only. [Le] Ch. X, Sec. 5, Example 12. 2.

1.421 [Le] Problem 18.

1.422 *Answer:* The equation $\mu = f(r^2) + r^{2(s+1)}$ determines periodic orbits in a neighborhood of the origin. These orbits are unstable. [M-MC] Remark to the Russian translation (Mir, Moskva, 1980) in Ch. 3B.

1.423 *Hint:* The multipliers of the associated equation in variations satisfy the equation $\rho^2 - \rho \, \text{tr} \, X(T) + \det X(T) = 0$, where X is the fundamental matrix of the linearized system around (ξ, η), $X(0) = I$. Obviously, $\rho_1 = 1$; hence $\rho_2 = \det X(T)$, which equals (from the Liouville formula) the exponent of the quantity in the Poincaré criterion. [De] Ch. 4, Sec. 4.

1.424 [HKW] Ch. 1, Sec. 7, Ex. 1, 5; Ch. 2, Sec. 2, Example 4.

1.425 *Hint:* In cylindrical coordinates this system takes the form $dr/d\varphi = ar(1 - r)$, $dz/d\varphi = ar(1 - z)$.
Answer: The equations $z = 1$, $r = 1$ describe the limit cycle.

1.426 *Hint:* The general solution is $x(t) = C_1 \exp(t \sin t)$, $y(t) = C_2 \exp(-t \sin t)$.
Answer: If $C_1 C_2 \neq 0$, then the lower Lipaunov exponent is 0, otherwise it is -1 (except for the case $x \equiv 0$, when it is not defined).

1.427 [De] Ch. 3, Sec. 12.

1.428 [De] Ch. 3, Ex. 22.

1.429 *Answer*: $-1, 1$ if $c = 0$; $-1, -(1 + e^{-4\pi})$ if $c \neq 0$.

1.430 [Ce] 6. 6; R. E. Vinograd, *Prikl. Mat. Mekh.* 17, 1953, 645–650.

1.431 *Answer*: The Perron example is $x' = p(t)x, y' = 2p(t)y$, where $p(t) = \sin(\log t) + \cos(\log t)$. A basis of solutions is given by $x_1 = e^{-q(t)}$, $y_1 = 0$; $x_2 = 0$, $y_2 = e^{2q(t)}$, where $q(t) = \int p = t \sin(\log t)$. The lower Liapunov exponents for (x_1, y_1), (x_2, y_2), and $(x_1 + x_2, y_1 + y_2)$ are $-1, -2$, and a positive number, respectively. [BVGN] Ch. I, Sec. 3. 3.

1.432 *Answer*: For

$$A(t) = \begin{bmatrix} -1 - 2\cos 4t & -2 + 2\sin 4t \\ 2 + 2\sin 4t & -1 + 2\cos 4t \end{bmatrix}$$

there is an eigenvalue -1 but the Liapunov exponent of the solution $(e^t \sin 2t, e^t \cos 2t)$ equals 1. [BVGN] Ch. IV, Sec. 9. 1.

1.433 *Answer*: The extremal Liapunov exponents are $-1, +1$ (while for the system with the matrix

$$\begin{bmatrix} 0 & 0 \\ 0 & \pi \sin \pi\sqrt{t} \end{bmatrix}$$

they are 0, 0). There exist arbitrarily small perturbations of the matrix in the problem such that the Liapunov exponent jumps from ± 1 to 0. [BVGN] Ch. V, Example 13. 5. 1.

1.434 *Answer*: For instance $x' = (x^2(2y - x) + 2y^5)/(x^2 + y^2)$ $(1 + (x^2 + y^2)^2)$, $y' = 8y^2(y - x)/(x^2 + y^2)(1 + (x^2 + y^2)^2)$. Here the Liapunov exponents are less than -0.1; all solutions approach the origin but the zero solution is unstable. To see this prove that $dy/dx > 2$ on the segment [(0,0), (1/4,1/2)]. The phase portrait can be sketched after a careful analysis of the sets where $x' > 0, <0, y' > 0, <0$. [BVGN] Ch. III, Example 6. 3. 1.

1.435 [H-S] Ch. 6, Sec. 6, Ex. 4.

1.436 *Hint*: $\Phi(x) = d^n u/dx^n$, $K(x,y) = \sum_{j=1}^n a_j(x)(x - y)^{j-1}/(j - 1)!$. [Tr2] Ch. I, Sec. 7.

1.437 [B1] Ch. 4, Ex. 9. 1. 4; [I] Ch. VIII, Ex. 3.

1.438 *Hint*: The system $a_0 y + \cdots + a_n y^{(n)} = 0$, $a_0 y' + \cdots + a_n y^{(n+1)} = 0, \cdots, a_0 y^{(n)} + \cdots + a_n y^{(2n)} = 0$ has a solution $(a_0, \ldots, a_n)$.

1.439 [I] Ch. VIII, Ex. 3. Compare with problem 1.187 concerning the Airy equation $(n = 2)$.

1.440 [Sa] Ch. V, Sec. 4. 5.

1.441 [Sa] Ch. IV, Sec. 1. 2.

1.442 *Hint*: Use the backward induction method. First prove that x', . . . , $x^{(n-1)}$ are also uniformly bounded on $\mathbb{R}$. [Sa] Ch. VI, Sec. 9. 5.

1.443 [Sa] Ch. VII, Sec. 3. 1.

1.444 [Ce] 3. 3.

1.445 *Answer*: $y(x) = ((n-1)!)^{-1} \int_1^x (x-t)^{n-1} g(t) t^{-n} \, dt$. *Amer. Math. Monthly* 71, 1964, 636, Putnam Competition 1963, I-3; [B-W] 5. 40; [I] Ch. VIII, Ex. 8.

1.446 G. Mammana, *Math. Zeit.* 33. 1931, 186–231; [Ra] Ch. V, Sec. 31; [P-S] V. 93.

1.447 [Ra] Ch. V, Sec. 30, Th. 29. 2; [P-S] V. 93 and related problems.
Remark: For $n = 2$ the positivity condition $y(x) \geq 0$ is equivalent to the conditions on Wronskians. For $n \geq 3$ it is no longer true. For instance, consider $L[y] = y^{(3)} + y'$ on the interval $[0,b]$ with $b > 2\pi$.

1.448 [Hi] Ex. 5. 2. 3, 5. 2. 4.

1.449 P. Schweitzer; G. N. Lewis, *SIAM Review* 28, 1986, 241–242, Problem 85-8.

1.450 [I] Ch. XV, Ex. 1.

1.451 [Hi] Ex. 3. 1. 6, 3. 1. 7.

1.452 [Hi] Ex. 3. 2. 2.

1.453 [Hi] Ex. 3. 2. 3.

1.454 [Bu] Ch. V, Ex. 3.

1.455 [Hi] Ex. 3. 3. 4.

1.456 [Hi] Ex. 4. 2. 5.

1.457 [Hi] Ex. 4. 1. 5.

1.458 [Hi] Ex. 4. 1. 6.

1.459 [Bu] Ch. VI, Ex. 1.

1.460 [Hi] Ex. 5. 1. 3. The result is due to G. Julia.

1.461 [Hi] Th. 5. 3. 1.

1.462 [V] Ch. VI, Sec. 95.

1.463 *Hint*: Consider as an example the equation $xy' - \lambda y = ax$. Its solution is (for $\lambda \neq 1$) $ax/(1 - \lambda) + Cx^\lambda$, hence for $\lambda \notin \mathbb{N}$ C must vanish in order to have an analytic solution. [V] Ch. VIII, Sec. 130; [I] Ch. 14. 4.

1.464 [Hi] Ex. 10. 2. 10.

1.465 [Hi] Ex. 11. 1. 3.

1.466 [Hi] 11. 1. 4.

1.467 *Hint*: In (vi) replace z by $1 + \varepsilon z$, d by d/ε^2, c by $c/\varepsilon - d/\varepsilon^2$, and let $\varepsilon \to 0$. The limiting form of this equation is (v). In (v) replace w by $1 + \varepsilon w$, b by $-b/\varepsilon^2$, a by $b/\varepsilon^2 + a/\varepsilon$, c by $c\varepsilon$, and d by $d\varepsilon$. In the limit $\varepsilon \to 0$, the equation becomes (iii). Similarly, in (v) replace w by $\varepsilon w\sqrt{2}$, z by $1 + \varepsilon z\sqrt{2}$, a by $1/(2\varepsilon^4)$, c by $-\varepsilon^{-4}$, d by $-(1/(2\varepsilon^4) + \delta/\varepsilon^2)$. In the limit equation (iv) arises. In (iii) replace z by $1 + \varepsilon^2 z$, w by $1 + 2\varepsilon w$, c by $1/(4\varepsilon^6)$, d by $-1/(4\varepsilon^6)$, a by $-1/(2\varepsilon^6)$, b by $1/(2\varepsilon^6) + 2b/\varepsilon^3$. In the limit the equation becomes (ii). Similarly, (ii) may be obtained from (iv) by replacing z by $\varepsilon z/2^{2/3} - 1/\varepsilon^3$, w by $2^{2/3}\varepsilon w + 1/\varepsilon^3$, a by $-1/(2\varepsilon^6) - a$, b by $-1/(2\varepsilon^{12})$ and taking the limit. Finally, in (ii) replace z by $\varepsilon^2 z - 6/\varepsilon^{10}$, w by $\varepsilon w + 1/\varepsilon^5$, a by $4/\varepsilon^{15}$, and in the limit the equation degenerates into (i). [Hi] Ex. 11. 1. 4, Ch. 12, Sec. 2; [I] Ch. 14. 4.

1.469 *Hint*: It is easy to compute $x(t) = \sin 2t + \cos 2t - (1000/9996) \sin 100t$, $y(t) = (2^{1/2} - 1) \sin t + \cos t - (1000/9999) \sin 100t$, but a direct comparison of x and y is a simpler method.
Answer: $x(t) \geq y(t)$ on $[0, \pi/4]$. [BSW] Ch. 5, Sec. 3, Example 5. 2.

1.470 *Answer*: $\lambda_n = 2n(2n - 1)$ for $n = 1,2,3, \ldots$ are the eigenvalues and Legendre polynomials P_{2n-1} are the eigenfunctions. [L] Ch. VI, Sec. 83.

1.471 *Hint*: This equation is the Euler equation for the minimum of the functional $J[y] = \int_0^a ((y')^2 - q(t)y^2 + 2r(t)y) \, dt$, $y(0) = y(a) = 0$. This extremum is realized by a unique function y. This minimizer cannot be positive (prove it a contrario). [Ra] Ch. III, Sec. 18, Th. 18. 1.

1.472 [KPPZ] Ch. 3, Sec. 14, Ex. 14. 14.

1.473 [KPPZ] Ch. 3, Sec. 14, Ex. 14. 15.

1.474 *Hint*: Change the variables $x(t) = y(t) \exp(\frac{1}{2} \int_a^t q(s) \, ds)$ and apply the Sturm comparison theorem. [V-K] Ex. 1. 2.

1.475 [V-K] Ex. 1. 5.

1.476 [B2] Part 1, Ex. 11. 3.

1.477 [V-K] Ex. 1. 7.

1.478 [V-K] Ex. 1. 8.

1.479 *Hint*: $(1 + t)^{ia} = \cos(a \log(1 + t)) + i \sin(a \log(1 + t))$.
Answer: $\lambda_n = 1/4 + (n\pi/\log 2)^2$, $x_n(t) = (1 + t)^{-1/2} \sin(n\pi \log(1 + t)/\log 2)$.

1.480 *Answer*: For $\lambda = i$, $x(t) = i + \cos t$ solves this problem. This example is due to Wielandt.

1.481 *Answer*: $y^{(4)} + y = -\lambda((34 - 9\varepsilon^{-1}e^{-ix})y')'$ with the boundary conditions $y^{(j)}(0) = y^{(j)}(2\pi)$, $j = 0,1,2,3$, has a solution $y_1(x) = e^{ix}$ for $\lambda_1 = 1/17$ and $y_2(x) = e^{ix} + \varepsilon e^{2ix}$ for $\lambda_2 = 1/8$.

1.482 *Answer:* $\sigma(L) = \mathbb{R}$. [Mt] Ch. V, Sec. 7, Ex. 16.

1.483 *Answer:* $\sigma(L) = \{\lambda : 4\,\text{Re}\,\lambda \geq (\text{Im}\,\lambda)^2\}$. [He] Ch. 5, Appendix, Ex. 4.

1.484 [M] Ch. III, Ex. 20.

1.485 *Answer:* (i) $\exp(-\,|\,x\,-\,y\,|\,)/2$; (ii) $(\exp(-\,|\,x\,-\,y\,|\,)\,-\,\exp(-x\,-\,y))/2$. [R-S] Ch. XI, Ex. 47.

1.486 [Z] Ex. 79.

1.487 [Z] Ex. 83.

1.488 *Hint:* $\|H_{ab}^{1/2}f\|^2 = a\,|\,f(1)\,|^2 + b\,|\,f(0)\,|^2 + \int_0^1 |\,f'(x)\,|^2\,dx$. [D] Ex. 4. 18.

1.489 *Corollary:* In the non(formally)selfadjoint case, the weak maximum principle does not imply the coercivity. [D-L] Ch. II, Sec. 8.

1.490 *Remark:* In general, a sequence of eigenvalues does not determine the potential q, but (the Borg-Levinson theorem, for instance: G. Borg, *Acta Math.* 78, 1946, 1–96; B. M. Levitan, *Sturm-Liouville Inverse Problems*, Nauka, Moskva, 1984, Th. 3. 3. 1) two increasing sequences of eigenvalues corresponding to two different selfadjoint boundary conditions do this. For multidimensional generalizations see, for instance, A. Nachman, J. Sylvester, G. Uhlmann, *Comm. Math. Phys.* 115, 1988, 595–605; S. Chanillo, *Proc. AMS* 108, 1990, 761–767.

1.491 *Answer:* ind $A_\pm = \pm 1$. A. A. Kirillov, A. D. Gvishiani, *Theorems and Problems in Functional Analysis*, Nauka, Moskva, 1988, 362; translation: Springer, New York, 1982; [B-W] 9. 113.

1.492 [B-S] Ch. II, Sec. 1, Example to Th. 1. 1.

1.493 *Hint:* Let $a < b$ and $W(x) = c < 0$ on $[a,b]$ and 0 elsewhere. Set $f(x) = \beta e^{kx} + \gamma e^{-kx}$ and choose the constants β, γ, k in $(-\infty,a)$, $[a,b]$, (b,∞) so that f is an eigenfunction of the operator $-d^2/dx^2 + W$. [Go] Ch. 2, Ex. 14. 23. 1.

1.494 [R-S] Ch. XIII, Ex. 21.

1.495 *Hint:* Let $N_1 = 100$ and x_1 denote the root of the equation $\int_0^x (x + 2)\log(x + 3))^{-1}\,dx = N_1$. Define $V(x) = (2\log 3)^{-2} - ((x + 2)\log(x + 2))^{-2}$ for $0 \leq x \leq x_1$. Then define recurrently $N_2, \ldots, N_k, x_1 < x_2 < \cdots < x_k$, so that $100(V(x_k) + 1)x_k \leq N_{k+1}$, $\int_{x_k}^{x_{k+1}} ((x + 2)\log(x + 3))^{-1}\,dx = N_{k+1}$ and put $V(x) = V(x_k) + 1 - ((x + 2)\log(x + 3))^{-1}$ for $x_k < x \leq x_{k+1}$.
Remark: Such a potential V may be modified to a smooth function. This example is due to M. Otelbaev. [K-S] Ch. IV, Sec. 7.

1.496 *Hint:* Compare $\lambda_n(c)$ with the eigenvalues of the operators $-d^2/dx^2 + x^2$ and $-d^2/dx^2 + cx^4$ and apply the minimax principle. [R-S] Ch. XIII, Ex. 9.

1.497 *Hint:* $\varphi_n(x) = (n!)^{-1/2}2^{n/2}i^{-n}\pi^{-3/4}\exp(x^2/2)\int u^n \exp(-u^2 + 2ixu)\,du$ are the eigenfunctions of H with the eigenvalues n. [D] Lemma 7. 12, Th. 7. 13.

1.498 [R-S] Ch. X, Sec. 9, Example 1, Th. X. 56.

1.499 *Answer*: The Hamiltonian is $H = \bar{p}p \exp(i(z - \bar{z})) + (k^2 - u)$ $\exp(-i(z - \bar{z}))$ and the canonical equations are $p' = -i \mid p \mid^2 \exp(i(z - \bar{z})) + i(k^2 - u) \exp(-i(z - \bar{z}))$, $z' = \bar{p} \exp(i(z - \bar{z}))$. [Pa] Ex. 3. 9. 1.

1.500 *Hint*: If $u(t)$, $v(t)$ are two linearly independent solutions with $\lambda = \lambda_0$ and (without loss of generality!) $(u(0)v'(0) - u'(0)v(0))p(0) = 1$, then for every λ the general solution x satisfies the relation $x(t) = c_1 u(t) + c_2 v(t) + (\lambda - \lambda_0) \int_a^t (u(t)v(s) - u(s)v(t))x(s) \, ds$ with arbitrary real a. Apply the Schwarz inequality and choose sufficiently large a to get $\mid \int_a^t (u(t)v(s) - u(s)v(t))x(s) \, ds \mid \leq C(\mid u(t) \mid + \mid v(t) \mid)(\int_a^t \mid x(s) \mid^2 ds)^{1/2}$ with $\int_a^\infty \mid u \mid^2 \leq C^2$, $\int_a^\infty \mid v \mid^2 \leq C^2$. To finish observe that $(\int_a^t \mid x \mid^2)^{1/2} \leq (\mid c_1 \mid + \mid c_2 \mid)C + 2 \mid \lambda - \lambda_0 \mid C^2 (\int_a^t \mid x \mid^2)^{1/2}$ and take a so large that $2 \mid \lambda - \lambda_0 \mid C^2 < 1/2$. [Ce] 5. 6; [R-S] Ch. X, Sec. 1, Appendix.

1.501, 1.502 The same reference as to the preceding problem.

1.503 *Hint*: Use the conservation of energy equation: $\frac{1}{2}(x')^2 + V(x) = $ const. [R-S] Ch. X, Th. X. 5.

1.504 [R-S] Ch. X, Th. X. 7.

1.505 [R-S] Ch. X, Th. X. 8.

1.506 The examples are due to J. Rauch, M. Reed, *Comm. Math. Phys.* 29, 1973, 105–111, and based on some ideas of E. Nelson. [R-S] Ch. X, Sec. 1, Appendix, Remarks to Ch. X.

1.507 *Hint*: Consider separately the cases $b < \pi$, $b = \pi$, $b > \pi$.
Answer: If $b < \pi$ the problem has a unique solution; if $b \geq \pi$, there is either one solution, no solution, or more than one solution, depending upon the number B. [BSW] Ch. 1, Sec. 4.

1.508 [KPPZ] Ch. 3, Sec. 15, Ex. 15. 16.

1.509 [KPPZ] Ch. 3, Sec. 15, Ex. 15. 11.

1.510 [PSV] Ch. I, Sec. 5. 8, Ex. 3.

1.511 [V-K] Ex. 6. 10.

1.512 *Answer*: There is (except for the case $\lambda = 0$) a nontrivial solution. However, it does not bifurcate from the zero solution. [A-K] M. Berger, Ch. VI, Sec. 3, Example 4.

1.513 *Hint*: Using the Green function of the equation $w'' + \lambda w = 0$ with suitable λ convert the problem into an integral equation. Apply the Banach contraction principle in the norm $\|x\| = \max_{a \leq t \leq b} \mid x(t) \mid /w(t)$. This result (the Picard theorem) is best possible: $x'' + Kx = 0$, $x(0) = x(b) = 0$, has the nontrivial solution $x(t) = \sin K^{1/2}t$ when $b = \pi/K^{1/2}$. [BSW] Ch. 3, Sec. 3, Th. 3. 2, compare with Th. 3. 5 (a generalization); see also [KPPZ] Ch. 3, Sec. 15, Ex. 15. 12.

1.514 [KPPZ] Ch. 3, Sec. 15, Ex. 15. 4, 15. 13. Compare with 1.543.

1.515 *Hint*: Assume (without loss of generality) that $A = B = 0$. Consider the operator $G(u)(t) = f(t,x(t),x'(t))$ from $C^1[a,b]$ into $C[a,b]$, the (compact) injection $j : C^2[a,b] \to C^1[a,b]$ and $(d^2/dt^2)^{-1} : C[a,b] \to C_0^2[a,b]$. A fixed point (which exists in virtue of the Schauder theorem) of $G \circ j \circ (d^2/dt^2)^{-1}$ provides a solution. For instance: [D-G] Ch. II, Sec. 5. 6; [BSW]; [V-K] Ch. 2, Sec. 3. 1; [H] Ch. XII, Sec. 4, Th. 4. 2; [Sa].

1.516 *Hint*: Define on the ball $B(r,0)$, $r = ma^2/8$, in the space $C^1(0,a)$, the operator T by $T(h) = x$, where $x(t)$ is the unique solution of $x'' = f(t,h,h')$ satisfying $x(0) = x(a) = 0$. Use the Tikhonov theorem. [H] Ch. XII, Sec. 4, Th. 4. 2.

1.517 *Answer*: For instance, the equation $x'' = 12t(x'/4)^{1/3}((\,|\,x\,|\,+\,|\,x'\,|\,)/(1 - t^4 + 4\,|\,t\,|^3))^{2/3}$ with homogeneous boundary conditions on $[-1,1]$ is solved by $c(1 - t^4)$ with an arbitrary constant c. Another example: the problem $x'' = x - 2\,|\,\tan t\,|^p\,|\,x\,|^q(\mathrm{sgn}\,x)|\,x'\,|^p$, $x(0) = x(\pi) = 0$, where $p, q > 0$, $p + q = 1$, has solutions of the form $c \sin t$. See also 1.509, 1.567.

Remark: Of course, in any example of this kind it is impossible to have an a priori bound on x and x'; compare with Scorza-Dragoni theorem in 1.515. [V-K] Introduction; Ch. 3, Sec. 3.

1.518 [BSW] Ch. 7, Sec. 4, Example 7. 4.

1.519 *Hint*: The result can be obtained by a direct integration: $t = \lambda^{-1/2} \int_0^x (c - e^s)^{-1/2}\,ds$, $c = \lambda^{-1}x'(0)^2$, and the condition $1 = 2\lambda^{-1/2}\int_1^c u^{-1}(c - u)^{-1/2}\,du$ should be satisfied.

Answer: The problem has no (smooth) solution for $\lambda > \beta$ (β a certain positive number), exactly one solution for $\lambda = \beta$, and two solutions for $0 < \lambda < \beta$. [Be] Ch. 1, Sec. 1. 2A; [BSW] Ch. 3, Sec. 6, Example 3. 2 and Ch. 7, Sec. 5, Example 7. 4.

1.520 *Hint*: Consider the integral form of this problem and use the Leray-Schauder theorem. Compare with 2.136. A. Krzywicki, T. Nadzieja, *Zastosowania Matematyki*, 21, to appear.

1.521 *Hint*: Put $v = x^{1/2}u$ and observe that v satisfies the equation $v'' - (\lambda^2 + q)v = 0$, where $q = (f(u)/u) - \lambda^2 - (4x^2)^{-1}$. A. Krzywicki, T. Nadzieja, *Math. Methods in the Appl. Sci.* 11, 1989, 403–408.

1.523 *Hint*: The cases $C \neq 0$ and $C = 0$ are different. [Ju] 15.

1.524 [KNK] Ch. 6, Ex. 66.

1.525 *Hint*: Let $\Phi(t)$ be the fundamental matrix of this equation such that $\Phi(0) = I$. Show that $I - \Phi(t)$ is invertible. Use the formula $y(T) = (I - \Phi(T))^{-1}\Phi(T)\int_0^T \Phi^{-1}(s)b(s)\,ds$. [So] Ch. III, Ex. 5.

1.526 *Hint*: It follows from the preceding exercise that for each continuous T-periodic function $b(t)$ there exists a unique T-periodic solution φ_b of the equation $x' = A(t)x + f(t,b(t))$. Prove that the mapping $b \mapsto \varphi_b$ is a contraction for sufficiently small $\delta > 0$. [So] Ch. III, Ex. 6.

1.527 *Hint*: Let $y(t)$ be the solution of the equation starting at y_0 for $t = 0$. The transformation $y_0 \mapsto y(T)$ is linear. [De] Ch. 3, Sec. 23; J. L. Massera, *Duke Math. J.* 17, 1950, 457–475, Th. 4.

1.528 *Hint*: Consider the bounded sequence $y(0)$, $U(y(0))$, $U^2(y(0))$, . . . , where U is the Poincaré(−Andronov) operator. If $y(0) = U(y(0))$, the theorem is proved; otherwise, show that the sequence in question is monotone and complete the proof. [R-M2] 3. 7. 1; [C-L] Ch. 16, Ex. 4; [P1] Ch. 2, Th. 2. 3; [PSV] Ch. IV, Sec. 1, Theorem; [KNK] Ch. I, Sec. 2, Ex. 4. The original paper is: J. L. Massera, *Duke Math. J.* 17, 1950, 457–475 (the example is on p. 458).
Remark: Observe that $\lim_{t\to\infty}(x(t) - y(t)) = 0$.

1.529 *Hint*: Compare with the Massera theorem in the preceding problem. [PSV] Ch. IV, Sec. 1, Ex. 3.

1.530 [PSV] Ch. IV, Sec. 4. 1, Ex. 4.

1.531 [P1] Ch. II, Sec. 5.

1.532 *Hint*: Consider $x' = y + (x^2 + y^2 - 1) \sin \lambda t$, $y' = -x$.
Remark: Such a phenomenon cannot be produced by any equation of the form $x' = f(x) + g(t)$. [P1] Ch. II, Sec. 1; [R-M2] Ch. 3, Ex. 7. 3.

1.533 [P1] Ch. II, Sec. 1, Th. 1. 8; J. Kurzweil, O. Vejvoda, *Czechoslovak Math. J.* 5, 1955, 362–370.

1.534 *Hint*: Observe that the map $x(0) \mapsto x(1)$ leaves invariant a suitable interval $[-k,k]$, $k > 0$. [Hr] Ch. B-III, Lecture 23.

1.535 *Hint*: Use the implicit function theorem for $\varphi(T,0,x,\lambda) - x = 0$, where $\varphi(t,t_0,x_0,\lambda)$ is the solution of $x' = f(t,x,\lambda)$, $x(t_0) = x_0$. [So] Ch. II, Ex. 11.

1.536 [De] Appendix, Ex. 9.

1.537 *Hint*: Prove that x is even and show that $x(t)$ and $y(t) = x(t + T)$ are solutions which satisfy $x(-T) = y(-T)$, $x'(-T) = y'(-T)$. [KNK] Ch. 6, Ex. 64.

1.538 *Hint*: Let $x(0) < x(T)$; if $x(2T) \le x(T)$ then for $y(t) = x(t) - x(t + T)$ one has $y(0) < 0$ and $y(T) \ge 0$. Hence there exists $t_0 \in [0,T]$ such that $x(t_0) = x(t_0 + T)$. [KNK] Ch. 6, Ex. 65.

1.539 [H] Ch. XII, Sec. 4, Ex. 4. 7 a).

1.540 [PSV] Ch. IV, Sec. 2, Theorem, Ex. 4.

1.541 [PSV] Ch. IV, Sec. 2, Ex. 1.

1.542 [G-H] Ex. 1. 5. 5.

1.543 [BSW] Ch. 6, Sec. 3, Example 6. 2; [Tr2]. Compare with 1.514.

1.544 *Hint*: Use the Poincaré-Bendixson theorem. [Fi] 1055.

1.545 [R-M2] 4. 7. 3.

1.546 *Answer*: There is a bifurcation of periodic orbits from (0,0,0) which are stable, $a > 0$. [M-MC] Ch. 4B, Ex. 4B. 1.

1.547 *Hint*: Under these symmetry assumptions the equation reduces to $dy/dx = -f(x) - g(x)/y$, where $y \neq 0$. [RSC] Th. 3. 2. 1 and 3. 2. 2; [S-C] Ch. VI, Complement 4; Z. Opial, *Ann. Polon. Math.* 5, 1958, 67–75.

1.548 [RSC] Th. 5. 1. 1.

1.549 [RSC] Th. 5. 1. 2.

1.550 *Example*: The Van der Pol equation is of the above form with $f(x) = \varepsilon(x^2 - 1)$, $\varepsilon > 0$. [RSC] Th. 4. 1. 3.

1.551 [Sb] Ch. 5, Sec. 5, Example 13; A. G. P. van der Burgh, Report 88-77, TU Delft; [A-P] Ch. 5, Ex. 5. 15.

1.552 [R-M2] Ch. 4, Ex. 9. 8.

1.553 **Remark:** This result can be generalized to the case of the equation $x'' + \mu h(x') + x = 0$, where $h \in C^1(\mathbb{R})$ is a T-periodic odd function, $h(s) > 0$ for $0 < s < T$ and $|\mu| < 2/\max(|h|, |h'|)$. There is exactly one limit cycle among orbits passing through the points $(\mu h(y), y)$, $y \in (nT, (n + 1)T)$, of the plane. For $\mu > 0$ the limit cycles corresponding to n even are stable and others unstable. For $\mu < 0$ the stability is exchanged. R. M. D'Heedene, *J. Differential Equations* 5, 1969, 564–571; J. Węgrzyn, *Zastosowania Matematyki* 12, 1972, 391–395.

1.554 [R-M1] Ch. IV, Ex. 12. 20.

1.555 *Answer*: $x = c \cos \tau + \mu(\cos 3\tau - \cos \tau)c^3/2 + \cdots$, where $\tau = 2\pi(1 + \mu c^2/4 + \cdots)^{-1}t$. [Du] Ch. III, Sec. 3.

1.556 *Hint*: Integrate over a period the expression $x''y + xy - y^4 - xy'' - xy - x^4 = 0$.
Answer: There is only trivial periodic solution $x = y = 0$ (in contrast to the behavior of the system linearized at the origin). [Be] Ch. 1, Sec. 1. 2B.

1.557 *Hint*: This solution can be obtained as the minimum of the (lower semicontinuous) functional $\int_0^T ((x')^2/2 + U(x,t))\, dt$ over all T-periodic C^1 vector functions $x(t)$. A natural functional setting is a larger (Hilbert) function space $H^1((0,T);\mathbb{R}^n)$. The growth condition on U guarantees the coercivity of the functional to be minimized. [Be] Ch. 6, Sec. 6. 1B.

1.558 [Be] Ch. 4, Sec. 4. 2, Th. 4. 2. 9 and Th. 4. 2. 16.

1.559 [KPPZ] Ch. 3, Sec. 15, Ex. 15. 4.

1.560 *Hint*: Let X be the Banach space of C^1 functions $h(t)$, $0 \leq t \leq a$, with the norm

$$\| h \| = \max\left(\max_{t \in [0,a]} | h(t) |, \frac{a}{4} \max_{t \in [0,a]} | h'(t) | \right)$$

and let $x(t)$ be the unique solution of $x'' = f(t,h(t),h'(t))$ satisfying $x(0) = x(a)$ $= 0$. Define the operator T on X by $T(h(t)) = x(t)$ and use the Banach contraction theorem. [H] Ch. XII, Sec. 4, Th. 4. 1.

1.561 [H] Ch. XII, Sec. 4, Ex. 4. 6.

1.562 [KPPZ] Ch. 3, Sec. 15, Ex. 15. 5.

1.563 [KPPZ] Ch. 3, Sec. 15, Ex. 15. 9.

1.564 *Hint*: First prove an a priori bound on possible solutions and then on their first derivatives. For instance: [D-G] Ch. II, Sec. 5. 7.

1.565 [V-K] Ex. 6. 19.

1.566 *Hint*: Let $\varphi(t,\gamma)$ be the solution of the problem $\varphi'' = f(t,\varphi,\varphi')$, $\varphi'(0,\gamma) = \gamma$, $\varphi(0,\gamma) = a$. Show that the function $u(t) = \partial\varphi/\partial\gamma$ exists on $[0,1]$ and satisfies the equation $u'' - (\partial f(t,x(t),x'(t))/\partial x')u' - (\partial f(t,x(t),x'(t))/\partial x)u = 0$, $u(0) = 0$, $u'(0) = 1$. Next, using $\partial f/\partial x > 0$, show that the equation $\varphi(1,\alpha)$ $-\beta = 0$ can be solved for α as a function of β for (α,β) in a neighborhood of $(x'(0),0)$. [C-L] Ch. 1, Ex. 5.

1.567 *Hint*: These solutions must satisfy $t = \int_0^x (c^4 - y^4)^{-1/2}\, dy$. $x(t)$ is of period $T = 2c^{-1}\int_{-1}^1 (1 - t^4)^{-1/2}\, dt$ and it suffices to have $T = A/n$ with an integer n. [Be], Ch. 1, Sec. 1. 2A.

1.568 *Hint*: It suffices to prove that $x_\lambda(t) < x_\mu(t)$ for $t \in (0,\pi/n)$. Denote by $T_+(A,\lambda)$ and $T_-(A,\lambda)$ the least positive t's such that $x_\lambda(T_+(A,\lambda)) = \max x_\lambda = A^{1/2}$, $x_\lambda(0) = 0$, $x_\lambda(T_+(A,\lambda) + T_-(A,\lambda)) = 0$. Prove that $T_\pm$ are strictly decreasing functions of λ for fixed A and $(T_+ + T_-)$ is a strictly increasing function for A for λ fixed. P. Biler, *Colloquium Math.* 52, 1987, 305–312; see also the next problem.

1.569 C. Chicone, *J. Differential Equations* 69, 1987, 310–321.

1.570 *Hint*: Rewrite the equation as the system $u' = v$, $v' + f(u) = 0$, and analyze the time T necessary for the solution starting at $(0,p)$ to arrive at $(\alpha(p),0)$. T has exactly one critical point. [S] Ch. 13, Sec. D.

1.571 *Hint*: It suffices to prove the a priori estimates $g(t) \le u(t) \le h(t)$ for any possible solution $u(t)$. [V-K] Ch. 3, Sec. 1. Compare with 1.540.

1.572 [V-K] Ex. 3. 3.

1.573 [V-K] Ex. 3. 5.

1.574 [V-K] Ex. 6. 9.

1.575 [V-K] Ex. 6. 15.

1.576 [Sa] Ch. XII, Sec. 7. For a generalization see the Mambriani theorem: If g is a continuous function for t, $x \ge 0$, $g(t,x) > 0$ for $t > 0$, $x > 0$, $g(t,0) = 0$, g increasing in x, $\liminf_{t\to\infty} g(t,c) > 0$ for $c > 0$, and if f is a positive continuous function, $\int_0^\infty f = \infty$, then there exists a unique solution of the equation $x'' = f(t)g(t,x)$ such that $x(0) = x_0 > 0$, $x(\infty) = 0$.

1.577 *Hint*: Consider a family of problems $t^{1/2}x'' = \lambda \mid x \mid^{1/2}x$ depending on $\lambda \in [0,1]$. Show the a priori estimates: $0 \le x(t) \le 1$, $\mid x'(t) \mid \le 3$, $\mid t^{1/2}x''(t) \mid \le 1$. Transform the dependent variable putting $u = x - (1 - t)$. For instance: [D-G] Ch. II, Ex. 13. 27.

1.578 [V-K] Ch. 7, Sec. 2.

1.579 *Hint*: The original solution used explicit formulas involving elliptic functions. A soft analysis approach uses rudiments of bifurcation theory. [Be] Ch. 4, Sec. 4. 3B.

1.582 *Answer*: If $q \ge 0$ is sufficiently large then the solution is unique. [PSV] Ch. IV, Sec. 2. [V-K] Ex. 8. 15.

1.583 [PSV] Ch. IV, Sec. 2.

1.584 [V-K] Ex. 9. 2.

1.585 [V-K] Ex. 9. 8.

1.586 *Hint*: Show that $\mid u_R' \mid \le \sigma/a$; next prove that the family (u_R) is uniformly bounded and separated from zero. Use the Arzelà-Ascoli theorem.
Remark: This problem arises in the theory of electrolytes. A. Krzywicki, T. Nadzieja, *Math. Methods in the Appl. Sci.* 11, 1989, 403–408; *ibidem* 12, 1990, 405–412.

1.587 *Hint*: $u(t) = F(t) - F(0) \cos(t/\varepsilon) - \int_0^t F'(s) \cos((t - s)/\varepsilon) \, ds$.
Answer: If $F' \in L^2$ and $F(0) = 0$, then the solutions converge in L^2, if $F(0) \ne 0$ then the convergence is only weak. [Ma] I. Ch. 2, Sec. 2, Example.

1.588 *Answer*: $u(x,\varepsilon) = c_1(\varepsilon) \exp(-x/\varepsilon) + c_2(\varepsilon) \exp(-(1 - x)/\varepsilon) + u_0(x,\varepsilon)$, where $\lim_{\varepsilon \to 0} c_1(\varepsilon) = c_1$, $\lim_{\varepsilon \to 0} c_2(\varepsilon) = c_2$, $\lim_{\varepsilon \to 0} u_0(x,\varepsilon) = u_0(x)$, and $-d^2u_0/dx^2 = p(x)$. This may be obtained from an explicit formula for the solution. [Lo] Ch. 1, Sec. 3.

1.589 *Hint*: Change the independent variable $x = \log(2t + 1)^{1/2}$ to get $\varepsilon \, du/dt + u = (2t + 1)^{3/2}$.
Answer: Formally $u(x,\varepsilon) = e^x - \varepsilon e^{-x} - \varepsilon^2(e^{-3x} + 3!!e^{-5x}\varepsilon + \cdots + (2n + 1)!!e^{-(2n+3)x}\varepsilon^n + \cdots)$ but this series diverges for $\varepsilon \ne 0$. [Lo] Ch. 2, Sec. 7. 2, Example 2.

1.590 [K-C] Ch. 3, Sec. 3, Problem 1; [PSV] Ch. II, Sec. 5. 2, Ex. 14.

1.591 *Answer*: $x(t) = 1/t + 3\mu + \mu^2(3/t^2 - 3t) = \mathcal{O}(\mu^2)$.

1.592 *Answer*: (i) $u(x,\varepsilon) = -1 + (\cosh(2\varepsilon)^{-1})^{-1} \cosh(\varepsilon^{-1}(x - 1/2)) = -1 + \exp(-x/\varepsilon) + \exp((x - 1)/\varepsilon) + \mathcal{O}(\varepsilon^N)$ for all N and $\varepsilon \to 0$. (ii) $u(x,\varepsilon) = x^2/2 - \varepsilon x - (1/2 - \varepsilon)(1 - \exp(-x/\varepsilon))(1 - \exp(-1/\varepsilon))^{-1} = x^2/2 - \varepsilon x - 1/2 + \varepsilon + (1/2 - \varepsilon) \exp(-x/\varepsilon) + \mathcal{O}(\varepsilon^N)$ for all N and $\varepsilon \to 0$.

1.593 *Hint*: This derivative satisfies the linear equation $y' = y + t + x^2$ for $\mu = 0$ and $y(0) = 0$; for $\mu = 0$, x solves $x' = x$, $x(0) = 1$.
Answer: $\partial x/\partial\mu|_{\mu=0} = e^{2t} - t - 1$.

1.594 [Kr] Ch. 5, Ex. 67.

1.595 *Hint*: $u_0(x) = \pm1$, $u_1(x) = (2k^2/(1 + k^2))^{1/2}$ sn$(K(k)x,k)$, where $1/\varepsilon = 2(1 + k^2)K(k) > \pi$, $K(k)$ is the quarter-period of the Jacobi elliptic function sn. As $\varepsilon \to 0$, $k \to 1$, $\delta(\varepsilon) = (1 - k^2)^{1/2} \simeq 4 \exp(-1/2^{3/2}\varepsilon)$, so $\tanh(x/\varepsilon(2 - \delta^2)^{1/2}) \le (1 + \delta^2/2(1 - \delta^2))u_1(x) \le 1$ for $0 \le x \le 1/2$. Thus as $\varepsilon \to 0$ and x is bounded away from 0 and 1, the function u_1 differs from 1 by terms exponentially small in ε and such that $u_1 < 1$. Near $x = 0$ and 1 there is a boundary layer of width $\mathbb{O}(\varepsilon)$ in which $u_1 \simeq \tanh(x/2^{1/2}\varepsilon)$. For instance: [Be] Ch. 4, Sec. 4. 4A. For a multidimensional generalization see Sec. 4. 4C.

1.596 *Answer*: For $|A| < 2^{1/2}$, $|B| < 2^{1/2}$, and every $n \ge 2$ there exist four solutions satisfying the limit relation $\lim_{\varepsilon\to0} x(t,\varepsilon) = 0$ for all $t \in [\delta, 1 - \delta]$, $0 < \delta < 1/2$, except for $t_j = j/n$, $j = 1, \ldots, n - 1$, where $\lim_{\varepsilon\to0} x(t,\varepsilon) = \pm2^{1/2}$. [Ch-H] Ch. 8, Sec. 1, Example 8. 3; R. E. O'Malley, *J. Math. Anal. Appl.* 54, 1976, 449–466.

1.597 [B2] Part 1, Ex. 28. 1. See also: [B3] Ex. 4. 27. 1.

1.598 [K-C] Ch. 2, Sec. 5, Problem 3.

1.599 [K-C] Ch. 2, Sec. 7, Problem 1.

1.600 [K-C] Ch. 3, Sec. 1, Problem 2.

1.601 [G-H] Ex. 4. 2. 1.

1.602 [G-H] Ex. 4. 2. 3.

1.603 *Answer*: $Z(t) = (1 - \varepsilon^2)^{-1/2} \exp(-\varepsilon t) \sin((1 - \varepsilon^2)^{1/2}t)$, which is a rather poor result as $x(t) = \exp(-3\varepsilon t/2) \sin t + \mathbb{O}(\varepsilon)$ on the time scale $1/\varepsilon$. [S-V] Sec. 2. 7. 6, Example.

1.604 *Hint*: The averaged equation is $z' = 0$ as $\lim_{T\to\infty} T^{-1}a(x) \int_0^T f(t)dt = 0$. [S-V] Sec. 3. 7. 5, Example.

1.605 [S-V] Sec. 5. 2. 5, 3. 8. 1, Examples; [K–C] Ch. 3, Sec. 3. 2.

1.606 *Answer*: $x(t) \simeq \frac{1}{2}(\alpha + \beta) \exp(-\varepsilon t/2)(\cos t + \sin t) + \frac{1}{2}(\alpha - \beta) \exp(-\varepsilon t/2)(\cos t - \sin t)$. [S-V] Sec. 2. 7. 2, Example. For more information see: [K-C] Ch. 3, Sec. 3. 1.

1.607 *Hint*: Observe that $u_\lambda(t,x) = u_1(t/\lambda,x)$ and $u(t) = u_1(t,x)$ satisfies the equation $u(t) = e^{-t}x + \int_0^t e^{-(t-s)}Cu(s) ds$. Then show $\Phi_n(t) \le e^{-t}n + \int_0^t e^{-(t-s)}\Phi_{n-1}(s) ds$, $\Phi_0(t) \le t$, where $\Phi_n(t) = \|u(t) - C^n x\|\, \|x - Cx\|^{-1}$. This can be done considering an integral inequality satisfied by $\Phi_n^2(t)$. For instance: [CHADP] Proposition 2. 12.

1.608 *Hint*: Consider the equation $\psi' = A\psi + f$ with $f = B\psi$. [D-K] Ch. II, Ex. 1.

1.609 *Hint*: Consider a selfadjoint nonpositive operator in an infinite-dimensional Hilbert space such that 0 is in its continuous spectrum. This phenomenon is not possible in finite-dimensional spaces. [D-K] Ch. II, Ex. 2, 3. Compare with 2.307.

1.610 *Answer*: Consider $X = \ell^2$ and the system given by $x_n' + \tanh(t - n)x_n = 0$ with a diagonal matrix $A, \|A\| = 1$. The general solution is $x_n(t) = (\cosh n/\cosh(t - n))x_n(0)$. Obviously, all $x(0)$ with a finite number of nonzero coordinates belong to the considered subspace but $x(0) = (1/n)$ fails to be there: $|x(t)| \geq |x_{[t]}(t)| \geq \cosh(t - 1)/t \cosh 1 \to \infty$ as $t \to \infty$.
Remark: There exist examples with a constant operator A (but not Hermitian). [M-S] Ex. 33. G.

1.611 *Answer*: (i) Let

$$A(t) = \frac{1}{2}\pi \begin{bmatrix} \sin 2\pi t & 1 - \cos t\, 2\pi t \\ -1 - \cos 2\pi t & - \sin 2\pi t \end{bmatrix}$$

hence

$$U(t) = \begin{bmatrix} \cos \pi t & \pi t \cos \pi t - \sin \pi t \\ \sin \pi t & \pi t \sin \pi t + \cos \pi t \end{bmatrix}$$

Observe that $U(1)$ does not have any (real) square root and, therefore, any logarithm.
Remark: $\int_0^1 |A(t)|\, dt = \pi$, so the result on the existence of the Floquet representation of order 1 when $\int_0^1 |A(t)|\, dt < \pi$ (even in infinite-dimensional Hilbert spaces) is optimal. [M-S] 111, Example F. (ii) Let $X = \ell^2$ ($\mathbb{Z}$) (real or complex space) with the standard basis (e_n) and the shift operator $Se_n = e_{n+1}$. Verify that $S = \exp W$, where the operator W is defined by $(We_m, e_n) = (-1)^{m-n}(m - n)^{-1}$ for $m \neq n$. The spectrum of W is the interval $[-\pi i, \pi i]$. Given $\varepsilon > 0$ consider the (symmetric or Hermitian) operator R defined by $Re_n = 0$ for $n < 0$, $Re_n = \varepsilon e_n$ for $n \geq 0$ and $A(t) = -W - 6t(1 - t) \exp(tW) R \exp(-tW)$ for $0 \leq t \leq 1$ (with 1-periodic extension). A direct computation shows that $U(t) = \exp(tW) \exp(t^2(3 - 2t)R)$ for $0 \leq t \leq 1$, so in particular $U(1) = S \exp R$. For each m, $U(1)^m$ does not have any logarithm, hence there is no Floquet representation. It may be computed that $\int_0^1 |A(t)|\, dt \leq \pi + \varepsilon$ (compare with the preceding remark). [M-S] 111, Ex. G.

1.612 *Hint*: Suppose that $x_m(0) \neq 0$. Then $x(n) = S^n x(0)$ and $x_m(n) = x_{m-n}(0) \to 0$ as $n \to \infty$. Show that even $x_m = (x, e_m)$ is not almost periodic.

Remark: This example shows that the hypothesis that each bounded solution of $x' = -Sx$ (in this situation!) is bounded away from the origin is essential in the (Amerio modification for uniformly convex Banach spaces) Favard theorem on almost periodic solutions. [M-S] 104, Th. A; 113, Ex. G.

1.613 *Hint:* Taking $u(t) = 1_{[t_0, t_0 + \tau]} y(t) / |y(t)|$ one proves $|x(t)| \le K\tau^{1/p}$ and $|y(t_0 + \tau)| \int_{t_0}^t |y(s)|^{-1} ds \le K\tau^{1/p}$. Then show $|y(t_0 + \tau)| \le \exp(\int_{t_0}^{t_0 + \tau} |A(s)| ds) |y(t_0)|$, which leads to $\int_{t_0}^{t_0 + \tau} |y(s)| ds \ge \tau(|y(t_0)| \exp(\int_{t_0}^{t_0 + \tau} |A(s)| ds))^{-1}$.

Remark: If for every $u \in C_0((0,\infty); E)$ the solution x of the first problem is uniformly bounded, then $|y(t)| \le N(t_0) \exp(-\alpha(t - t_0)) |y(t_0)|$ for some positive α and $N(t)$. In fact, $x \in C_0((0,\infty); E)$. [Bn1] Ch. III, Sec. 5, Th. 5. 4, Th. 5. 5; J. L. Massera, J. J. Schäffer, *Ann. Math.* 57/3, 1958.

1.614 *Answer:* $p_0(t) = \int_0^t q_1(y) \exp(-(\lambda + \mu)y) dy + \delta_{0a}$, $p_n(t) = \mu^{-1} \exp(-(\lambda + \mu)t) \sum_{k=1}^n q_k(t)(\lambda/\mu)^{n-k} + (\lambda/\mu)^n p_0(t)$, where $q_n(t) = \mu\beta^{n-a}(1 - \delta_{0a})(I_{n-a}(\alpha t) - I_{n+a}(\alpha t)) + \lambda\beta^{n-a-1}(I_{n+a+1}(\alpha t) - I_{n-a-1}(\alpha t))$, I_n is the modified Bessel function: $\exp((\lambda s + \mu/s)t) = \sum_{n=-\infty}^{\infty} (\beta s)^n I_n(\alpha t)$, $\alpha = 2(\lambda\mu)^{1/2}$, $\beta = (\lambda/\mu)^{1/2}$. P. R. Parthasarathy, *Adv. Appl. Prob.* 19, 1987, 997–998; R. Syski, *ibidem* 20, 1988, 693; W. A. Massey, *SIAM Review*, to appear.

1.615 *Answer:* The classical example due to Dieudonné is $x' = f(x)$ where $f: c_0 \to c_0$ defined by $f((x_n)) = (|x_n|^{1/2} + 1/(n + 1))$ is a continuous map. There is no solution of this differential equation with the initial condition $x(0) = 0$. [Di] Ch. 10, Sec. 5, Ex. 5. For more information on the (non)existence of solutions of ordinary differential equations with continuous right-hand side in Banach spaces see A. N. Godunov, *Mat. Zametki* 15, 1974, 467–477; *Vestnik MGU* 29, 1974, 31–39; *Funkc. Anal. Pril.* 9/1, 1974, 59–60.

1.616 J. J. Schäffer, *J. Differential Equations* 56, 1985, 426–428; B. M. Garay, J. J. Schäffer, *ibidem* 64, 1986, 48–50.

1.617 [K] Ch. V, Sec. 3, Ex. 5.

1.618 [Fa] Ch. X, Ex. 2; [B-W] 5. 48, 7. 88.
Hint: For the question recall the spectral synthesis theorem of L. Schwartz.

1.619 [B-V] 91.

1.620 For instance, [HKW] Ch. 4, Sec. 3; [M-MC] Ch. 4C, Example 4C, (1); [Ha1] Sec. 11. 4; [Ha2] Example 4. 1. 3; [I-J] Ch. VIII, Sec. 4, Example VIII. 2.

1.621 *Answer:* (i) $\exp(z_n t)$ where $z = z_n$ satisfies the quasicharacteristic equation $z + ae^{-z} = 0$, that is, $z_n = -\log |2\pi n/a| \pm (\frac{1}{2} \operatorname{sgn} a + 2n)\pi i + \mathbb{O}(\log n/n)$; (ii) $z_n = \log(-a) \pm 2n\pi i + \mathbb{O}(1/n)$ if $a < 0$ and $z_n = \log a \pm (2n + 1)\pi i + \mathbb{O}(1/n)$ if $a > 0$. [E] Ch. II, Sec. 3, Examples 1, 2.

1.622 *Answer*: 0 is an asymptotically stable solution of this equation.
Hint: The trivial solution of the linear approximation $x'(t) + 3x(t) + 2x(t - \tau) = 0$ is stable for all $\tau > 0$. [E] Ch. III. Sec. 7, Example 1.

1.623 *Answer*: If $f(t) = a_0/2 + \sum_{n=1}^{\infty}(\alpha_n \cos nt + \beta_n \sin nt)$ and the quasicharacteristic equation $z + a + b \exp(-\tau z) = 0$ does not have roots of the form mi, m an integer, then

$$x(t) = \frac{\alpha_0}{2(a + b)} +$$

$$\sum_{n=1}^{\infty} \frac{(a_n(a + b \cos n\tau) - b_n(n - b \sin n\tau)) \cos nt}{+ (b_n(a + b \cos n\tau) - a_n(n - b \sin n\tau) \sin nt}{(a + b \cos n\tau)^2 + (n - b \sin n\tau)^2}$$

is the unique periodic solution. In the resonant case $\alpha_m = \beta_m$ should vanish and the solution is as above except for the mth coefficients which are arbitrary. [E] Ch. IV, Sec. 3, Example 1.

1.624 *Answer*: The example given by F. Oliva is: $x'(t) = -2t \exp(1 - 2t) x(t - 1)$ has the solution $x(t) = \exp(-t^2)$ on $[-1,\infty)$. [Ha1] Property 3. 3. 1.

1.625 *Answer*: For example, $x'(t) = f(\mid x_t \mid)$, where $x_t(s) = x(t + s)$, $-1 \le s \le 0, f(s) = 0$ for $0 \le s \le 1, f(s) = -3(s^{1/3} - 1)^2$ for $s > 1$, has two solutions coinciding for $s \ge 0$: $x(t) \equiv 0$ and $x(t) = -t^3$ for $t < 0$ and $x(t) = 0$ for $t \ge 0$. [Ha1] Property 3. 4. 2.

1.626 *Answer*: Consider the Popov system $x'(t) = 2y(t)$, $y'(t) = -z(t) + x(t - 1)$, $z'(t) = 2y(t - 1)$. For $t \ge 1$ the vector $(x,y,z)(t)$ is orthogonal to $(1,-2,-1)$. [Ha1] Property 3. 5. 2.

1.627 *Hint*: There exists a nontrivial solution which tends to zero faster than any exponential function e^{-ct} as $t \to \infty$. [Ha1] Property 3. 3. 2, Theorem 8. 1. 2.

1.628 *Hint*: The characteristic equation is $\lambda(1 - e^{-\lambda}) + c = 0$, hence Re $\lambda \to 0$ as $\mid \lambda \mid \to \infty$. [Ha2] Ch. 1.

1.629 J. Roe, *Math. Proc. Cambridge Phil. Soc.* 87, 1980, 69–73. For a generalization see: R. Howard, *Proc. AMS* 105, 1989, 658–663.
Remark: In the multidimensional case the "proper" generalization is: If $\Delta f_k = f_{k+1}$ and $\mid f_k \mid_\infty$ are uniformly bounded, then $\Delta f_0 + f_0 = 0$. See R. Strichartz, *Characterization of eigenfunctions of the Laplacian by boundedness condition*, preprint Cornell University, 1989.

1.630 *Answer*: For example, $f(x) = \sum_{n=-\infty}^{\infty}(-1)^n c^{-n(n+1)/2} \exp(-c^n x)$ satisfies these conditions.

Remark: It is interesting to compare with the case $c < 1$. T. Kato, J. B. McLeod, *Bull. AMS* 77, 1971, 891–937; Ll. G. Chambers, *Quart. Applied Math.* 32, 1975, 445–456; *ibidem* 45, 1987, 695–696; L. Fox, D. F. Mayers, J. R. Ockendon, A. B. Tayler, *J. Inst. Math. Appl.* 8, 1971, 271–307; J. D. Love; P. O. Frederickson, *SIAM Review* 22, 1980, 503–504, Problem 79–20; [B-W] 5. 46.

1.631 [L] Ch. II, Ex. 13.

1.632 *Hint*: If u is twice differentiable, then this equation is equivalent to $u'' + (2\lambda - 1)u = 0$. [Gr] Ch. XXXI, Ex. 5; [C-H] III. 10. 2; [KKM1] Ch. II, Sec. 23.

1.633 [V12] 5. 55.

1.634 [V12] 5. 53.

1.635 [Ho] 2. 1, 2. 2.

1.636 [Ho] 2. 7.

1.637 [Ho] 7. 2.

1.638 *Hint*: One has $\int_0^\infty e^{-as} \sin(xs)\, ds = x/(a^2 + x^2)$, so $\int_0^\infty s/(a^2 + s^2) \sin(xs)\, ds = \pi e^{-ax}/2$ (integrate the function $z/(a^2 + z^2)e^{izx}$ along a closed curve). [Gr] Ch. XXXI, Ex. 4; see also: [C-H] III. 10. 2; [Y].

1.639 *Answer*: $\int_a^b K(x,y)v(y)\, dy = C_1\, u(x)$, $\int_a^b K(x,y)u(x)\, dx = C_2\, v(y)$.
Hint: Replacing u by $u + hU$, v by $v + kV$ with arbitrary U, V, and observing that the considered integral has a minimum for $h = k = 0$, one obtains $\int_a^b U(x) \int_a^b (K(x,y)v(y) - u(x)v^2(y))\, dy\, dx = 0$ and a similar identity with V. [Gr] Ch. XXXII, Ex. 5.

1.640 *Answer*: The Urysohn example is $K(x,t) = t \exp(x^{-2}-1)$ for $0 \le t \le x \exp(1 - x^{-2})$, $= x$ for $x \exp(1 - x^{-2}) \le t \le x$, $= 0$ for $t > x$, hence $u(x) = C/x$ with an arbitrary constant C. An example with a kernel analytic off the origin is $K(x,t) = 2\pi^{-1}xt^2/(x^6 + t^2)$. The inhomogeneous equation $v(x) = \int_0^x K(x,t)v(t)\, dt - 2\pi^{-1}x^{-2} \arctan x^2$ has an integrable solution and $v(x) + x^{-2}$ is a nonintegrable one. The equation $u(x) = \int_0^x t^{x-1}u(t)\, dt$, $0 \le x$, $t \le 1$, has a unique continuous solution $u(x) \equiv 0$. The function $u(x) = Cx^{x-1}$ is a discontinuous solution of the above equation. See 1.635. [KKM1] Ch. 1, Sec. 3.

1.641 *Hint*: The iterated kernel is $K^{(m)}(s,t) = \pi^{m-1} \sum_{n=1}^\infty \sin ns \sin(n + m)t/(n^2(n + 1)^2 \cdots (n + m - 1)^2)$ and therefore the Neumann series converges for all values of λ. [C-H] III. 10. 7.

1.642 [V11] Sec. 2. 4; Ex. 17. 5, Ex. 29. 7.

1.643 [V12] 5. 54.

1.644 *Answer*: (i) If $h < 0$, then $u(t) \equiv 0$ (under the solvability condition $f = 0$); if $0 < h < 2$, $u(t) = if/(t - ih)$; and if $h > 2$, $u(t) = -if/(t + i(h - 2)) +$

$c((t + i(h - 2))^{-1} + (t - ih)^{-1})$. (ii) If $h < 0$, then $v(t) = f/(1 - h)$; if $0 < h < 2$, $v(t) = u(t) + c(t - i)/(t - ih)$; and if $h > 2$, $v(t) = u(t) + c_1(t - i)/(t - ih)$. [Gv] Ch. III, Ex. 8.

1.645 [KKM1] Ch. II, Sec. 15.

1.646 *Hint*: To prove the global result use the Bielecki renorming trick: modify the usual sup norm in $C[0,T]$ putting $\|u\| = \max_{0 \leq t \leq T} e^{-Lt} |u(t)|$. In this norm, a suitable operator in $C[0,T]$, whose fixed point is the solution of the original problem, is a contraction—even if L is large. [D-G] Ch. I, Sec. 2. 1, Th. (2.1).

1.647 [M1] Ch. II.

1.648 [M1] Ch. I, Problem 15.

1.649 [M1] Ch. I, Th. 10. 1.

1.650 [M1] Ch. I, (10.2).

1.651 [M1] Ch. I, (10.9).

1.652 [M1] Ch. I.

1.653 [M1] Ch. II, Problem 14.

1.654 P. J. Bushell, *Math. Proc. Cambridge Phil. Soc.* 79, 1976, 329–335.

1.655 W. R. Schneider *Z. Angew. Math. Phys.* 33, 1982, 140–142.

1.656 P. J. Bushell, W. Okrasiński, *Math. Proc. Cambridge Phil. Soc.* 106, 1989, 547–552.

1.657 *Answer*: This initial value problem is equivalent to the following integral equation: $n!u(x) = \int_0^x (x - s)^n d(g \circ u)(s)$, $0 < x < \varepsilon$. Using the Jensen inequality for a nontrivial solution u of this equation, one obtains $u'(x)/g(u(x)) \leq (n!)^{-1/n}(u(x)/g(u(x)))^{(n-1)/n}$, from which the necessary condition of the theorem follows. The sufficient condition: a nontrivial solution u of the equation can be constructed as the limit of the following sequence: $v_0(x) = 2F(cx)$, $v_{k+1}(x) = \int_0^x (x - s) d(g \circ v_k)(s)$, $k = 0,1,2, \ldots , 0 < x < \varepsilon$, where F is the inverse function to $\int_0^s (s/g(s))^{1/n} s^{-1} ds$, and $c > 0$ is a sufficiently small number. G. Gripenberg, *Math. Scand.* 48, 1981, 59–67; W. Okrasiński, *Math. Methods in the Appl. Sci.* 13, 1990, 273–279; W. Mydlarczyk, *Math. Scand.*, 68, 1991, 83–88.

1.658 [Ap] Ch. X, Ex. 11.

1.659 [Ap] Ch. X, Sec. 219.

1.660 [Ga] Ch. 2, Sec. 2. 7, Ex. 8.

1.661 For instance: [Ga] Ch. 2, Sec. 2. 7, Ex. 6. Compare a generalization in 1.365.

1.662 [Ga] Ch. 2, Sec. 2. 7, Ex. 7.

1.663 [Ga] Ch. 2, Sec. 2. 7, Ex. 10.

1.664 [Ga] Ch. 2, Sec. 2. 7, Ex. 11.

1.665 *Answer*: The cycloid $x = r(\varphi - \sin \varphi)$, $y = r(1 - \cos \varphi)$, is the only solution of this problem.
Remark: The problem has an obvious application to the isochronic clock problem.

1.666 [Ga] Ch. 2, Sec. 2. 10, Observation 2, Proposition 15.

1.667 [Ap] Ch. X, Ex. 5.

1.668 [Ga] Ch. 2, Sec. 2. 10, Corollary 16.

1.669 [A-M] 3. 7. E; [Gd] Ch. 3, Sec. 3.

1.670 [Ap] Ch. XI, Ex. 2.

1.671 [Ap] Ch. XI, Ex. 10.

1.672 [PTKY] 8. 2.

1.673 *Answer*: $F(r) = -\alpha/r^2 + \beta/r^3$, where $p = \omega^2 k_0^2/(m\alpha)$, k_0 is the initial value of the (kinetic) momentum, $\omega = (1 + \beta m/k_0^2)^{1/2}$. [PTKY] 8. 19, 8. 20.

1.674 *Hint*: $U = T - \text{const}$, where T is the kinetic energy calculated using the relation $x' \, \partial\omega/\partial x + y' \, \partial\omega/\partial y = 0$. In the Bertrand problem (see the next problem) this is obtained from the area integral $x'y - xy' = \text{const}$. [G1] Ch. 1, Sec. 1. 3.

1.675 *Hint*: Use the result of the preceding problem. Another method: Differentiate the equation of conical sections $(x^2 + y^2) = ex + p$, where e is the eccentricity and p is the focal parameter. Supposing that $x'' = X(x,y)$, $y'' = Y(x,y)$, one gets $(xy' - x'y)^2/r^2 = y(y'X - x'Y)/x'$ and $Y_0/X_0 = y_0/x_0$. The area integral $x'y - xy' = \text{const}$ leads to $X = -c^2 p^{-1} r^{-3} x$, $Y = -c^2 p^{-1} r^{-3} y$. [G1] Ch. 1, Sec. 1. 2; [Ap] Ch. XI, Sec. 233.

1.676 [A3] Ch. 2, Sec. 8; [Ap] Ch. XI, Sec. 233.

1.677 [A] Ch. XI, Sec. 225.

1.678 [Ap] Ch. XI, Ex. 3.

1.679 [Ap] Ch. XI, Sec. 233.

1.680 For instance: [Li] Ch. III, Sec. 3. 7.

1.681 [Po] Ch. 1, Sec. 5, Ex. 5. 5.

1.682 *Hint*: It suffices to prove that the equation below for the function $u(t) = x(t) - bt - a$, $u(t) = \int_t^\infty \int_s^\infty f(u(\tau) + b\tau + a, \tau) \, d\tau \, ds$ has a unique solution defined for t in some neighborhood of infinity. The autonomous case $f = f(x)$ is only slightly simpler. M. A. Astaburuaga, C. Fernández, V. H. Cortés, *J. Math. Anal. Appl.* 134, 1988, 471–481. Compare R. G. Newton, *Scattering Theory of Waves and Particles*, Springer, New York, 1982.

1.683 [Po] Ch. 2, Sec. 1, Ex. 1. 3.

1.684 *Hint:* $I'' = 2T + \sum (x_k, \partial U/\partial x_k)$ so $-U = \sum (x_k, \partial U/\partial x_k)$, $U = \sum Gm_i m_j / |x_i - x_j|$. [Po] Ch. 2, Sec. 2, Ex. 2. 1.

1.685 *Hint:* $I'' \to \infty$, first exclude total collapse in $t \to \infty$. Then from the Sundman inequality $c^2 := |\sum m_k x_k \times v_k|^2 \leq 4I(I'' - E)$ infer that $c^2/4 \leq (EI + \text{const})/\log I^{-1}$, hence $c = 0$. [Po] Ch. 2, Sec. 4.

1.686 *Hint:* From $I'' = T + E$ one gets $t^{-1} I' = t^{-1} \int_0^t T + E + \text{const} \cdot t^{-1}$, so another equivalent condition in the theorem is $\lim_{t \to \infty} t^{-1} I'(t) = 0$. For the implication $\lim_{t \to \infty} t^{-2} I'(t) = 0 \Rightarrow \lim_{t \to \infty} t^{-1} I'(t) = 0$ observe that $I'' \geq E = \text{const}$. [Po] Ch. 2, Sec. 5; H. Pollard, *Bull. AMS* 70, 1964, 703–705.

1.687 *Hint:* $I'' = T + E$, where T is the kinetic energy and E is the total energy. [Po] Ch. 2, Sec. 2, Ex. 2. 2.

1.688 G. F. Hilmi, *Doklady AN SSSR* 79, 1951, 419–422.

1.689 *Hint:* There exists $R > 0$ such that if $|Q| \geq R$, then $r^2(t) \geq |Q| + r'(0)t + 2ht^2$ where $r(t) = (x^2(t) + y^2(t) + z^2(t))^{1/2}$. If $|Q| < R$, use the Maupertuis-Lagrange variational principle. V. A. Antonov, E. I. Timoshkova, K. V. Kholshevnikov, *Introduction to Theory of Newtonian Potential*, Nauka, Moskva, 1988, Ch. 1, Sec. 1. 4, Th. 1. 4. 2.

1.690 *Answer:* In general, no. [PTKY] 22. 23, 22. 24.

1.691 [Pa] Ex. 2. 3. 2.

1.692 *Hint:* Consider the Hamiltonians $H = pF$ and $H = p_0 F^{-1}$. *Answer:* In (x,p) variables for $H = pF$ the canonical equations are $x' = [x,H] = F$, $p' = [p,H] = -p \nabla F$. In (t,p_0) variables for $H = p_0 F^{-1}$ the equations take the form $dt/dx = [t,H] = 1/F$, $dp_0/dt = [p_0,H] = p_0(\partial F/\partial t)F^{-2}$. [Pa] Ex. 3. 3. 1.

1.693 *Answer:* If $c_n = (k_n)^{1/2} a_n$ then the system takes the form $a'_n = [a_n, H]$ for $H = k(\overline{a_1} \overline{a_2} a_3 \exp(-i\gamma(t)) + a_1 a_2 \overline{a_3} \exp(i\gamma(t))$, $k = (k_1 k_2 k_3)^{1/2}$. The action-angle variables are constructed as follows. Let $b_n = a_n \exp(i\gamma) = (I_n)^{1/2} \exp(-i\varphi_n)$. Then apply the canonical transformation with the generating function $\varphi_1 I'_1 + \varphi_2 I'_2 + (\varphi_3 - \varphi_1 - \varphi_2) I'_3$, that is, $I_1 = I'_1 - I'_2$, $I_2 = I'_2 - I'_3$, $I_3 = I'_3$, $\varphi_1 = \varphi'_1$, $\varphi_2 = \varphi'_2$, $\varphi_3 = \varphi'_1 + \varphi'_2 + \varphi'_3$. The new Hamiltonian is $H = -(d\gamma/dt)(I'_1 + I'_2 + I'_3) + 2k((I'_1 - I'_3)(I'_2 - I'_3))^{1/2} \cos \varphi'_3$. The first integrals are $I'_1 I'_2$, and (if $\gamma = qt$) $-q(I'_1 + I'_2 - I'_3) + 2k((I'_1 - I'_3)(I'_2 - I'_3))^{1/2}(I'_3)^{1/2} \cos \varphi'_3$. [Pa] Ex. 1. 5. 5.

1.694 *Hint:* Consider the canonical transformation $x_1 = (q_1 + q_2 + q_3)/3$, $x_2 = q_2 - q_1$, $x_3 = q_3 - q_1$, $p_1 = y_1/3 - y_2 - y_3$, $p_2 = y_1/3 + y_2$, $p_3 = y_1/3 + y_3$, generated by the function $(p_1 + p_2 + p_3)y_1/3 + (q_2 - q_1)y_2 + (q_3 - q_1)y_1$. In these variables the Hamiltonian takes the form $h = y_1^2/6 + y_2^2 + y_3^2 + \exp(x_3) + \exp(-x_2) + \exp(x_2 - x_3)$.

Answer: In the variables introduced in the above hint y_1 and h are obvious first integrals, $I = y_2 y_3 (y_2 + y_3) + y_3 \exp(-x_2) + y_2 \exp(x_3) - (y_2 + y_3) \exp(x_2 - x_3)$ is a specific integral with $[I,h] = 0$. [Pa] Ex. 1. 5. 3.

1.695 *Answer*: $H = -(p, A^{-1} \partial f / \partial t) + \Psi(q_i, t)$, where $A = (\partial f_i / \partial q_s)$ and Ψ is an arbitrary function of q and t. [PTKY] 20. 32, 20. 33.

1.696 [PTKY] 20. 34.

1.697 [A-M] 3. 4. E.

1.698 *Answer*: $\sum_{i=1}^n q_i'(aq_i + \sum_{j=1}^n b_{ij} q_j + c_i) - (\frac{1}{2} \sum_{i=1}^n q_i'^2 + V(q_i)) \times (2at + d)$ for some constants a, c_i, d and a skew symmetric matrix (b_{ij}). [PTKY] 20. 40, 20. 39.

1.699 *Answer*: $W(x_1, x_2, \alpha, E) = \frac{1}{2}(x_1(2a - \omega_1^2 x_1^2)^{1/2} + 2(\alpha/\omega_1) \arcsin (\cos \omega_1 x_1/(2\alpha)^{1/2}) + x_2(2(E - \alpha) - \omega_2^2 x_2^2)^{1/2} + 2(E - \alpha)/\omega_2 \arcsin(\omega_2 x_2/(2(E - \alpha))^{1/2})$. [B-V] 146.

1.700 *Hint*: If $\partial S / \partial t + H(q, \partial S / \partial q, t) = 0$, then $S(q,t) = S_0(q)$,

$$S(q,t) = S_0(q_0(q,t)) + \int_0^t \left(p \frac{\partial H(q,p,\tau)}{\partial q} - H(q,p,\tau) \right) \Big|_{\substack{q = \varphi(q_0(q,t),\tau) \\ p = \psi(q_0(q,t),\tau)}} d\tau$$

where $q = \varphi(q_0,t)$, $p = \psi(q_0,t)$. Solve the system of canonical equations $q' = \partial H / \partial p$, $p' = -\partial H / \partial q$ and let $\partial \varphi / \partial q \neq 0$.
Answer: $S = c(q - 1/2ct)$, $S = \frac{1}{2} q^2 / (t - 1)$, no solutions, $S = (-q_0 \arctan q_0 + \frac{1}{2} \log(1 + q_0^2) + \frac{1}{2} t \arctan^2 q_0)$, where $q = q_0 + t \arctan q_0$, $0 \leq t < 1$. [B-V] 93.

1.701 *Hint*: In general situation, f is a first integral and the Poisson bracket of f and the Hamiltonian vanishes. Test the functions $\varphi = q^2, p^2, qp$. [PTKY] 20. 56, 20. 53, 20. 54, 20. 55.

1.702 For instance: [B-P] Ch. III, Example 3; [A3] Ch. 5, Sec. 25, D; [Gb] Ch. II, Sec. 6.

1.703 [BNF] Ch. 2, Sec. 2, Example 1.

1.704 *Answer*:

$$\alpha = C_1 \sin\left(\left(\frac{(2-\sqrt{2})g}{\ell}\right)^{1/2} t + \alpha_1\right)$$
$$- C_2 \sin\left(\left(\frac{(2 + \sqrt{2})g}{\ell}\right)^{1/2} t + \alpha_2\right) + 2\left(\frac{a}{\ell}\right) \sin\left(\left(\frac{g}{\ell}\right)^{1/2} t\right)$$
$$\beta = C_1\sqrt{2} \sin\left(\left(\frac{(2 - \sqrt{2})g}{\ell}\right)^{1/2} t + \alpha_1\right)$$

$$+ C_2\sqrt{2}\,\sin\!\left(\!\left(\frac{(2+\sqrt{2})g}{\ell}\right)^{1/2}\! t + \alpha_2\right) + \left(\frac{a}{\ell}\right)\sin\!\left(\!\left(\frac{g}{\ell}\right)^{1/2}\! t\right)$$

where α and β are the angles of rotation of the pendulums. [PTKY] 18. 36.

1.705 *Hint*: The angle of rotation of jth pendulum is $\varphi_j \approx \sin\varphi_j = (x_j - x_{j+1})/\ell$. Determine the amplitudes in the solution $x_j = u_j\sin(\omega t + \alpha)$ using the recurrent formula for the Chebyshev polynomials $(k-1)L_{k-2}(p) + (p - 2k + 1)L_{k-1}(p) + kL_k(p) = 0$, $L_k(p) = (1/k!)e^p d^k(p^k e^{-p})/dp^k$.
Answer: $\varphi_j = \sum_{i=1}^{n} C_i(L_{j-1}(\omega_i^2\ell/g) - L_j(\omega_i^2\ell/g))\sin(\omega_i t + \alpha_i)$. [PTKY] 16. 98.

1.706 *Answer*: $\langle x.y \rangle = \sum_{j=1}^{2} c_j\langle(b^2 - \omega_j^2)^{1/2}\cos(\omega_j t + \varphi_j),\ (a^2 - \omega_j^2)^{1/2}\sin(\omega_j t + \varphi_j)\rangle$, where $a^2 = 2g\alpha - \Omega^2$, $b^2 = 2g\beta - \Omega^2$, $\omega_j^2 = \frac{1}{2}(a^2 + b^2 + 4\Omega^2) \pm \frac{1}{2}((a^2 + b^2 + 4\Omega^2)^2 - 4a^2b^2)^{1/2}$ and c_j, φ_j are arbitrary constants. [PTKY] 16. 54.

1.707 For instance, [Da] Ch. 7, Sec. 5.

1.708 *Answer*: For purely horizontal motion, $r\cos\theta = k = $ const. Letting $u = k\tan\theta$ leads to $u'' + cu'/m + gu/k = 0$, $u(0) = k\tan\theta_0$, $u'(0) = 0$. If $\beta = |\alpha|^{1/2}$, where $\alpha = 1 - 4gm^2/kc^2$ and $\gamma = c/2m$, then $u = (k/\beta)\exp(-\gamma t)(\sinh(\gamma\beta t) + \beta\cosh(\gamma\beta t))\tan\theta_0$ for $\alpha > 0$, $u = (k/\beta)\exp(-\gamma t)(\sin(\gamma\beta t) + \beta\cos(\gamma\beta t))\tan\theta_0$ for $\alpha < 0$, $u = k\exp(-\gamma t)(1 + \gamma t)\tan\theta_0$ for $\alpha = 0$, and $r^2 = k^2 + \omega^2$. M. A. Abdelkader; D. F. Lockhart, *SIAM Review* 26, 1984, 436–437, Problem 83-14.

1.709 *Hint*: The energy is $m\ell^2(\alpha'^2 + \omega^2\alpha^2)$ where $\omega = (g/\ell)^{1/2}$ and the increments of the energy over time intervals of length $\approx 1/\varepsilon$ are not negligible. Suppose that ℓ varies so that $\omega' = \varepsilon\omega h(\omega)$ where h is a sufficiently smooth function. The first-order approximation for the new variables $\varphi = \alpha' + i\omega\alpha$, $\overline{\varphi} = \alpha' - i\omega\alpha'$ is given by the equations $\varphi' = (i\omega + \frac{5}{2}\varepsilon h(\omega))\varphi$, $\overline{\varphi}' = (-i\omega + \frac{5}{2}\varepsilon h(\omega))\overline{\varphi}$. It follows from the above system that for $\varphi\overline{\varphi} = \alpha^2 + \omega^2\alpha^2$, $d(\varphi\overline{\varphi})/d\omega = 5\omega^{-1}\varphi\overline{\varphi}$, hence $E/\omega = $ const (the so-called adiabatic invariant). [B-P] Ch. III, Example 1. Compare with 1.605.

1.710 *Answer*: The equations are $x'' = -x/r^3 + f$, $y'' = -y/r^3$, $z'' = -z/r^3$, where $r^2 = x^2 + y^2 + z^2$. The integrals include the energy $\frac{1}{2}((x')^2 + (y')^2 + (z')^2) - 1/r - fx = h$, the area $zy' - yz'$, and the (generalized) Laplace integral $x'rr' - x/r - \frac{3}{2}fx^2 - \frac{1}{2}fr^2 - 2hx$.
Remark: This system is integrable in quadratures, for instance, in parabolic coordinates. The variety of possible trajectories is extraordinarily rich, having in mind the relatively simple physical model. [B1] Ch. 3.

1.711 [A-M] 3. 7. E.

1.712 *Answer*: $T = 85$ min. [T-P] 28. 2.

1.713 *Answer:* 0.405. [T-P] 28. 39.

1.714 *Hint:* The potential energy is $k(x_2 - x_1 - b)^2 + k(x_3 - x_2 - b)^2$ where $2k$ is the stiffness of the junctions, b is the distance between the atoms, and x_1, x_2, x_3 are the coordinates on the line.
Answer: $\omega_1 = 0$ (which corresponds to motion as a rigid body), $\omega_2 = (2k/m)^{1/2}$, and $\omega_3 = (2km^{-1}(1 + 2m/M))^{1/2}$. [Gd] Ch. 10, Sec. 4.

1.715 *Answer:* $a(1 + 3\sin^2\theta)(\theta')^2 - 6g\cos\theta(1 - \cos\theta) = $ const, where the angle *BAD* equals 2θ and $\cos\alpha > 1/4$. The length is $13a/18$ and the period of small oscillations is $2\pi(13/18g)^{1/2}$. [At] Ch. XVI, Example 1.

1.716 [PTKY] 7. 25.

1.717 [PTKY] 3. 21.

1.718 *Answer:* The trajectory is a cycloid. *Amer. Math. Monthly* 64, 1957, 23, Putnam Competition 1956, I-3; [B-W] 5. 127.

1.719 [B-W] 5. 128.

1.720 [At] Ex. V. 5.

1.721 [At] Ch. V, Example 4.

1.722 [T-P] 36. 17.

1.723 *Answer:* $y = h\cosh((w/H)^{1/2}x)$. Since the tension at each point of the bridge is tangent to the curve, it follows that no mortar will be needed if the bridge has the shape of a catenary. [T-P] 36. 22.

1.724 J. S. Lew, *SIAM Review* 21, 1979, 565–566, Problem 78-17.

1.725 [At] Ex. VI. 8.

1.726 *Hint:* Show that after the particle has fallen a distance x, $kv^2 = g(1 - \exp(-2kx))$. [At] Ex. X. 42.

1.727 [At] Ex. X. 38.

1.728 [At] Ex. VI. 13.

1.730 *Answer:* $x(t) = x_0(t) + kx_1(t) + k^2x_2(t) + \cdots$, where $x_1(t) = A^2(1/2 - (1/3)\cos t - (1/6)\cos 2t)$, $x_2(t) = A^3(-1/3 + (29/144)\cos t + (5/12)t\sin t + (1/9)\cos 2t + (1/48)\cos 3t)$.
Remark: x_2 contains a resonance term $t\sin t$. If one replaces the above expansion with one of the form $x(t) = y_0(\omega t) + y_1(\omega t)k + y_2(\omega t)k^2 + \cdots$, where $\omega = 1 + \omega_1 k + \omega_2 k^2 + \cdots$ is to be a periodic function of $\tau = \omega t$ of period 2π, then the solution will be periodic. Here this Lindstedt procedure gives $y_1(\tau) = A^2(1/2 - (1/3)\cos\tau - (1/6)\cos 2\tau)$, $y_2(\tau) = A^3(-1/3 + (29/144)\cos\tau + (1/9)\cos 2\tau + (1/48)\cos 3\tau)$, ... [Sb] Ch. I, Sec. 2, Ex. 9; Ch. 3, Sec. 2, Ex. 5; Ch. 3, Sec. 3, Example 3.

1.731 For instance, [G-H] Ch. 2, Sec. 4 and Ex. 5. 2. 9; L. D. Pustylnikov, *Uspekhi Mat. Nauk* 23:4, 1968, 251–252; *Trans. Moscow Math. Soc.* 14, 1978,

1–101; G. M. Zaslavskiĭ, *Stochasticity of Dynamical Systems*, Nauka, Moskva, 1984, Ch. 2, Sec. 2; R. Douady, *J. Math. pures appl.* 68, 1989, 297–318.

1.732 *Answer:* For instance, $M_1 = -\lambda Ax_1$, $M_2 = -\lambda Bx_2$, $M_3 = -\lambda Cx_3$ (this solution is not unique). [G1] Ch. 1, Sec. 7.

1.733 *Answer:* $p = \exp(-\lambda^2 t)b \sin((C - A)/(A - \mu^2)\beta \exp(-\mu^2 t) + \alpha)$, $q = \exp(-\lambda^2 t)b \cos((C - A)/(A - \mu^2)\beta \exp(-\mu^2 t) + \alpha)$, $r = \beta \exp(-\mu^2 t)$, where α, β, b are arbitrary constants. [PTKY] 11. 51.

1.734 [R-M2] 2. 7. 10. See 1.418.

1.735 [At] Ex. VII. 25.

1.736 [At] Ex. VII. 21.

1.737 [At] Ex. III. 5.

1.738 *Hint:* The attraction of such a plate at a point on the axis distant r from the center is $2GMa^{-2}(1 - r(r^2 + a^2)^{-1/2})$. [At] Ex. I. 5.

1.739 *Hint:* If x_k denotes the wine concentration in the kth cup, then $x_1' + x_1 = 1$, $x_k' + x_k = x_{k-1}$, $k \geq 2$.
Answer: The final concentration in the kth cup is $1 - e^{-1} \sum_{j=1}^{k-1} 1/j!$. [B-W] 1. 120; C. Gilping, R. E. Gaines, *Amer. Math. Monthly* 28, 1921, 143–145, 2791.

1.740 *Hint:* The force on a particle carrying a charge e moving with velocity v in an electric field E and a magnetic field H is $e(E + c^{-1}v \times H)$. [At] Ex. IV. 13.

1.741 [B-W] 1. 124.

1.742 *Answer:* This problem reduces to the study of extremum for the functional $\int_A^B (u(x')^2 + v(y')^2)^{1/2} dt$ under the constraint $\int_A^B ((x')^2 + (y')^2)^{1/2} dt$ fixed, where u, v are positive constants. [Fa] Ch. XIII, Ex. 5. See also the Vieille problem: [Gr] Ch. XXXIV, Ex. 12.

1.743 [A-M] 2. 6. F.

1.744 [Sg] 3. 10. C3.

1.745 [T-P] Ex. 17. 4, 36. 6.

1.746 *Answer:* Let $x_i(t)$ denote the position of car A_i at time t, $i = 1, 2$. The condition leads to the first-order differential equation $x_1(t) + t_0 x_1'(t) = x_2(t)$. Solving this equation, one obtains $x_1(t) = \exp(-t/t_0)(x_1(0) + t_0^{-1} \int_0^t \exp(s/t_0)x_2(s) ds)$. For example, if $x_2(t) = bt$ for $t \leq 0$, $x_2(t) = bt + ct^2$ for $t \geq 0$, then $x_1(t) = b(t - t_0)$ for $t \leq 0$, $x_1(t) = (b - 2ct_0)(t - t_0) + c(t^2 - 2t_0^2 \exp(-t/t_0))$ for $t \geq 0$. V. Salmon; S. C. Pinault, *SIAM Review* 30, 1988, 507–508, Problem 87-14.

1.747 *Answer:* Let $\varphi(x) = 2 \exp(\int_0^x 2\rho(y) dy)$, $I = \int_0^\infty \varphi(\lambda)(\lambda + 1)^{-2} d\lambda$. (i) Escape will not occur if $\rho(x) \geq (2x + 2)^{-1}$; in that case $I = \infty$. (ii) The finiteness of the integral I is a necessary and sufficient condition for escape to occur. (iii) The minimum initial velocity which a missile must possess in order to escape is $I^{1/2}$. D. J. Newman; J. E. Wilkins, Jr., *SIAM Review* 7, 1965, 423–424, Problem 64-3.

1.748 *Answer*: The problem here is to minimize $\int dt$ subject to the constraints $(x')^2 + (y')^2 + 2gy = at$, $\int x' \, dt = X$, $\int y' \, dt = Y$. Defining $\tan \gamma = y'/x' = dy/dx$, the problem becomes: minimize $\int dt$ subject to the constraints $(y')^2/\sin^2 \gamma + 2gy = at$, $\int y' \cot \gamma \, dt = X$, $\int y' \, dt = Y$. The usual Lagrange multiplier technique yields the variational problem: min $\int \{\lambda(t)((y')^2/\sin^2 \gamma + 2gy - at) - Cy' \cot \gamma\} \, dt$. The resulting Euler equations are $(2\lambda y'/\sin^2 \gamma - C \cot \gamma)' = 2gy$, $2\lambda y' = C \tan \gamma$. Finally, $2g^2 t/a = \log(\sec \gamma)/(\sec \gamma_0) - \gamma \tan \gamma_0 + \gamma_0 \tan \gamma_0$, $16g^3 x/a^2 = 2(1 + k^2)(\gamma - \gamma_0) + (k^2 - 1)(\sin 2\gamma - \sin 2\gamma_0) - 2k(\cos 2\gamma - \cos 2\gamma_0)$, $16g^3 y/a^2 = 4 \log(\sec \gamma)/(\sec \gamma_0) + 4k(\gamma - \gamma_0) - (k^2 - 1)(\cos 2\gamma - \cos 2\gamma_0) - 2k(\sin 2\gamma - \sin 2\gamma_0)$. If γ (final) and γ_0 are chosen properly, one can ensure that $x = X$, $y = Y$ (finally). So one has the parametric equations for the optimal path plus the minimum time of flight. D. J. Newman, *SIAM Review* 6, 1964, 311–313, Problem 59-7.

1.749 [Sg] 3. 10. C4.

1.750 *Hint*: $d(t)$ satisfies the inequality $d' \geq 1/d - 1$ if the mouse's velocity is opposite to the cat's direction. Suppose that $d(T) \leq 1$ and T is the smallest t satisfying the inequality $\int_{d(0)}^{d(T)} (1/s - 1)^{-1} \, ds \leq T$ or $\int_1^{d(0)} s(s - 1)^{-1} \, ds$, hence $T = \infty$.

1.751 [Br] Ch. I, Sec. 1. 5.

1.752 *Hint*: The differential equation is a Bernoulli equation.
Answer: $M = \exp(kNt + krt^2/2)(N_0^{-1} + k \int_0^t \exp(kNv + krv^2)/2) \, dv)^{-1}$. [Sg] 3. 13. 2. C2.

1.753 [Br] Ch. 4, Sec. 4. 10.

1.754 [Da] Ch. 14, Sec. 3, Ex. 5.

1.755 *Answer*: These are conical sections. [PTKY] 21. 35, 21. 36.

1.756 *Hint*: Apply the Poincaré lemma to the one-form $\alpha(x)v = (A(x),v)$. [A-M] 2. 5. C.

1.757 *Hint*: The extremals are determined by the equation $x^c y'(1 + (y')^2)^{-1/2} = $ const. [V] Ch. XI, Sec. 192.

1.758 [V] Ch. XI, Sec. 192.

1.759 *Answer*: Each strict relative minimizer of g is a local attractor, whereas each strict relative maximizer is a local repeller. Moody T. Chu, *SIAM Review* 30, 1988, 375–387; for the steepest descent method see: [Be] Ch. 3, Sec. 3. 2.

1.760 *Answer:* The solution satisfies the equation $f(x(t)) = e^{-t}f(x(0))$. The flow always moves in the direction along which the magnitude of $f(x)$ is reduced exponentially. The solution $x(t)$ either diverges to infinity, converges to a local minimum of f, or converges to a zero of f. Moody T. Chu, *SIAM Review* 30, 1988, 375–387.

1.761 *Answer:* If $y = b_1$, $y = b_2$ are the parametrizations of ℓ_1, ℓ_2, respectively, then for γ described by $y = y(x)$ one obtains the differential equation $x - yx' = \pm a^2 y/((b_1 - y)(b_2 - y))$. Hence $x = y(C + a^2 \log(\pm y)/(b_1 b_2) - a^2 \log(\pm(y - b_1))/(b_1(b_2 - b_1)) + a^2 \log(\pm(y - b_2))/(b_2(b_2 - b_1))$. Discuss also the case $b_1 = b_2$. [Ju] 2.

1.762 *Answer:* $y^{(3)} (1 + (y'')^2) - 3y'(y'')^2 = 0$. [I] Ch. I Ex. 5.

1.763 *Answer:* $((y'')^{-2/3})^{(3)} = 0$. Observe that $((y'')^{-2/3})'' = 0$ describes all coplanar parabolas. [I] Ch. I, Ex. 6.

1.764 *Hint:* Take $y = ax + b + (cx^2 + 2dx + e)^{1/2}$, so $y'' = (ce - d^2)(cx^2 + 2dx + e)^{-3/2}$ and finally $((y'')^{-3/2})^{(3)} = 0$.
Answer: In the Halphen form this equation reads $40(y^{(3)})^3 - 45y^{(2)} y^{(3)}y^{(4)} + 9(y^{(2)})^2 y^{(5)} = 0$. [V] Sec. 75.

Partial Differential Equations

2.1 *Answer*: The famous Hadamard example is: $\Delta u = 0$ in $\mathbb{R}^2$, $u(x,0) = \exp(-k^{1/2}) \cos kx$, $u_y(x,0) = 0$ with the solution $u(x,y) = \exp(-k^{1/2}) (\cos kx) (\cosh ky)$. As $k \to \infty$, the Cauchy data and their derivatives of each order tend uniformly to zero, but the solutions are unbounded if $y = 0$. For example: [M] Ch. 1, Sec. 1, Ex. 3.

2.2 *Answer*: This is the famous Kowalevskaya example. The only solution analytic at the origin would be $u(x,t) = \sum_{m,n=0}^{\infty} u_{m,n} x^m t^n$, where $u_{2m,n} = (-1)^{m+n}(2m + 2n)!/((2m)!n!)$, $u_{2m+1,n} = 0$. The series is plainly divergent in $(0, t)$ for all $t > 0$. [M] Ch. I, Sec. 2, Example 2.

2.3 K. Keller, A. Schneider, *Manuscripta Math.* 39, 1982, 31–37; W. Walter, *Amer. Math. Monthly* 92, 1985, 115–126.

2.4 *Hint*: First observe that there exists $\delta = \delta(m,n) > 0$ such that for a polynomial P of degree m in $\mathbb{R}^n$ $|P(\xi)|^{-\delta} \in L^1_{\text{loc}}$ (a proof follows, for instance, from the Weierstrass preparation theorem). To show that lemma infer (from the above fact) that there exist $\delta, M > 0$ such that $\int |P(\xi)|^{-\delta}(1 + |\xi|^2)^{-M} d\xi < \infty$. Then $\int \sum_{k \in \mathbb{Z}^n} |P(k + \alpha)|^{-\delta} (1 + |k|^2)^{-M} d\alpha \leq C \int_{\mathbb{R}^n} |P(\xi)|^{-\delta} (1 + |\xi|^2)^{-M} d\xi < \infty$, where the first integral is taken over $0 \leq \alpha_j \leq 1$, $j = 1, \ldots, n$. [Ty] Ch. I, Sec. 7; see also [Fo] Ch. 1, Sec. F.

2.5 *Hint*: Take a positive function $f \in C_0^\infty(\mathbb{R}^2)$, $f(x,y) = f(-x,y)$ and f vanishing outside the union of disjoint disks D_n in the half-plane $x > 0$ which converge to the origin as n tends to infinity, $f > 0$ in D_n. Suppose that u is a solution and put $v(x,y) = \frac{1}{2}(u(x,y) + u(-x,y))$, $w(x,y) = \frac{1}{2}(u(x,y) - u(-x,y))$. Hence $\partial v/\partial x + ix\ \partial v/\partial y = 0$, $\partial w/\partial x + ix\partial w/\partial y = f$, and the change of variables $s = x^2/2$ transforms the first equation into the Cauchy-Riemann equation $\partial v/\partial s + i\partial v/\partial y = 0$. Observe that v is smooth and therefore analytic as the function of $s + iy$ as well as w. w should vanish in (a connected set!) $\mathbb{R}^2 \setminus \cup_n D_n$. An application of the Green formula to the function f on D_n leads to a contradiction: w vanishes on ∂D_n while f is strictly positive on D_n. The same result is true in the case of the Mizohata equation $\partial u/\partial x + ix^{2k+1}\ \partial u/\partial y = f$, $k \in \mathbb{N}$.

Remark: The first example of a linear partial differential equation without solutions given by H. Lewy was slightly more complicated: $\partial u/\partial x + i\ \partial u/\partial y + i(x + iy)\ \partial u/\partial z = f$. [Eg] Ch. VI, Sec. 1. 1.

2.6 *Hint:* $L(\omega,s) = \{f \in \mathscr{S}\ (\Omega) : u \in H^s,\ Pu = f \text{ in } \omega\}\ (\neq \mathscr{S}(\Omega))$ is the projection onto $\mathscr{S}(\Omega)$ of $K(\omega,s) = \{(u, f) \in H^s \times \mathscr{S}(\Omega): Pu = f \text{ in } \omega\}$. This set is, of course, closed, hence $L(\omega,s)$ is closed and nowhere dense in $\mathscr{S}(\Omega)$. Now Ω can be represented as a countable union of ω_j and $\cup_j L(\omega_j,\ s)$ is of the first category.

2.7 *Hint:* The Cauchy-Riemann operator $\partial/\partial\bar{z}$ is analytic-hypoelliptic. [T] Ex. 3. 1.

2.8 [T] Ex. 17. 9.

2.9 [Ty] Ch. III, Sec. 3, Ex. 1.

2.10 [T] Ex. 19. 6.

2.11 *Hint:* The function $1/\pi z_j$ is a fundamental solution of $\partial/\partial\bar{z}_j$. Take $|z_2| + \cdots + |z_n| \to \infty$ in the (inhomogeneous) Cauchy formula with respect to z_i. [T] Ex. 5. 6.

2.12 *Hint:* Observe that $\partial h/\partial\bar{z}_j = 0, j = 1, \ldots, n$. Use a cutoff function $g \in C_0^\infty\ (\mathbb{R}^{2n})$, equal to 1 in a neighborhood of K, and solve the system $\partial u/\partial\bar{z}_j = f_j$ where $f_j = (\partial/\partial\bar{z}_j)((1 - g)h)$.

Remark: This extension result is the classical Hartogs theorem. The same result would not be true if $n = 1$. [T] Ex. 5. 7.

2.13 [T] Ex. 21. 6.

2.14 *Hint:* It suffices to do it for $(0, a, 0)$ as the operator in translation invariant with respect to x and z.

Answer: $u(x,y,z) = 3^{1/2}\ (2\pi)^{-1}\ ((2x + y - a)\ (a^3 - y^3) - 3z^2)^{-1/2}$ if $x > 0$, $y < a$, and the expression under the square root is positive; otherwise, $u = 0$. [Hö] Ch. I, 7. 6. 14.

2.15 [T] Ex. 2. 10.

2.16 *Hint*: Apply the maximum modulus principle to the holomorphic function $w(z) - r(z)$, where $r(z) = \pi^{-1} \int \int_G \partial w/\partial \bar{z}(\zeta)\,(\zeta - z)^{-1}\,d\xi\,d\eta$, $\zeta = \xi + i\eta$. Observe that $\sup_G |r| \leq 2R\sup_G |\partial w/\partial \bar{z}| \leq 2RK\sup_G |w|$. [BJS] Ch. II, 2. 7.

2.17 *Hint*: It suffices to prove $w \equiv 0$ in the small disk $\{\,|z| \leq R\}$ for some $R > 0$. The functions $z^{-n}w(z)$, $n = 1, 2, \ldots$, vanish at $z = 0$ and satisfy the inequality from the assumption. The result of the preceding problem implies $\max_{|z| \leq R} |z^{-n}w(z)| \leq kR^{-n}\max_{|z| \leq R} |w(z)|$, hence w must vanish everywhere in $\{\,|z| \leq R\}$. [BJS] Ch. II, 2. 8. For recent results on the unique continuation property see: [R-S] Ch. 13, Appendix to Sec. 13; [Hö] Ch. 17, Sec. 2; C. E. Kenig, *Proceedings of the International Congress of Mathematicians, 1986, Berkeley*, AMS, 1987, pp. 948–960; L. Hörmander, *Comm. Partial Differential Equations* 8, 1983, 21–64; D. Jerison, C. Kenig, *Annals Math.* 121, 1985, 463–494; J.-C. Saut, B. Scheurer, *J. Differential Equations* 43, 1982, 28–43; *ibidem* 66, 1987, 118–139.

2.18 *Hint*: Assume contrarily that there exists a sequence $(u_k) \subset C_0^\infty(\Omega)$ with the supports of u_k shrinking to a point. (u_k) is compact in $H^{s'}$, the limit function u can be represented as $u = P(D)\delta$ with some constant-coefficient operator P of order m. Then $\hat{u} = P \neq 0$ and $P(\xi)(1 + |\xi|^2)^{s/2} \in L^2(\mathbb{R}^n)$, which is possible only if $m + s < -n/2$. For an example consider an approximative identity.
Remark: If $s > s'$, then for every $\varepsilon > 0$ there exists c that for all $u \in H^s (\mathbb{R}^n)$ $\|u\|_{s'} \leq \varepsilon \|u\|_s + c\varepsilon^{1/(s' - s)} \|u\|_{s' - 1}$. This follows easily from the inequality $\xi^{s'} \leq \varepsilon \xi^s + c\varepsilon^{1/(s' - s)}\xi^{s' - 1}$ valid for all $\xi \geq 1$, $\varepsilon > 0$. [Eg] Ch. I, 3. 14, 3. 15.

2.19 [BJS] Ch. 6, Sec. 3. The Bers-Bojarski-Nirenberg theorem.

2.20 *Answer*: $f(x_1, \ldots, x_n) = \int_0^1 h(tx_1, \ldots, tx_n)\,t^{p-1}\,dt$. [Ca] Ex. 3, 10; [B-W] 5. 114.

2.21 *Hint*: Introduce new independent variables $t = x + y$, $z = x - y$. The general solution of the equation $u_t = \frac{1}{2}w(z)$ (equivalent to the aforementioned equation) is $u(t, z) = \frac{1}{2}w(z)\,t + c(z)$ with an arbitrary differentiable function c of z.
Remark: Each (generalized) continuous solution function of this equation is of the above form. [P1] Appendix, Sec. 52, Remark.

2.22 *Hint*: Show that $\max_D u \leq 0$, $\min_D u \geq 0$, using conditions for a maximum at a boundary point. [J1] Ch. 1, Sec. 6, Ex. 2.

2.23 [T] Ex. 1. 10.

2.24 *Hint*: Rewrite $\int |f(x^2 + y^2)|^2\,dx\,dy$ by introducing u, assumed to satisfy the equation. [T] Ex. 2. 1.

2.25 [Hö] Example 13. 6. 5.

2.26 [G] Ch. 2, Sec. 1, Ex. 3.

2.27 *Answer*: In the Fourier variables

$$\begin{bmatrix} \exp(-i\xi t) & -i\xi t \exp(-i\xi t) \\ 0 & \exp(-i\xi t) \end{bmatrix}$$

is the matrix of $S(t)$, hence it is not bounded in $L^2(\mathbb{R})$. However, an estimate of the form $\| \langle u(t), v(t) \rangle \|_2 \le C(1 + |t|)(\| \langle u_0, v_0 \rangle \|_2 + \|A \langle u_0, v_0 \rangle\|_2$ holds, where

$$A = -i\xi \begin{bmatrix} 1 & 1 \\ 0 & 1 \end{bmatrix}$$

[F] Example 8. 6. 1; R. Beals, *J. Functional Analysis* 10, 1972, 281–299.

2.28 *Answer*: The solution operator is given by

$$\exp(-it\xi) \begin{bmatrix} \cosh((-i\xi)^{1/2} t) & (-i\xi)^{1/2} \sinh((-i\xi)^{1/2} t) \\ -(i\xi)^{-1/2} \sinh((-i\xi)^{1/2} t) & \cosh((-i\xi)^{1/2} t) \end{bmatrix}$$

The solution satisfies an estimate of Gevrey type $\| \langle u(t), v(t) \rangle \|_2 \le$

$$c\sum_{n=0}^{\infty} C^n (n!)^{-\alpha} \|A^n \langle u_0, v_0 \rangle\|_2,$$

where

$$A = \begin{bmatrix} -i\xi & -i\xi \\ 1 & -i\xi \end{bmatrix} \quad \text{and} \quad \alpha > 1$$

[F] Ex. 8. 6. 6; and the reference to the preceding problem.

2.29 *Hint*: By a linear transformation reduce to the case $a = e = 1, b = d = 0$. [J1] Ch. 2, Sec. 4, Ex. 4.

2.30 A. Ungar, W. Allegretto, *SIAM Review* 20, 1978, Problem 77-4, 190–191.

2.31 [P-W] Ch. 2, Sec. 4, Ex. 3.

2.32 [G] Ch. 7, Sec. 1, Ex. 2.

2.33 *Hint*: Seek for the solution in the form $\sum_{n=1}^{\infty} u_n(x) \sin n\omega y$, $\omega = \pi/b$. *Answer*: For every $\lambda \in \mathbb{R} \setminus \{-k\omega \coth k\omega a : k \in \mathbb{N}\}$, there exists the unique solution obtained by the above separation of variables method. If $\lambda = -k\omega \coth k\omega a$, then the necessary compatibility condition is $\int_0^b \sin k\omega y\, g(y)\, dy = 0$ and $u_k(x) = c \sinh k\omega(x - a)$ with an arbitrary constant c shows non-uniqueness of solutions. [D-L] Ch. II, Sec. 4, Example 23.

2.34 *Hint*: Represent u as $v + w$, where v and w are harmonic, $v(x,y,0) = f(x,y)$, $v_z(x,y,0) = 0$, $w(x,y,0) = 0$, $w_z(x,y,0) = g(x,y)$. *Answer*: $u(x,y,z) = \sum_{k=0}^{\infty} (-1)^k 4^k z^{2k}((2k)!)^{-1} \partial^{2k} f/\partial x^k \partial y^k + \sum_{k=0}^{\infty} (-1)^k 4^k z^{2k+1} ((2k + 1)!)^{-1} \partial^{2k} g/\partial x^k \partial y^k$. [Ya] Ch. 2, Sec. 1.

2.35 For instance: [W] Ex. 14. 2.

2.36 *Hint*:   Use the Taylor expansion. [T] Ex. 10. 2; [PPJ] 16. 9.

2.37 For instance: [Tr1] Ch. IV, Ex. 10.

2.38 *Hint*:   Let $n = 3$ and $u(\xi,\eta,\zeta)$ be harmonic if $u(x,y,z)$ is, ξ, η, $\zeta = \xi, \eta,$ ζ (x, y, z). A necessary and sufficient condition for this reads $\Delta\xi = \Delta\eta = \Delta\zeta$ $= 0$, $|\nabla\xi| = |\nabla\eta| = |\nabla\zeta|$, $\nabla\eta \cdot \nabla\zeta = \nabla\zeta \cdot \nabla\xi = \nabla\xi \cdot \nabla\eta = 0$. Then show that for $V = (\xi, \eta, \zeta)$ $|V_x| = |V_y| = |V_z|$ $(=: \rho)$, $V_y \cdot V_z$ $= V_z \cdot V_x = V_x \cdot V_y = 0$ hold. Differentiating the relations above one gets $V_{xx}V_x$ $= \rho\rho_x$, $V_{xx}V_y = V_xV_y = -\rho\rho_y$, $V_{xx}V_z = -\rho\rho_z$, and then $\Delta V = 0$, which implies that ρ is a constant. [Fa] Ch. X, Ex. 1.

2.39 [G-T] 4. 7.

2.40 *Hint*:   Use the mean value property of harmonic functions. [G-T] 2. 13.

2.41 *Hint*:   Use comparison functions.
Remark: As is known from 2.40, the gradient of a harmonic function at $x \in$ D may grow as $1/\text{dist}(x,\partial D)$, so the assumption on C^2 extendability of the boundary data g onto D is essential. [P-W] Ch. 2. Sec. 14, (2).

2.42 *Hint*:   Prove the following lemma: Under the aforementioned assumptions on G and G', let the function v be harmonic in G, vanish on a, and satisfy the inequality $0 \le |v| \le 1$ on b. Then there exists a positive constant $q <$ 1, depending only on the configuration of the domains G and G', such that v satisfies the inequality $|v| \le q$ on b'. [C-H] IV. 4. 2.

2.43 *Hint*:   Show that the weak maximum principle applies to the function with restricted mean value property. Then use the assumption on the solvability of the Dirichlet problem. R. B. Burckel, *Amer. Math. Monthly* 87, 1980, 819–820. The original paper is: O. D. Kellogg, *Trans. AMS*, 36, 1934, 227–242. See also: L. Zalcman, *Arch. Rational Mech. Anal.* 47, 1972, 237–254; [B-W] 5. 122.

2.44 *Answer*:   Yes; for example, $f(x_1, \ldots , x_n) = \text{sgn } x_1$ satisfies all these conditions but the continuity.
Remark: A similar result holds for the spherical means. T. Radó, *Subharmonic Functions,* Ergebnisse der Mathematik, Springer, Berlin, 1937. More generally: M. Brelot, *Théorie classique du potentiel,* Centre de documentation universitaire, Paris, 1969, Ch. 2; A. G. O'Farrell, R. L. Cooke, *Amer. Math. Monthly* 86, 1979, 229–230, 6135; [B-W] 5. 123.

2.45 [Sn] Ch. 4, Sec. 3, Problem 1.

2.46 [BST] IV. 161.

2.47 [K] Ch. VII, Sec. 6, Ex. 5.

2.48 *Hint*:   Simply take $\rho(x) = c |x - x_0|^\lambda$, $0 < \lambda < 1$, in a small disk on a (flat) Liapunov surface. [KGS] Ch. XX, Sec. 10.

2.49 [Ep] Ch. 7, Sec. 7, Ex. 34.

2.50 *Answer:* Yes, let $u(x,y) = \mathrm{Re}\, \exp(z^{-4})$ for $z = x + iy$, $z \neq 0$, and $u(0,0) = 0$. This function does not define even a distribution near the origin. [S] Ch. 8; [P2] Ch. 3, Sec. 27. For some discontinuous generalizations of harmonic functions see: P. Mikusiński, *Proc. AMS* 106, 1989, 447–449.

2.51 [G-T] 2. 2.

2.52 *Hint:* Observe that $u(x,y) = \mathrm{Re}(1 - 2/z)$, hence u does not tend to any limit when x, y tend to the origin. [Gr] Ch. XXVII, 512; [P-W] Ch. 2, Sec. 3, Ex. 2.
Remark: If the condition $\lim \inf_{\rho \to 0} (\sup\{u(x,y) : (x - 1)^2 + y^2 = \rho^2\})/(\log 1/\rho) \leq 0$ were satisfied, then $u \equiv 0$ would follow. [P-W] Ch. 2, Sec. 9 (9).

2.53 *Hint:* This trace equals $\cos(1/2) \exp(\cos \vartheta/2(1 - \sin \vartheta))$ and its singularity at $\vartheta = \pi/2$ is of infinite order. [D-L] Ch. II, Sec. 6, Example 4. See also: [C-H] VI, App. 3. 5; [T] Ex. 27. 3.

2.54 *Answer:* Let $E \subset S^1$ be of measure zero. Consider open sets G_n such that $G_n \supset E$ and the Lebesgue measure of G_n is $\leq 2^{-n}$. Finally take $u(r \exp(i\vartheta)) = \sum_{n=0}^{\infty} (2\pi)^{-1} \int_{G_n} (1 - r^2)(1 + r^2 - 2r \cos(\vartheta - \varphi))^{-1} d\varphi$. For $\vartheta \in E$ $\lim \inf_{r \to 1} u(\exp(i\vartheta)) \geq \sum_{n=0}^{N} 1 = N + 1$, for all $N \in \mathbb{N}$. L. Carleson, *Selected Problems on Exceptional Sets,* Van Nostrand, Princeton, 1967, Ch. V, Sec. 5.

2.55 [M] Ch. IV, Problems 19, 20, 21.

2.56 *Answer:* $u = g$ on the circle $x = 0$, $y^2 + z^2 = 1$.
Hint: $v = \partial u/\partial x$ satisfies the Laplace equation in the ball and $v = f$ on the sphere, so v is uniquely determined. The supplementary condition can be read: $\partial^2 u/\partial y^2 + \partial^2 u/\partial z^2 = v - \partial v/\partial x$, $u = g$ on the circle $x = 0$, $y^2 + z^2 = 1$. [Eg] IV. 1. 3.

2.57 *Hint:* Apply the Riesz representation theorem for the superharmonic function $v(x) = u(x) + 1/|x|$. [W] Th. 18. 2.

2.58 *Hint:* Apply the Kelvin transform and use the Bôcher theorem. [W] Sec. 18, Corollary 1. See also 2.39.

2.59 [Gr] Ch. XXVII, 514; S. Zaremba, *Bull. Acad. Sci. Cracovie,* 1909.

2.60 [K] Ch. X, Sec. 2, Ex. 5.

2.61 [Ty] Ch. XI, Sec. 2, Ex. 2.

2.62 For instance, [Ty] Ch. XI, Sec. 5, Ex. 3.

2.63. *Hint:* Observe that $v(x_1,0,0) \to 0$ as $x_1 \to 0^-$. Each bounded harmonic function which coincides with v on the boundary of the domain except for the origin is identically equal to v. For related examples see [C-H] IV. 4. 2, IV. 4. 4; [T] Ex. 29. 10.

2.64 *Hint*: Use the Poisson representation formula or the estimates for derivatives of harmonic functions $\max_{B(x,R)} |D^\alpha u| \le (n|\alpha|/\rho)^{|\alpha|}$ $\max_{B(x,R+\rho)} |u|_u$. For $m = 0$ consider $u + c$ and show, using the mean value property, that $\nabla(u + c) = 0$. [PPJ] 16. 40, 16. 42.

2.65 [G] Ch. 9, Sec. 2, Ex. 17.

2.66 *Hint*: Consider comparison functions $\delta \sin((1 - \varepsilon)^{1/2} (x_n/h + \rho))$ $\sum_\beta \exp(h^{-1}\sum_{j=1}^{n-1}\beta_j x_j)$, where $\delta, \rho > 0$, $2\varepsilon = 1 - \sum_{j=1}^{n-1} a_j^2$, $|\beta_j| = b_j$, $\sum_{j=1}^{n-1} b_j^2 = 1 - \varepsilon$.
Remark: The assumptions in this theorem are the weakest possible; consider for instance $u(x) = \sin(x_n/h)\exp(h^{-1} \sum_{j=1}^{n-1} a_j x_j)$ with $\sum_{j=1}^{n-1} a_j^2 = 1$. [PPJ] 16. 43.

2.67 *Answer*: The solution exists if $B = aA/2$ and this is determined up to an additive constant: $u(r) = Ar^2/4 + \text{const.}$ [BST] IV. 31.

2.68 *Hint*: $u(x) = (x_1^2 - x_2^2) (-\log |x|)^{-1/2}$ solves this equation in the classical sense outside the origin. If v were a classical solution then $u - v$ would be harmonic and bounded outside the origin, hence harmonic in the whole ball. [M] IV, Sec. 3.

2.69 *Hint*: Let $0 < r_1 < r_2 < R$ and h be an affine function of $\log r$ or r^{2-n} which coincides with M for $r = r_1$ and $r = r_2$. Using the maximum principle, compare $u(x)$ and $h(|x|)$. [PPJ] 17. 10.

2.70 *Hint*: Apply the maximum principle for $u - (a + b \log r)$ with suitably chosen constants a and b. [PPJ] 17. 13.

2.71 *Answer*: First show that $\Delta w = \exp(\sum_{j=1}^n a_j x_j) (\Delta u + 2\sum_{j=1}^n a_j \partial u/\partial x_j + \sum_{j=1}^n a_j^2 u) \ge 0$. Then $u \Delta u - |\nabla u|^2 \ge 0$, which means that $\log u$ is subharmonic. [PPJ] 17. 29.

2.72 [PPJ] 17. 34.

2.73 [PPJ] 20. 11.

2.74 *Hint*: Verify that $\nabla(ru_r) = -2$, hence $2u - ru_r = -u\Delta(ru_r) + ru_r \Delta u$. Using the Green's formula show that $\int_\Omega (2u - ru_r) \, dx = c^2 \int_{\partial\Omega} r \, \partial r/\partial \nu \, ds = nc^2|\Omega|$. On the other hand $\int_\Omega ru_r = -n\int_\Omega u \, dx$ and $(n + 2) \int_\Omega u \, dx = nc^2|\Omega|$. Furthermore $1 = (\Delta u)^2 \le n |\nabla u|^2 \le n\sum_{i,j=1}^n (\partial^2 u/\partial x_i \, \partial x_j)^2$ and $\Delta (|\nabla u|^2 + 2u/n) = 2\sum_{i,j=1}^n (\partial^2 u/\partial x_i \partial x_j)^2 - 2/n \ge 0$. Then it follows from the strong maximum principle that $|\nabla u|^2 + 2/n < c^2$ or $\equiv c^2$ in Ω, as this function is equal to c^2 on $\partial\Omega$. The former relation is not possible as it implies $(1 + 2/n) \int_\Omega u \, dx < c^2|\Omega|$. So the latter gives $\partial^2 u/\partial x_i \partial x_j = -\delta_{ij}/n$ and $u = (2n)^{-1} (A - r^2)$. The boundary condition for u identifies Ω as a ball of radius $A^{1/2}$. [PPJ] 16. 30. The original paper is: H. Weinberger, *Arch. Rational Mech. Anal.* 43, 1971, 319–320. For more general setting see P. Aviles, *Amer. J. Math.* 108, 1986, 1023–1036; N. Garofalo, J. L. Lewis, *ibidem* 111, 1989, 9–33.

2.75 *Hint*: Use the Harnack inequality for harmonic functions. [P-W] Ch. 2, Sec. 10, Ex. 6.

2.76 [G-T] Th. 3. 9.
Remark: This theorem is valid also for functions $u \in H^1(\Omega)$ satisfying in weak sense the equation $\Delta u = f$. The function $u(x,y) = |xy| \log(|x| + |y|)$ in the unit disk demonstrates its sharpness. [G-T] 8. 4.

2.77 *Answer*: From the formula $\partial^2 u/\partial x_i\, \partial x_j\,(x) = x_i x_j f(x)/|x|^2 + (\delta_{ij}/n - x_i x_j/|x|^2)\, (\omega_n |x|^n)^{-1} \int_{B(0,|x|)} f(y)\, dy$ valid for all radial functions u one obtains $\|\partial^2 u/\partial x_i \partial x_j\,(x)\|_\infty \le (2 - \delta_{ij}/n)\|f\|_\infty$. [D-L] Ch. II, Sec. 3, Proposition 7, Remarque 5.

2.78 *Hint*: Construct such examples as Newtonian potentials of certain continuous densities.
Answer: (i) For instance $u(x,y) = xy(-\log(x^2 + y^2))^{1/2}$ if $x^2 + y^2 > 0$, $u(0,0) = 0$. Original papers: H. Petrini, *Acta Math.* 31, 1908, 127–332; *J. Math. pures appl.* 5, 1909, 127–233. [D-L] Ch. II, Sec. 3; [B-W] 5. 95.

2.79 [D-L] Ch. II, Sec. 3, Remarque 6; [Se] Ch. V.

2.80 [T] Ex. 34. 7.

2.81 [P-W] Ch. 2, Sec. 8, Ex. 2.

2.82 [J1] Ch. 6, Sec. 2, Ex. 1.

2.83 [D-L] Ch. II, Sec. 8, Remarque 3; H. Garnir, *Les problèmes aux limites de la physique mathématique*, Birkhäuser, Basel, 1958; I. A. Shishmariov, *An Introduction to the Theory of Elliptic Equations*, MGU, Moskva, 1979, Ch. 3, Th. 1. 1.

2.84 [Fa], Ch. X, Ex. 3b).

2.85 *Hint*: Represent u in terms of a specific fundamental solution that fulfills the radiation condition. [G] Ch. 7, Sec. 1, Ex. 3; Sec. 2, Ex. 20.

2.86 [V12] 18. 27.

2.87 *Hint*: $s = s(x,y)$ and $t = t(x,y)$ should satisfy the system $s_x^2 + s_y^2 = a^2$, $t_x^2 + t_y^2 = a^2$, $s_x t_x + s_y t_y = 0$, that is, $s_x = a\cos\varphi$, $s_y = a\sin\varphi$, $t_x = a\sin\varphi$, $t_y = -a\cos\varphi$ for a function $\varphi = \varphi(x,y)$. R. Bellman, *Bol. Unione Mat. Ital.* 13, 1958, 535–538; [B3] Part 3, Ex. 3. 1–3. 5.

2.88 *Answer*: For instance, $\exp(i\ell\varphi) \exp((a + \ell^2)^{1/2} \log r)$, $\exp(i\ell\varphi) \exp(-(a + \ell^2)^{1/2} \log r)$, where $\varphi = \arctan(y/x)$, $r^2 = x^2 + y^2$. [Ya] Ch. 4, Sec. 4.

2.89 [P2] Ch. 2, Sec. 25.

2.90 [G] Ch. 11, Sec. 2, Ex. 10.

2.91 *Hint*: This and three subsequent results follow easily from the Courant minimax principle characterizing eigenvalues. [C-H] VI. 2. Th. 3.

2.92 [C-H] VI. 2. Th. 5.

2.93 [C-H] VI. 2. Th. 6.

2.94 [C-H] VI. 2. Th. 7.

2.95 *Hint*: The equality sign holds in the Faber-Krahn inequality for every circular membrane. The use of the Steiner symmetrization is crucial in the proof. [G] Ch. 11, (11.60), Ch. 11, Sec. 3, Ex. 3; G. Pólya, G. Szegö, *Isoperimetric Inequalities in Mathematical Physics*, Ann. Math. Studies 27, Princeton University Press, Princeton, 1951.

2.96 *Hint*: Begin with the case of a smooth boundary. This is a deep result. H. Levine, H. F. Weinberger, Arch. Rational Mech. Anal. 94, 1986, 193–208.

2.97 *Hint*: Observe that $u(y) = -\int_{\partial D} u(x) \, \partial K(x,y)/\partial \nu \, d\sigma = -\int_{\partial D} K(x,y) \, \partial u(x)/\partial \nu \, d\sigma$. [G] Ch. 7, Sec. 3, Ex. 5.

2.98 [G] Ch. 7, Sec. 3, Ex. 6.

2.99 *Answer*: For instance, $u(r,\varphi) = \sum_{k=1}^{\infty} r^k k^{-2} \sin(k^4 \varphi)$, $0 \le r \le 1$, $0 \le \varphi < 2\pi$. [G] Ch. 8, Sec. 1, Ex. 5, Ex. 11; another examples: [Ep] Ex. 7. 27; [Fa] Ch. XIII, Sec. 15; [Gr] Ch. XXVII, Ex. 5.

2.100 *Answer*: Consider in any finite domain G a point P and a small ball of radius $a < 1$ centered at P and contained in G. Let $\varphi_a(r) = 0$ for $r > a$, $\varphi_a(r) = 1$ for $r < a^2$, and $\varphi_a(r) = \log(r/a)/\log a$ for $a^2 < r < a$ (of course, $\varphi_a \equiv 0$ on the boundary of G). The Dirichlet integral $D(\varphi_a) = 2\pi(\log a)^2 \int_{a^2}^{a} r \, dr = -2\pi/\log a$. Now if $a \to 0$, $D(\varphi_a) \to 0$ for the corresponding functions φ_a but $\varphi_a(P) \equiv 1$ while the unique solution φ is identically zero. [C-H] IV. 2. 4.

2.101 [T] Ex. 37. 8.

2.102 [T] Ex. 23. 8.

2.103 [G-T] 3. 9; Lemma 3. 4 (this lemma is false in general for strictly and uniformly elliptic equations in divergence form, but it is true if the coefficients are Hölder continuous).

2.104 *Hint*: Any such solution u has a Fourier series expansion $\sum f_n(y) \sin nx$, in which the coefficients $f_n(y)$ satisfy the ordinary differential equation $y^2 f_n'' - n^2 f_n = 0$. This equation has the independent solutions $\exp(b_n \log y)$, $\exp(c_n \log y)$, where $b_n = \frac{1}{2}(1 + (1 + 4n^2)^{1/2}) > 0$, $c_n = \frac{1}{2}(1 - (1 + 4n^2)^{1/2}) < 0$. The fact that u is bounded at $y = 0$ requires that $f_n(y) = \text{const} \cdot \exp(b_n \log y)$, and hence $f_n(0) = 0$. It follows that the only continuous solutions on $\overline{R}$ satisfying the prescribed boundary conditions must have zero boundary values on $y = 0$. [G-T] 6. 6.

2.105 *Answer*: Each function u of the form $(1 - |z|^2) f(z)$, where f is an analytic function, is a solution of the Bitsadze equation. [PPJ] 20. 2; [BJS] II. 1, App. 1; [Fo] Ch. 7. A.

2.106 *Answer:* If $w = u_1 + iu_2$, $z = x + iy$, then the equation becomes $\partial^2 w / \partial \bar{z}^2 = 0$. Hence the general solution is $w(z) = \bar{z}\varphi(z) + \psi(z)$ with arbitrary analytic functions φ and ψ of z. [Bi] Ch. II, Sec. 1. 1.

2.107 *Remark:* This problem is not of Fredholm type since the system degenerates at the origin. [Bi] Ch. II, Sec. 4. 5.

2.108 [G] Ch. 5, Sec. 1, Ex. 11.

2.109 *Hint:* Apply the Harnack inequality in suitable annuli extending to infinity. [G-T] 3. 3.

2.110 [T] Ex. 9. 2–9. 4; [Fr2] Problem 9. 8.

2.111 [Mi] I. 8.

2.112 *Hint:* In the new variables $x = \tan(\vartheta/2) \cos \varphi$, $y = \tan(\vartheta/2) \sin \varphi$, the equation reads $\Delta u + 4n(n + 1) (1 + x^2 + y^2)^{-2} u = 0$ or, in the complex form, $\partial^2 u / \partial z \partial \zeta + n(n + 1)(1 + z\zeta)^{-2} u = 0$, where $z = x + iy$, $\zeta = x - iy$. If $U(z, \zeta)$ solves the latter equation then $U(\zeta(z - t)/(1 + \tau z), \xi^{-1}(\zeta - \tau)/(1 + t\zeta))$ is also a solution; t, τ, ξ are arbitrary complex numbers such that $\xi \neq 0$, $1 + t\tau \neq 0$. Take $\cos \vartheta = (1 - z\zeta)/(1 + z\zeta)$.
Answer: The functions $P_n(\cos \vartheta)$ where P_n is the nth Legendre function, satisfying the equation $(1 - z^2)P_n''(z) - 2zP_n'(z) + n(n + 1)P_n(z) = 0$, $n \in \mathbb{R}$, are solutions of the considered equation. [Ve] Ch. 1, Sec. 5, Example 4; Sec. 12.

2.113 [G] Ch. 5, Sec. 1, Ex. 10.

2.114 *Answer:* The fundamental solution matrix has the elements $-(4\pi\mu)^{-1}$ $(r^{-1}\delta_{jk} - (\lambda + \mu)/2(\lambda + 2\mu) \partial^2 r/\partial x_j \partial x_k)$. [J1] Ch. 6, Sec. 1, Ex. 3, 4.

2.115 *Hint:* The easiest way to prove it is (iv) $\Rightarrow$ (iii) $\Rightarrow$ (ii) $\Rightarrow$ (i) $\Rightarrow$ (iv). [D-L] Ch. V, Sec. 5. 3, Proposition 5.

2.116 *Hint:* Prove the following implications (v) $\Rightarrow$ (iv) $\Rightarrow$ (iii) $\Rightarrow$ (ii) $\Rightarrow$ (i) $\Rightarrow$ (v) and (vi), (vi) $\Rightarrow$ (ii). [D-L] Ch. V, Sec. 5. 2, Th. 1.

2.117 [G-T] 3. 8, see also 9. 4, where a similar construction is used to demonstrate sharpness of a maximum principle for quasilinear equations.

2.118 *Answer:* The function $u(x,y) = 1 + 1/\log r$, where $r^2 = x^2 + y^2$, solves the uniformly elliptic equation $(1 + 2x^2 / (r^2(-\log r - 2)))u_{xx} + (4xy/ (r^2(-\log r - 2)))u_{xy} + (1 + 2y^2 / (r^2 (-\log r - 2)))u_{yy} = 0$ for $0 < r \leq 0.1$. The coefficients of this equation are continuous everywhere and smooth off the origin. The solution attains a maximum at the origin. [P-W] Ch. 2, Sec. 9, Ex. 1; D. Gilbarg, J. B. Serrin, *J. Analyse Math.* 3, 1954–56, 309–336.

2.119 [Fa] Ch. X, Ex. 3. 2.

2.120 *Hint:* Show that there exists a harmonic function u_1 satisfying $\Delta(u - x_1 u_1) = 0$. [PPJ] 20. 23; [Tr1] Ch. IV, Ex. 14; [Fa] Ch. XII, Ex. 8.

2.121 [Ve] Ch. 5, Sec. 37.

2.122 [Fr1] Ch. 10, Problem 6.

2.123 [Bi] Ch. II, Sec. 4. 2.

2.124 *Hint*: $\Delta u = (I - \Delta)^{-1} (\Delta - \Delta^2)u \geq 0$ in $\mathscr{D}'(\mathbb{R}^n)$, hence u is subharmonic. [D-L] Ch. V, Sec. 5. 3, Remarque 2.

2.125 [G] Ch. 12, Sec. 3, Ex. 6; A. Pliś, *J. Math. Mech.* 9, 1960, 557–562.

2.126 [PPJ] 20. 8.

2.127 *Hint*: Use the maximum principle. [PPJ] 16. 22.

2.128 *Hint*: If L is large enough, then $\lambda_1 < a$, where λ_1 is the first eigenvalue of the Laplacian with a positive eigenfunction φ. Take $\delta\varphi$ as a subsolution of the considered problem, with $\delta > 0$ sufficiently small. If $z = z(x)$ is a nonconstant positive solution of the ordinary differential equation $z'' + az(1 - z) = 0$, $z(-L) = z(L) = 0$, then z is a supersolution of the problem. [S] Ch. 10, Sec. C.

2.129 *Hint*: Observe that $-\int_\Omega |\nabla u|^2 + \int_\Omega(\mu u^2 - u^4) = 0$. Use the Poincaré inequality to conclude $0 \geq -\int_\Omega u^4 \geq (\lambda - \mu) \int_\Omega u^2 > 0$ if μ is less than the first eigenvalue of $-\Delta$.
Remark: This solution is stable as a solution of the corresponding parabolic problem. It loses its stability as $\mu > \lambda$. [S] Ch. 10, Sec. C, Ch. 12, Sec. B, Example 2.

2.130 *Hint*: Consider the substitution $u = e^v$.
Answer: If $u > 0$ in D, then there exists a unique solution with a harmonic function v (this is true also in n-dimensional domains). If u changes its sign, then the solutions are not unique. To see this consider the family of functions $u(z) = |F(z)|^2 \exp(-\pi^{-1} \int_C (1 - |z|^2)|t - z|^{-2} \log|F(t)| dt)$ with an arbitrary analytic function F and D which is continuous up to the boundary and $F \neq 0$ on C. [Bi] Ch. V, Sec. 2. 2.

2.131 *Hint*: Consider $v(x) = \exp(-u(x))$.
Answer: A unique solution of the Dirichlet problem is of the form $u(x) = -\log v(x)$ for some harmonic function v. The Neumann problem can be reduced to the existence of a harmonic function satisfying $\partial v/\partial \nu + fv = 0$ on the boundary of the considered domain. It is instructive to solve completely the above problem in the one-dimensional case: $u'' - (u')^2 = 0$, $-1 < x < 1$, $u'(-1) = A$, $u'(1) = B$. A nontrivial solution exists for $B - A - 2AB = 0$ and it is real for $B > -1/2$. [Bi] Ch. V, Sec. 2. 4.

2.132 *Hint*: Use the Schauder fixed point theorem.
Remark: The boundedness condition on F can be slightly relaxed. For instance: [D-G] Ch. II, Sec. 5. 6.

2.133 *Hint*: Show the Pohozhaev-type identity $(2n/(n - 2) - 1) \int u^2 dx = (2n/(n - 2) (p + 2) - 1) \int |u|^p u^2 dx$ for the solutions of this equation. [Be] Ch. 1, Sec. 1. 2A.

2.134 *Hint*: Use the Rellich identity obtained by taking the scalar product in L^2 of the equation with $x \cdot \nabla u$. [Ba] Ch. IV, Sec. 2. 5.

2.135 [Ba] Ch. I, Sec. 3. 2, Proposition 1.6, Problem 3.

2.136 *Hint*: Observe that $\sigma^2 = (\int_{\partial\Omega} \partial u/\partial v)^2 \le \int_{\partial\Omega}(\partial u/\partial v)^2(x,v) \times \int_{\partial\Omega}(x,v)^{-1}$ and use the Pohozhaev identity. A. Krzywicki, T. Nadzieja, *Zastosowania Matematyki* 21, to appear.

2.137 *Hint*: Consider integrated form of this problem and then use the Leray-Schauder theorem. A. Krzywicki, T. Nadzieja, *Zastosowania Matematyki* 21, to appear; T. Suzuki, K. Nagasaki, *Proc. Japan Acad. Sci.* 65(A), 1989, 1–3 and 74–76.

2.138 [Ba] Ch. II, Sec. 3. 2, Proposition 2. 3.

2.139 *Hint*: Consider the function $F(x) = |\nabla u(x)|^2 + 2\int_0^{u(x)} f(s)\, ds$. A computation yields that $\Delta F - (2|\nabla u|^2)^{-1} \sum_{k=1}^n (\partial F/\partial x_k - 4f(u)\, \partial u/\partial x_k)\, \partial F/\partial x_k \ge 0$, so F satisfies an elliptic differential inequality and, hence, the maximum principle at all points where $|\nabla u| \ne 0$. Complete the argument in the case when either F assumes its maximum on the boundary $\partial\Omega$ or at a point where $|\nabla u| = 0$. L. Payne, I. Stakgold, *J. Appl. Anal.* 3, 1973, 295–306.

2.140 *Hint*: Use a real integral equation of the form $u(s,t) = U(s,t) + (2\pi)^{-1} \int \int F(x,y,u,u_x,u_y) \log((x-s)^2 + (y-t)^2)^{1/2}\, dx\, dy$, where U stands for a harmonic function. What can be said when F is not analytic but does possess continuous partial derivatives of some specific order? [G] Ch. 16, Sec. 3, Ex. 7.

2.141 *Answer*: Consider the equation $\sum_{|\alpha| \le 1} (-1)^{|\alpha|} D^\alpha a_\alpha(x, Du, D^2u) = 0$ where $a_\alpha = \partial F/\partial \xi_\alpha$, $|\alpha| \le 2$, and $F(x,Du,D^2u) = \sum_{i,j=1}^n (\partial^2 u/\partial x_i\, \partial x_j)^2 + (\sum_{i,j=1}^n (\delta_{ij} + 4(n-2)^{-1} (\partial u/\partial x_i \cdot \partial u/\partial x_j)(1 + |Du|^2)^{-1})\, \partial^2 u/\partial x_i\, \partial x_j)^2$. The function $u(x) = (x_1^2 + \ldots + x_n^2)^{1/2}$ is a weak solution of the Dirichlet problem (on the unit ball in $\mathbb{R}^n$, $n \ge 3$, with the boundary condition $u = 1$), $u \in H^2$, but $u \notin C^1$. A, Kufner, S. Fučik, *Nonlinear Differential Equations*, Elsevier, Amsterdam, 1980, Sec. 17. 7.

2.142 *Hint*: If u_1 and u_2 are two solutions with the same boundary values and if $u = \lambda u_1 + (1 - \lambda)u_2$, then the area integral $I = \int \int (1 + u_x^2 + u_y^2)^{1/2}\, dx\, dy$ is a convex function of the parameter $\lambda \in [0, 1]$, yet $I'(0) = I'(1) = 0$. In fact, the minimal surfaces equation is the Euler equation for the Plateau problem of minimizing the above area integral where the values of u are prescribed along the boundary curve. [G] Ch. 7, Sec. 1, Ex. 15.

2.143 *Hint*: (iii) without loss of generality, Φ is a convex function, hence $\xi = x + \Phi_x(x,y)$, $\eta = y + \Phi_y(x,y)$ is a diffeomorphism of the plane $\mathbb{R}^2$. Therefore $w(\xi + i\eta) = x - \Phi_x(x,y) - i(y - \Phi_y(x,y))$ is an entire function. Observe that $|w'(\xi + i\eta)| < 1$ and apply the Liouville theorem. E. Giusti,

Minimal Surfaces and Functions of Bounded Variation, Birkhäuser, Boston, 1984, 17. 1, 17. 2.

2.144 *Hint*: It suffices to show that Ker$(-\Delta + V + I)^* = \{0\}$. The Kato inequality $\Delta|\ u\ | \geq \mathrm{Re}((\mathrm{sgn}\ u)\Delta u)$ (for u, $\Delta u \in L^1_{loc}$, the inequality is interpreted in the sense of $\mathscr{D}'$) may be useful. [R-S] Ch. X, Sec. 4, Th. X. 28, X. 27; T. Kato, *Israel J. Math.* 13, 1973, 135–148.
Corollary: The operator $-d^2/dx^2 + x^2 + x^4$ is essentially selfadjoint on C_0^∞ ($\mathbb{R}$). The analogous statement is true for $-\Delta + P$ where $P(x)$ is a polynomial bounded from below on $\mathbb{R}^n$. [R-S] Ch. X, Sec. 4, Ex. 2.

2.145 *Hint*: $H_0 + mV$ increases monotonically with the quadratic form limit equal to $H_\infty : \lim_{m\to\infty}(\sup_{|\ t\ | \leq a} \|\exp(-i(H_0 + mV)t)f - \exp(-itH_\infty)f\|) = 0$ for all $f \in L^2(U)$ and $0 \leq a < \infty$. As one would expect, for large m the potential mV provides an insurmountable barrier and the wave function remains within U for all time. [D] Ex. 4. 33.

2.146 *Hint*: Take $\psi \in C^\infty(\mathbb{R}^n)$ such that $\psi(x) = |\ x\ |^{2-n}$ for $|\ x\ | \geq 1$ and $\psi(x) > 0$ everywhere. For $n \geq 5$, $\psi \in L^2(\mathbb{R}^n)$ and one may take $V = \psi^{-1}$ as the potential (and ψ as an eigenfunction). [B-S] Ch. III, Sec. 4, Remark.

2.147 [R-S] Ch. XIII, Sec. 13, Ex. 1.

2.148 *Hint*: It suffices to show that for every $\lambda > 0$ there exists κ such that $|\ V\ |^{1/2}(-\Delta + \kappa^2)^{-1}|\ V\ |^{1/2}$ has an eigenvalue greater than $1/\lambda$. The above operator is a positive, selfadjoint Hilbert-Schmidt operator, hence it suffices to prove $\lim_{\kappa\to 0} \|\ |\ V\ |^{1/2}(-\Delta + \kappa^2)^{-1}|\ V\ |^{1/2}\| = \infty$. To show this take $f \in L^2(\mathbb{R}^n)$ so that $u(x) = |\ V(x)\ |^{1/2} f(x) \geq 0$ does not vanish a.e. and estimate $(u, (-\Delta + \kappa^2)^{-1} u) = \int |\ \hat{u}(\xi)\ |^2 (\ |\ \xi\ |^2 + \kappa^2)^{-1} d\xi$. [R-S] Ch. XIII, Sec. 3, Th. XIII. 11.

2.149 *Answer*: Let $n = 4$, $u_0(x) = \exp(-|\ x\ |^2)$ and $u_k(x) = k^{1-\varepsilon}$ $\exp(1 - 1/(1 - 4k|\ x-x_k\ |^2))$ if $|\ x - x_k\ | < (2k^{1/2})^{-1}$ and 0 elsewhere, $x_k = (k, 0, \ldots)$, $\varepsilon > 0$. Consider $u = u_0 + \Sigma_{j=0}^\infty u_{k_j}$ where (k_j) is strictly increasing and $k_j \geq 2^{j/2\varepsilon}$. Put $V = u^{-1} \Delta u$. Observe that $V(x) \geq -c|\ x\ |$ and $\lim_{|\ x\ |\to\infty} |\ x\ |^{-1} u(x) = 0$ is the best possible result in this situation since $u(x_{k_j}) \geq |\ x_{k_j}\ |^{1-\varepsilon}$. A. M. Hinz, *J. reine ang. Math.* 370, 1986, 82–100. For results on the decay of eigenfunctions of Schrödinger operators with more regular potentials, see, for instance, [B-S] Ch. 3, Sec. 2; [R-S] Ch. XIII, Sec. 11; S. Agmon, *Lectures on Exponential Decay of Solutions of Second-Order Elliptic Equations,* Math. Notes 29, Princeton University Press, Princeton, 1982; *Schrödinger Operators,* Lecture Notes in Math. 1159, Springer, Berlin, 1985, 1–38.

2.150 *Hint*: Represent the solution as a regular wave packet $\psi(x,t) = \pi^{-1/2}$ $\int_0^\infty C(k)\varphi(k,x) \exp(-ik^2t)\ dk$, where $\varphi(k,x)_{xx} + k^2\varphi(k,x) = V(x)\varphi(k,x)$ are the (generalized) eigenfunctions of the stationary problem. Denote by $R(k)$ and $T(k)$

the reflection and the transmission coefficients of the scattering matrix associated with V.

Answer: $\psi(x,t) = (2t)^{-1/2}C(-x/2t)R(-x/2t)\exp(i(x^2/t - \pi)/4 + \mathbb{O}(1/t)$ if $x < -a$, $\psi(x,t) = (2t)^{-1/2}C(x/2t)T(x/2t)\exp(i(x^2/t - \pi)/4) + \mathbb{O}(1/t)$ if $x > a$.

Remark: It suffices to assume that V is a short-range potential: $V(x) = \mathbb{O}((1 + |x|)^{-1-\alpha})$, $|x| \to \infty$, and $\int (1 + |x|^{2+\alpha}) |V(x)| \, dx < \infty$ for some $\alpha > 0$. For instance: P. Biler, *Lett. Math. Phys.* 8, 1984, 1–6.

2.151 [R-S] Ch. IX, Th. IX, 31; for far-reaching generalizations see: R. Strichartz, *J. Functional Analysis* 40, 1981, 341–357.

2.152 *Hint:* The first inequality follows from the explicit formula $u(x,t) = (4\pi it)^{-n/2} \int \exp(|x - y|^2/4it)u_0(y) \, dy$. Then use the Riesz-Thorin interpolation theorem. The number $2 + 4/n$ in the third inequality is unique from the dilation invariance of the Schrödinger equation $x \mapsto \lambda x$, $t \mapsto \lambda^2 t$. The fourth inequality follows from the third by an interpolation argument. For instance: R. Strichartz, *Duke Math. J.* 44, 1977, 705–714.

2.153 [B-V] 183.

2.154 [B-V] 193.

2.155 *Hint:* $u(x,t) = (4\pi it)^{-1} \int_{\mathbb{R}^n} \exp(i|x - y|^2/4t)u_0(y) \, dy$. [Hr] Ch. A-IV, Lecture 15. For far-reaching generalizations of this result on smoothing effect for the Schrödinger equation see P. Constantin, J.-C. Saut, *Journal AMS* 1, 1988, 413–439.

2.156 *Hint:* Show that if $(1 + |x|)^\ell u_0 \in L^2$, then $(x - 2it\nabla)^k u \in L^2$ for all $k \le \ell$, hence u is locally in $H^\ell(\mathbb{R}^n)$. [St] Ch. 2. For applications of such techniques to local smoothness of solutions of nonlinear equations, see G. Ponce, *J. Differential Equations* 78, 1989, 122–135.

2.157 *Hint:* Show the identity $(d^2/dt^2) \int r^2|u|^2 \, dx = 16E(u) + 4 \int (nf(u)\bar{u} - 2(n + 2)F(u)) \, dx$. After integrating one gets $\int r^2|u|^2 \, dx \le 16E(\varphi) t^2 + c_1 t + c_0$, which is negative for sufficiently large t—a contradiction. [St] Ch. 4, Th. 1; R. Glassey, *J. Math. Phys.* 18, 1977, 1794–1797.

2.158 *Answer:* The charge $\int |u|^2 \, dx$ is a conserved quantity. The energy integral is $\int(\frac{1}{2}|\nabla u|^2 + F(u)) \, dx$, where $F(u) = \frac{1}{2}G(|u|^2)$, $G(0) = 0$, $G' = g$. From the dilation transforms one obtains

$$\frac{d}{dt} \int \left(\frac{1}{2} |xu - 2it\nabla u|^2 + 4t^2 F(u)\right) dx$$

$$= -2t \int (nf(u)\bar{u} - 2(n + 2)F(u)) \, dx$$

so the equation is conformally invariant for $f(u) = C|u|^{p-1}u$ with $p = 1 + 4/n$ (or $C = 0$). Then also $\int(\frac{1}{2} \text{Im } ru_r \bar{u} + t|\nabla u|^2 + 2tF(u)) \, dx$ is conserved. [St] Ch. 2. For more general equations see: G. Ponce, *J. Differential Equations* 78, 1989, 122–135.

2.159 [F] Ex. 1. 4. 1, Ex. 1. 6. 1: P. Brenner, *Math. Scand.* 19, 1966, 27–37.

2.160 *Hint*: $u(x,t) = \int_{\mathbb{R}^n} f(\xi) \exp(i \mid \xi \mid^2 t) \exp(ix\xi) \, d\xi$. Use techniques of maximal functions.
Remark: The best known results are: (i) sufficient conditions: $f \in H^{1/4}(\mathbb{R}), f \in H^s(\mathbb{R}^n)$ with $s > 1/2$ for $n > 1$; (ii) necessary conditions: $f \in H^s(\mathbb{R}^n)$, $s \geq 1/4$. L. Carleson, *Euclidean Harmonic Analysis,* Lecture Notes in Math. 779, Springer, Berlin, 1979, 5–45; L. Vega, *Proc. AMS,* 102, 1988, 874–878; E. Prestini, *Mh. Math.* 109, 1990, 135–143 for $n = 2$: J. Bourgain, preprint 1990.

2.161 [Fa] Ch. XI, Ex. 1.

2.162 [C] Ex. 1. 5.

2.163 [G] Ch. 12, Sec. 3, Ex. 2.

2.164 W. Fulks, *Proc. AMS* 17, 1968, 6–11.

2.165 *Hint*: First prove that u satisfies the parabolic maximum principle. Then compare u with an obvious candidate. W. Fulks, *Proc. AMS* 17, 1968, 6–11.

2.166 [Ty] Ch. III, Sec. 2, Ex. 2.

2.167 *Hint*: Consider $u(x,t) = \sum_{k=0}^{\infty} f^{(k)}(t)x^{2k}/(2k)!$ where $f \not\equiv 0$ is a smooth function supported in [0, 1] such that $\|f^{(m)}\|_{\infty} \leq C^m m^{m(1+d)}$ for some $C > 0$ and $0 < d < 1$. Apply the Denjoy-Carleman theorem on quasianalytic classes (see, for instance, W. Rudin, *Real and Complex Analysis,* McGraw-Hill, New York, 1974, Ch. 19) to establish the existence of such a function f. [Fr2] Ch. 1, Sec. 9; [J1] Ch. 7, Sec. 1; another examples can be found in: P. C. Rosenbloom, D. V. Widder, *Amer. Math. Monthly* 65, 1958, 607–609; [S] Ch. 9.

2.168 *Answer*: No. Such inequalities are satisfied by solutions of the heat equation in any compact contained in the domain of definition of the solution. Consider for instance $u(x,t) = \int_0^1 g(s) \exp(-x^2/4(t - s))(t - s)^{1/2} \, ds$ with two different functions g on [0, 1] but coinciding on [1/3, 2/3]. [V] Ch. XVI, 299.

2.169 *Answer*: $u(x,t) = \sum_{m=0}^{\infty} a^m f^{(qm)}(t)x^{pm}/(pm)!$ is formally a solution of the equation which vanishes (together with its derivatives) for $t = 0$. u satisifies the required properties if, for instance, f is a smooth function with its support in [0, 1] such that $\mid f^{(m)}(t) \mid \leq C^m m^{(1+\varepsilon)m}$ for some $\varepsilon > 0$. Taking $\varepsilon > 0$ such that $(1 + \varepsilon)q < p$ one has $\mid u(x,t) \mid \leq C_1 \exp(C_2 \mid x \mid^{\alpha})$, $\alpha = p/(p - (1 + \varepsilon)q)$, while it can be proved that solutions of the Cauchy problem are unique in the class of functions growing not faster than $C_1 \exp(C_2 \mid x \mid^{p/(p-q)})$. [Fr1] Ch. 7, Sec. 3, Remark 2 and Th. 7. 9.

2.170 *Hint*: Let $L = \Delta - \partial/\partial t$, hence the formally adjoint operator is $L^* = \Delta + \partial/\partial t$. Write the Green identity for $vLu - uL^* v$ with $v = h(y)\Gamma(x,t;y,s)$

where h is a suitable smooth cutoff function and Γ is a fundamental solution. Remember that $|\Gamma| + |\nabla\Gamma| \leq C(t - s)^{-(n+1)/2} \exp(-c|x - y|^2/(t - s))$ for some positive constants c, C. Estimate $u(x,t)$ with small $t > 0$ using the above information and assumptions.

Remark: The Hölder continuity of the coefficients is needed in the construction of Γ. [Fr2] Ch. 1, Th. 16.

2.171 *Hint:* First show the following lemma: If $Lu \leq 0$, $u(x,t) \geq -B \exp(\beta|x|^2)$, $u(x,0) \geq 0$, then $u(x,t) \geq 0$. To prove it consider $v = u \exp(-k|x|^2/(1 - \mu t) - \nu t)$ for some large positive μ, ν, and observe that $\lim \inf_{|x|\to\infty} v(x,t) \geq 0$ for $0 \leq t < 1/2\mu$. Apply the maximum principle to v which satisfies a certain parabolic equation. [Fr2] Ch. 2, Th. 10.

2.172 *Hint:* Use approximative identities but with the parabolic dilations that are appropriate to the problem; see, for example, [Se].

Remark: An analogous result can be proved for positive solutions of linear parabolic equations in divergence form (one uses the fundamental solution which is comparable to that of the heat equation, see the Aronson estimate). There is also a generalization of the abovementioned result for the porous medium equation $u_t = \Delta(u^m)$ which is a (possibly) degenerate nonlinear parabolic equation. B. E. J. Dahlberg, E. B. Fabes, C. E. Kenig, *Proc. AMS* 91, 1984, 205–212.

2.173 [J1] Ch. 7, Sec. 1, (d), Theorem; [Wi]; D. G. Aronson, *SIAM J. Math. Anal.* 12, 1981, 639–651; C. H. Wilcox, *Amer. Math. Monthly* 87, 1980, 183–186.

2.174 *Hint:* It is quite easy to construct selfsimilar solutions such that $u(x,t) = (T - t)^{-\gamma}w(y)$, where $x = (T - t)^{1/2}y$ and $\gamma > 0$. They blow up for $t \to T$, that is, $\lim_{t\to T} u(x,t) = \infty$. However, it is not hard to construct initial data of the form $\sum_{k=-\infty}^{\infty} c_k\delta_k$ such that for the solution u either
(i) $u(x, T)$ is finite for all x, or
(ii) $u(x,T) < \infty$ for $x < x_0$, $u(y,T) = \infty$ for $y \geq x_0$, or
(iii) $u(x,T) < \infty$ for $x \leq x_0$, $u(y,T) = \infty$ for $y > x_0$
and u cannot be continued beyond T.
Answer: The selfsimilar solutions with $T = 1$ satisfy the equation $\nabla \cdot (\nabla w - \frac{1}{2}yw) = (\gamma - 2/n)w$. Evidently, for $\gamma = 2/n, w = \exp(|y|^2/4)$, but for every $\gamma > 0$ there is a radial solution of the above equation, namely $w(y) = 1 + \sum_{k=1}^{\infty} |y|^{2k}\gamma(\gamma + 1) \cdots (\gamma + k - 1)/(k!4^k(n/2)(n/2 + 1) \cdots (n/2 + k - 1))$. P. Biler, *Recent Advances in Nonlinear Elliptic and Parabolic Problems,* ed. P. Bénilan, M. Chipot, L. C. Evans, M. Pierre, Research Notes in Math. 208, Pitman, Harlow, 1989, 28–38; *Colloquium Math.* 58, 1989, 85–110.

2.175 *Hint:* Use of Poisson-Fourier formula. For an example see M. Krzyżański, *Ann. Polon. Math.* 6, 1957, 288–299.

Remark: The condition with the existence of the limit of integral means is necessary and sufficient for the stabilization of u as t tends to ∞. This can be proved using the Wiener Tauberian theorem. V. D. Repnikov, *Doklady AN SSSR* 148, 3, 1963, 527–530; see also: I. V. Suveĭka, *Tauberian Theorems*, Shtinca, Kishinev, 1982, 38–41.

2.176 *Hint*: $u_0 \mapsto u(.,t)$ is a semigroup of contractions in $L^1(\mathbb{R}^n)$ generated by the convolution with a kernel satisfying certain homogeneity properties with respect to t. The same would be true assuming analyticity of the corresponding semigroup for general parabolic equation and/or the Aronson-type estimates for the associated fundamental solution (see for instance [Fr2] Ch. 1, Sec. 6, (6.12), (6.13)). Z. Brzeźniak, B. Szafirski, *Bull. Pol. Acad. Sci. ser. Math.*, to appear.

2.177 *Hint*: Derive the representation $u(y,s) = \int_{-a}^{a} u(x, 0) \, G(y,s;x,0) \, dx \pm \int_0^s u \partial G/\partial x \mid_{x=\mp a} dt$, where $G(x,t;y,s) = \sum_{m=-\infty}^{\infty} \Gamma (x + 4ma, t; y, s) - \sum_{m=-\infty}^{\infty} \Gamma (2a - x + 4ma, t; y, s)$ and Γ is the fundamental solution of the heat equation. [Fr2] Ch. 3, Problem 4.

2.178 **Remark:** Together with analyticity of u in x this implies that u is of Gevrey class 2. [J1] Ch. 7, Sec. 1, Problem, (1.46).

2.179 [Ty] Ch. XI, Sec. 4, Ex. 7.

2.180 *Hint*: This is a nontrivial fact which follows from the so-called maximal regularity result for interpolation semigroups.
Remark: An analogous property is not satisfied when the inhomogeneous heat equation is considered in the space $C^2(\mathbb{T} \times [0, 1])$. Let Φ be a C^∞ function on $\mathbb{R}^2$ such that $\Phi(x,t) = 0$ if $x^2 + |t| \geq 1$ and $\Phi(x,t) = 1$ if $x^2 + |t| \leq 3/4$. Consider the function $u_N(x,t) = \sum_{k=1}^{N} \Phi(2^k x, 2^{2k}t)(x^2 + 2t)$. The function $f_N = (u_N)_t - (u_N)_{xx}$ satisfies $\max_{\mathbb{T} \times [0,1]} |f_N| \leq M$ independently of N, whereas $\max_{\mathbb{T} \times [0,1]} |(u_N)_{xx}| \geq (u_N)_{xx}(0,0) = N$. The heat semigroup is analytic on $C(\mathbb{T})$ but it does not have the maximal regularity property. [CHADP] Ex. 6. 20, 6. 21, Th. 6. 10.

2.181 *Hint*: The function $w = \varepsilon(|x|^2 + kt)^p \exp(\alpha t)$ ($\varepsilon > 0, k > 0, \alpha > 0, 2p > q$) satisfies $Lw < 0$ for appropriate k, α. Consider $w + u$. [Fr2] Ch. 2, Problem 2.

2.182 *Hint*: Choose a constant $\varepsilon > 0$ so that $v_t - \Delta v \leq 0$ in G_{R+r} for $v = ((R + r)^2 - |x|^2 - |t|^2)^2 |\nabla u|^2 + \varepsilon u^2$. In fact, it suffices to take $\varepsilon = (2n + 10)(R + r)^2$. Apply the maximum principle to get $v \leq \varepsilon \max u^2$.
Remark: The estimate can be generalized to $|D_x^\alpha u|^2 \leq (ck^2 r^{-2})^k \max_{\overline{G_{R+r}}} |u|^2$, where $c = 2n + 10$, $|\alpha| = k$, and even $|(\partial^p/\partial t^p) D_x^\alpha u|^2 \leq n^{2p}(c(k + 2p)r^{-2})^{k+2p} \max_{\overline{G_{R+r}}} |u|^2$. [PPJ] 21. 16–21. 18.

2.183 *Hint*: Use the Bernstein method to obtain the estimate $\max |(\partial^p/\partial t^p) D_x^\alpha u| \leq cr^{-(k+2p)}(1 + (R + r)^2 + |T|)^{q/2}$ in parabolic regions $\{|x|^2 + |t| \leq R^2, t < 0\}$, $R > 0$, with $|\alpha| + 2p \geq [q] + 1$ and a constant c independent of r. Compare with the estimate in the remark to the preceding problem. [PPJ] 21. 21.

2.184 *Answer:* Consider $f(t) = f_n(t) = n^{-2} \cos(2n^2t)$ and $g(t) = n^{-1}$ $(\cos(2n^2t) - \sin(2n^2t))$. Then the solution is $n^{-2} \exp(nx) \cos(nx + 2n^2t)$ and it does not tend to zero as n goes to infinity. [C] Ch. 2, Sec. 5.

2.185 *Hint:* Construct a solution of the heat equation in $0 < x < \infty$, $0 < t$, with $u(0,t) = f(t) = \exp(-t^{-2})$ and $u_x(0,t) = 0$.
Answer: $u(x,t) = \sum_{j=0}^{\infty} f^{(j)}(t) x^{2j+1}/(2j+1)!$ is a locally bounded solution and $(4\pi t)^{-1/2} \exp(-x^2/4t)$ is another one, unbounded (in fact, this is not a distribution solution in $0 \le x < \infty$, $0 \le t$). [C] Ch. 5, Sec. 3.

2.186 *Answer:* A singular solution is $v(x,t) = xt^{-1}(4\pi t)^{-1/2} \exp(-x^2/4t)$. For $x \ne 0$, $v(x,0) = 0$ in the sense of L^1_{loc} convergence (even uniform convergence on compacts of $(0, 1]$). There exists also a unique regular solution continuous up to the boundary. For instance: [Ep] Ch. 8, Sec. 1.

2.187 *Answer:* $u \equiv 0$ and $v(x,t) = \exp(-n^2t) \sin \pi x$ solve the same problem of the considered type.
Hint: For the uniqueness proof represent a solution of the homogeneous problem as a series of solutions with separated variables. The existence is constructive, too. [C] Ch. 9, Sec. 1.

2.188 [C] Ch. 11, Sec. 6.

2.189 *Hint:* Assume that $0 \in \Omega$ and $b > 0$ is such that $\overline{\Omega} \subset D = \{ \,|\, x_j \,| \le \pi/4b, \ j = 1, \ldots, n\}$. Compare u with the function $M \exp(-nb^2 \, t)$ $\Pi_{j=1}^{n} \cos(bx_j)$ for sufficiently large M. [PPJ] 21. 14.

2.190 *Hint:* Use the Fourier separation of variables method. Observe that in the case $c < 0$ there is an eigenfunction of the stationary problem of the form $\cosh \lambda x$.
Answer: The solution decays exponentially if $c > 0$ and it grows exponentially if $c < 0$ (this case has no physical meaning).

2.191 *Hint:* Apply the maximum principle. [Fr2] Ch. 5, Problem 3.

2.192 *Answer:* Let $P_k(L,T) = \sum_{j=0}^{k}(-1)^j L^{k-j} T^j$ for $k = 0, 1, 2, \ldots$, and $L = -\Delta$, $T = \partial/\partial t$. A necessary and sufficient condition for the solution of the heat equation is $[P_k(L,T) f]_{x \in \partial G, \ t=0} = 0$. Similarly, for the wave equation $P_k(L,T)f = 0 = P_k(L,T)(\partial f/\partial t)$ for $x \in \partial G$, $t = 0$, and all $k = 0, 1, 2, \ldots$, where $T = \partial^2/\partial t^2$.
Remark: These conditions can be interpreted as compatibility conditions for f guaranteeing that u, obtained via the Fourier method, is smooth. S. Smale, *Comment. Math. Helvetici* 55, 1980, 1–12.

2.193 [C] Ch. 15, Sec. 4, Th. 15. 4. 1.

2.194 [C] Ch. 16, Sec. 4, Lemma 8. See also: L. C. Evans, R. Gariepy, *Arch. Rational Mech. Anal.* 78, 1982, 293–314; G. Lieberman, *Applicable Analysis* 33, 1989, 25–43.

2.195 [He] Ch. 1, Sec. 3, Ex. 14; Ch. 3, Sec. 3, Ex. 10.

2.196 [He] Ch. 3, Sec. 2, Ex. 1.

2.197 *Hint*: The Fourier transform U of u satisfies the equation $-\xi^2 U - \eta\, \partial U/\partial \xi = \partial U/\partial t$.
Answer: $u(t,x,y - tx) = 3^{1/2}(2\pi t^2)^{-1} \exp(-(x - x_0)^2/t + 3(x - x_0)(y - y_0)/t^2 - 3(y - y_0)^2/t^3)$. [Hö] (7. 6. 13).

2.198 *Hint*: Observe that $u(x,t) = \exp(-\pi^2 \int_0^t a(y)\, dy) \sin \pi x$.
Answer: $a(t) = -h'(t)/\pi^2 h(t)$. [C] Ch. 13, Sec. 2, Problem 1.

2.199 *Hint*: Represent the solution using the semigroup $u(x,t) = \exp(t(\Delta - q))u(x,0)$; apply the positiveness of $\exp(t\Delta)$ and the Trotter formula $L^2 - \lim_{n\to\infty} (\exp(tn^{-1}\Delta) \exp(-tn^{-1}q(x)))^n f = \exp(t(\Delta - q))f$. Use the analyticity of the semigroup $\exp(t(\Delta - q))$, the inequality $\| \nabla \exp(t\Delta) \| \leq Ct^{-1/2}$, $t > 0$, and the Sobolev embedding theorem. [R-S] Ch. X, Sec. 8, Examples 4, 5, Ex. 53, 54.

2.200 [Go] Ex. 9. 21. 6.

2.201 *Answer*: $u_1(x, t) = A_1 + B_1\Phi(x/2a_1 t^{1/2})$, $u_2(x,t) = A_2 + B_2\, \Phi(x/2a_2 t^{1/2})$, where $A_1 = U_1$, $B_1 = -U_1/\Phi(\alpha/2a_1)$, $A_2 = -U_2\Phi(\alpha/2a_2)/ (1 - \Phi(\alpha/2a_2))$, $B_2 = U_2/(1 - \Phi(\alpha/2a_2))$, and α is the root of the equation $k_1 U_1 \exp(-\alpha^2/4a_1^2)/ a_1\Phi(\alpha/2a_1) + k_2 U_2 \exp(-\alpha^2/4a_2^2)/a_2(1 - \Phi(\alpha/2a_2)) = -\rho_2^{\frac{1}{2}}\pi^{1/2}\alpha$, $\Phi(z) = 2\pi^{-1/2} \int_0^z \exp(-s^2)\, ds$. [BST] III. 117.

2.202 *Answer*: The system $u_{xx} - u_t \geq 0$, $v_{xx} - 9u_x - v_t \geq 0$ is satisfied in $\{0 \leq x \leq 1, 0 \leq t \leq 1\}$ by the functions $u = -\exp(x + t)$, $v = t - 4(x - 1/2)^2$. However, u and v are nonpositive on the parabolic boundary, while v is positive on the line $x = 1/2$. [P-W] Ch. 3, Sec. 8, Remark (ii).

2.203 *Hint*: Use the inverse transform (the Hopf-Cole transform) $u = -2\mu v_x /v$. [C] Ex. 1. 6, 1. 7.

2.204 *Hint*: The substitution $v = \exp(-ct)u$ reduces the above problem to another one with f strictly increasing in u and an operator L satisfying the maximum principle. [Fr2] Ch. 7, Th. 5.

2.205 *Answer*: Consider the equation $u_t = u_{xx} + f(x,t,u)$ where $x \in [-\pi/2, \pi/2]$, $t \in [0, \pi/4]$, and $f(x,t,u) = u - (\cos^2 x - u^2)^{1/2}$ if $| u | \leq \cos x$, $f(x,t,u) = u$ otherwise. The functions $u_1(x,t) = \cos x \cdot \cos t$, $u_2(x,t) = \cos x$ satisfy the equation and the same boundary conditions on the parabolic boundary of the considered region, but $u_1 \neq u_2$. [Sz] Sec. 67; W. Mlak, *Ann. Polon. Math.* 13, 1963, 101–103.

2.206 *Hint*: Such a solution φ is positive by the maximum principle. Observe that $\varepsilon\varphi$ is a subsolution of the stationary problem if $\varepsilon > 1$ and a supersolution if $0 < \varepsilon < 1$. The corresponding solutions u_ε of the parabolic problem satisfy $\partial u_\varepsilon/\partial t > 0$, < 0, respectively. Hence φ is not stable. [S] Ch. 10, Sec. C.

2.207 *Hint*: Show that $u(x,t) \geq 0$ on its interval of existence. Then use a continuation argument.
Remark: If $u_0(x) \leq 0$, then $u(x,t) \leq 0$, but it may blow up in a finite time. [He] Ch. 3, Sec. 3, Ex. 9; Ch. 4, Sec. 3, Ex. 4.

2.208 *Hint*: Show that $\int_\Omega |u(x,t)|^6 dx$ is a nonincreasing function. [He] Ch. 3, Sec. 6, Ex. 1; Ch. 4, Sec. 1, Example 4 (for the Dirichlet boundary conditions).

2.209 *Hint*: Consider the Liapunov function $\int_0^\pi (u_x^2 - au^2 + bu^4/2) \, dx$. [He] Ch. 1, Sec. 1; Ch. 4, Sec. 3, Example 1; Ch. 6, Sec. 2, Example 1; [Ha2] Ch. 4, Sec. 3; N. Chafee, E. Infante, *J. Appl. Anal.* 4, 1977, 17–37.

2.210 [He] Ch. 5, Sec. 4, Ex. 5.

2.211 *Hint*: Consider a smooth positive initial data such that $\int_0^\pi \sin x \, u_0(x) \, dx > 2$. By the maximum principle $u(x,t) \geq 0$, whenever the solution exists. The function $s(t) = \int_0^\pi \sin x \, u(x,t) \, dx$ satisfies $s' = -s + \int_0^\pi \sin x \, u^3(x,t) \, dx \geq -s + \frac{1}{4}s^3$. Since $s(0) > 2$, $s(t) \to \infty$ before $t = \frac{1}{2} \log(s(0) + 2)/(s(0) - 2)$.
Remark: If $\int_0^\pi (u_0^2 + (u_0)_x^2) \, dx$ is sufficiently small, then the problem has a global solution which tends to 0 as $t \to \infty$. [He] Ch. 3, Sec. 1.

2.212 Y. Giga, R. V. Kohn, *Comm. Pure Appl. Math.* 38, 1985, 297–319; *Indiana Univ. Math. J.* 36, 1987, 1–40; A. Friedman, *Nonlinear Diffusion Equations and Their Equilibrium States I,* ed. W.-M. Ni, L. A. Peletier, J. Serrin, MSRI Publ. 12, Springer, New York, 1988, 301–318.

2.213 [SGKM] Ch. I, Sec. 3, Example 3.

2.214 *Hint*: Use the parabolic maximum principle. [He] Ch. 5, Sec. 1, Ex. 8.

2.215 *Hint*: Let $v(t)$ be the solution of $v'(t) = Q(v(t))$, $t > 0$, $v(0) = \max(\sup u_0, \sup u_1)$, $v(t)$ is defined on $(0, \infty)$. Using the maximum principle one obtains $u(x,t) \leq v(t)$. [SGKM] Ch. I, Sec. 2.

2.216 *Hint*: Consider $H(t) = \int_\Omega u(x,t) \, dx$ and show that $H'(t) \geq |\Omega| Q(H(t)/|\Omega|)$, where $|\Omega|$ is the volume of Ω. [SGKM] Ch. I, Sec. 2, Example 3.

2.217 *Hint*: Consider $E(t) = \int_\Omega u(x,t)\psi_1(x) \, dx$ and show that $E'(t) \geq -\lambda_1 E(t) + Q(E)$. [SGKM] Ch. I, Sec. 2, Example. 2.

2.218 *Hint*: $u(x,t) = a(t) \sin x + b(t) \sin 2x$ is a solution of this equation provided $a' = -a + g(a,b)$, $b' = -4b + h(a,b)$. It may be proved that each solution of the parabolic problem converges as $t \to \infty$ to a solution of the ordinary differential equation $u_{xx} + f(x,u) = 0$. [He] Ch. 3, Sec. 1. For realization of any vector field in a scalar parabolic equation see P. Poláčik (*Ann. Scuola Norm. Sup. Pisa*) 18, 1991, 83–102, and the references therein.

2.219 *Hint*: Show that all the norms $\|u(.,t)\|_q$, $q = p2^m$, $m = 0, 1, 2, \dots$, are uniformly bounded in q and $t \geq 0$. To prove this consider the inequality

$d(\int_\Omega |u|^s \, dx)/dt \le -4(s-1)s^{-1}\int_\Omega |\nabla|u|^{s/2}|^2 \, dx + sB\int_\Omega |u|^s \, dx + 2A(\int_\Omega |\nabla|u|^{s/2}|^2 \, dx)^{1/2} (\int_\Omega |u|^s \, dx)^{1/2}$. Using the Gagliardo-Nirenberg inequalities estimate $\sup \|u(.,t)\|_s$ by $\sup \|u(.,t)\|_{s/2}$ and $\|u(.,0)\|_s$. [He] Ch. 3, Sec. 5, Ex. 4; N. Alikakos, *J. Differential Equations* 33, 1979, 201–225.

2.220 *Answer*: There is a connected global attractor. To study the flow on the attractor, first consider bifurcations of equilibria, that is, the equation $u_{xx} + \lambda^2 f(u) = 0$ or, after rescaling, $u_{xx} + f(u) = 0$, $u(0) = u(\lambda) = 0$. If $0 < b < 1/2$, then there exist one, two, or three solutions as $0 < \lambda < \lambda_0$, $\lambda = \lambda_0$, $\lambda > \lambda_0$, respectively, for certain λ_0. If $b = -1$ all solution bifurcate from $u = 0$ at $\lambda_n = n\pi$, $n = 1, 2, 3, \dots$. Only the first branch of stationary solutions is stable. [Ha2] Ch. 4, Sec. 3. 4.

2.221 *Hint*: The function φ satisfies the equation $\varphi'' - V\varphi' + f(\varphi) = 0$. There is a solution φ such that $\lim_{s \to -\infty} \varphi(s) = 0$, $\lim_{s \to +\infty} \varphi(s) = 1$. *Answer*: An explicit solution is $\varphi(s) = 1/(1 + \exp(-s2^{-1/2}))$, $V = 2^{1/2}(1/2 - a)$. [He] Ch. 5, Sec. 4; see also: [S] Ch. 14, Ex. 1. 2.

2.222 *Answer*: $u = -c - a\tanh(a(x + ct)/2\nu)$. [He] Ch. 5, Sec. 4, Ex. 2.

2.223 For instance: [Te] Ch. VII, Sec. 4. 3.

2.224 *Hint*: Consider this problem in the spaces $(L^2(\Omega))^n \supset \{u \in (H^2(\Omega))^n : (u, v) = 0, (\partial u/\partial v) \times v = 0 \text{ on } \Gamma\} =: D$. The difficulty is that D is not mapped into itself by the nonlinear operator $u \mapsto (u, h)^2 u - (u, h)h$. This can be overcome using differentiability properties of the nonlinearity. [Hr] Ch. A-I, Th. 14, Application; Ch. A-IV, Th. 6, 7.

2.225 [Ha2] Ch. 4, Sec. 3. 4; A. Acker, W. Walter, *Nonlinear Analysis* 2, 1978, 449–505 and Lecture Notes in Math. 564, Springer, New York, 1976; H. Levine, J. T. Montgomery, *SIAM J. Math. Anal.* 11, 1980, 842–847; H. A. Levine, *Trends in the Theory and Practice of Non-Linear Analysis,* Arlington 1984, ed. V. Lakshmikantham, Elsevier, North-Holland, 1985, 275–286.

2.226 *Hint*: Write a differential inequality satisfied by $\Lambda(t)$. The exponent σ may be determined by considering special solutions of the equation. For related equations see: [Te] Ch. III, Sec. 5, J.-M. Ghidaglia, A. Marzocchi, *Applicable Analysis*, to appear. See also J. E. Saá, *J. Math. Anal. Appl.* 155, 1991, 345–363.

2.227 *Hint*: Introduce the new independent variable $z = xt^{-1/2}(A/4P_0)^{1/2}$ and the dependent variable $w(z) = \alpha^{-1}(1 - u^2(z)/P_0^2)$, where $\alpha = 1 - (P_1/P_0)^2$. In these variables, the problem takes the form $w'' + 2z(1 - \alpha w(z))^{-1/2} w' = 0$, $w(0) = 1$, $w(+\infty) = 0$. [BSW] Ch. 7, Sec. 5, Example 7. 3; Ch. 8, Sec. 3, Example 8. 4.

2.228 *Answer*: $\alpha = \beta n$, $\beta = 1/(2 + (m-1)n)$, $M = \int_{\mathbb{R}^n} U(x,t;M) \, dx$ for all t, $B = \beta(m-1))/2m$, and A is determined from the condition on M. Observe that $\lim_{t\to 0} U(x,t;M) = M\delta_0$. The limit of $U(x,t;1)$ as $m \to 1$ is equal to the fundamental solution of the heat equation in $\mathbb{R}^n$.

Remark: $U(x,t;M)$ is a classical solution where $U > 0$. Observe that $V = m(m-1)^{-1}u^{m-1}$ corresponding to the Barenblatt solutions, that is, $V(x,t;M) = \beta(2t)^{-1}(r^2 |t| - |x|^2)_+$ where $r(t) = (AB^{-1})^{1/2}t^\beta$, is continuous and V_t, ∇V have only jump discontinuities across $|x| = r(t)$. D. G. Aronson, *Nonlinear Diffusion Problems*, ed. A. Fasano, M. Primicerio, Lecture Notes in Math. 1224, Springer, Berlin, 1986, 1–45. The importance of the Barenblatt solutions is clearly shown by the following result: If $M = \int_{\mathbb{R}^n} u(x,0)\, dx$ for a positive solution u, then $\lim_{t\to\infty} t^{n/(2+n(m-1))}| u(x,t) - U(x,t;M)| = 0$ locally uniformly in x. A. Friedman, S. Kamin, *Trans. AMS* 262, 1980, 551–563.

2.229 *Answer:* Let $p = \Delta v$. Formally one has $v_t = (m-1)\, v\, \Delta v + |\nabla v|^2$, hence $p_t = (m-1)v\, \Delta p + 2m\, \nabla v \cdot \nabla p + (m-1)p^2 + 2\sum_{i,j=1}^n (\partial^2 v/\partial x_i\, \partial x_j)^2$. Estimating $\sum_{i,j}(a_{ij})^2$ by $\sum_i(a_{ii})^2 \geq n^{-1}(\sum_i a_{ii})^2$ one obtains $L(p) \equiv p_t - (m-1) v\Delta p - 2m\, \nabla v \cdot \nabla p - C^{-1}p^2 \geq 0$. On the other hand, $L(-C/t) = Ct^{-2} - C^{-1}C^2 t^{-2} = 0$. Thus $p \geq -C/t$ and the estimate for v_t follows from $v_t = (m-1)\, v\, \Delta v + |\nabla v|^2 \geq (m-1)\, v(-C/t)$.

Corollary: If $v(x_0, t_0) > 0$, then $v(x_0, t) > 0$ for all $t > t_0$. D. G. Aronson, the reference to the preceding problem, p. 9; D. G. Aronson, Ph. Bénilan, *C. R. Acad. Sci. Paris* 288, 1979, 103–105.

2.230 *Hint:* Using the Aronson-Bénilan inequality show that $v(x + y, t) + x^2/(2(m + 1)t)$ is continuous, nonnegative, and convex for all y, $t > 0$. Then $|\varphi'(x)| \leq h^{-1}\sup_{[x-h,x+h]} |\varphi|$ and $|v_x(y,t)| \leq h^{-1}(\|v(.,t)\|_\infty + h^2/(2(m + 1)t))$. Minimize over $h > 0$. D. G. Aronson, the reference to preceding problems, p. 10.

2.231 *Hint:* Observe that $T(t + h)f - T(t)f = \lambda^{-1/(m-1)}(T(t)(\lambda^{1/(m-1)} f) - T(t)f) + (\lambda^{-1/(m-1)} - 1) T(t)f$ with $\lambda = 1 + h/t$.

Remark: If X is an ordered space, $T(t)f \geq T(t)g$ whenever $f \geq g, f, g \in X$, then moreover $(m - 1)(T(t + h)f) - T(t)f) \geq (m - 1)((1 + h/t)^{1/(m-1)} - 1) T(t)f$. For the porous medium equation this yields the Aronson-Bénilan inequality $(m - 1)u_t \geq -u/t$ in $\mathcal{D}'((0, \infty) \times \mathbb{R})$. [CHADP] Th. 4. 33, 4. 34.

2.232 *Hint:* Look for selfsimilar solutions $u(x,t) = (T - t)^{-\gamma}w(y)$ with $x = (T - t)^{1/\alpha}y$.

Answer: For instance, the Barenblatt solutions $u(x,t) = (T - t)^{-k/(m-1)}(AT^k + BT| x|^2/(T - t)^{1-k})^{1/(m-1)}$, where $k = n(m - 1)/(n(m - 1) + 2)$, $u_0(x) = (A + B| x|^2)^{1/(m-1)}$, $c = k/(2mn)$, $T = c/B, A > 0$. These are radial solutions of the equation $\gamma w + \alpha^{-1}y\cdot\nabla w = \Delta(w^m)$ for selfsimilar blowing up solutions, where $\gamma = k/(m - 1)$, $\alpha = 2/(1 - k) = n(m - 1) + 2$. Hence $z = w^{m-1}$ satisfies $\nabla \cdot (z^{1/(m-1)}(\nabla z - 2cy)) = 0$. It can be proved that the Barenblatt solutions are in a certain sense stable solutions of the latter equation. The initial values of these solutions have critical growth in x; compare the existence condition for positive solutions of the porous medium equation. P. Bénilan, M. G. Crandall, M. Pierre, *Indiana Univ. Math. J.* 33, 1984, 51–87; B. E. J. Dahlberg, C. E. Kenig, *Comm. Partial Differential Equations* 9, 1984, 409–437; and the references to the problem 2.174.

2.233 *Hint*: Use the d'Alembert formula. Conversely, set $h = 0$, divide by k^2, and add $-2u(x,t)$. Then use the Taylor expansion. [S] Ch. 3, Lemma 3. 2.

2.234 [Fa] Ch. IX, Ex. 4.

2.235 *Hint*: The general solution of the above equation is $u(x,y) = F_1(x) + F_2(y)$, so this problem is not solvable for all choices of φ's and ψ's. If the solution exists, it is not stable. [Mn] Ch. 8, Sec. 5, Example 2.

2.236 *Answer*: No. An example is given by $u(x,y,0) = \exp(-\alpha(r - 2)^2/ (r - 1)(3 - r)$ for $1 \le r = (x^2 + y^2)^{1/2} \le 3$ and 0 elsewhere. From an explicit formula $u(0,0,t) = -t\int_1^3 (t^2 - r^2)^{-3/2} \exp(-\alpha(r - 2)^2/(r - 1)(3 - r))\, r\, dr$ for $t > 3$. Thus for small $\alpha > 0$, $-u(0,0,t)$ will be arbitrarily large.
Remark: A proper formulation (the Weinstein theorem) of the maximum principle is: If $\partial(u_t - \Delta u)/\partial t \le 0$, $u_t(x,y,0) \le 0$, $u_{tt}(x,y,0) \le 0$, then $u(x,y,t) \le u(x,y,0)$, so the maximum of u is attained on the initial plane $t = 0$. [P-W] Ch. 4, Sec. 11, Th. 14.

2.237 *Hint*: Consider, for instance, polynomials f and g. Recall the Huygens principle or prove directly divergence of the series for f, $g \ne 0$ with compact support. [B-K] 364.

2.238 *Answer*: Conservation of energy $\frac{1}{2}\int(|u_t|^2 + |\nabla u|^2)\, dx$ follows from time invariance. Conservation of angular momenta $\int |x_k\, \partial u/\partial x_j - x_j\, \partial u/\partial x_k|^2\, dx$, j, $k = 1, \ldots, n$, follows from rotational invariance in $\mathbb{R}^n$. Genuinely Lorentz transformations lead to conservation of $\int |xu_t + t\nabla u|^2\, dx$. The space-time dilation invariance gives conservation of $\int |tu_t + x \cdot \nabla u + (n - 1)u|^2\, dx$. See for instance: J. Ginibre, G. Velo, *Ann. Institut Henri Poincaré, Physique th.* 47, 1987, 221–261; S. Klainerman, *Comm. Pure Appl. Math.* 38, 1985, 321–332; [St] Ch. 2.

2.239 [G] Ch. 6, Sec. 3, Ex. 11.

2.240 *Hint*: Use the Holmgren uniqueness theorem. [T] Ex. 21. 3.

2.241 *Answer*: Here the mathematical setting is the wave equation $c^{-2}\partial^2 u/\partial t^2 = \Delta u$, $z \ge 0$, supplemented with the boundary conditions $u\,|_{z=0} = \varphi(x,y,t)\, \delta(t - a^{-1}(x^2 + y^2)^{1/2})$, $u_t\,|_{z=0} = \psi(x,y,t)\, \delta(t - a^{-1}(x^2 + y^2)^{1/2})$. If u is represented as the sum $u_1 + u_2$ of solutions of the above problem with vanishing second (respectively first) condition, then the discontinuous component of u_1

(found by applying the method of characteristics) equals $v_1(x,y,z,t) = f_+(x,y,z,t)$ $\exp(S_+(x,y,z)\ \partial/\partial t)\delta(t) + f_-(x,y,z,t)\ \exp(S_-(x,y,z)\ \partial/\partial t)\delta(t)$. Here $S_\pm$ satisfy the equations $\partial S_\pm/\partial z \pm (c^{-2}(\partial S_\pm/\partial x)^2 - (\partial S_\pm/\partial y)^2)^{1/2} = 0$ and the initial conditions $S_\pm\,|_{z=0} = -a^{-1}(x^2 + y^2)^{1/2}$. For $a < c$ the derivative $\partial S_\pm/\partial z\,|_{z=0}$ is imaginary, hence no wave propagates in $\mathbb{R}^3$. A similar argument works for u_2. [B-V] 206.

2.242 T. Morley, *SIAM Review* 27, 1985, 69–71.

2.243 *Answer*: Such a differential equation is equivalent to an equation of the form $2u_{xy} + Bu_x + Cu = 0$ where B, C are functions of x, y, where $x + y$, $x - y$ represent the time and space coordinates, respectively, and where $x = $ const, $y = $ const are the characteristics. The existence of the wave family $u = g(x,y)S(y)$ requires that $g_x = 0$ holds as well as $2g_{xy} + Bg_x + Cg = 0$ and hence $C = 0$. If, in addition, a wave family $u = h(x,y)S(x)$ advancing in another direction is to exist, then $2h_y + Bh = 0$ must be satisfied together with $2h_{xy} + Bh_x = 0$ so that $B_x = 0$ follows. But the equation $2u_{xy} + B(y)u_x = 0$ is equivalent to the equation $u_{xy} = 0$. [C-H] VI. 18. 1.

2.244 [M] V. 1, V. 5.

2.245 *Hint*: Use the stationary phase method for the Fourier transform of the solution.
Remark: (ii) is the optimal result if n is even (take $u(x,0) = 0$). If $n \geq 3$ is odd then the estimate (i) holds also for $|x| \leq (1 - \varepsilon)|t|$ (the Huygens principle). Describe the situation in full details in the one-dimensional case. [R-S] Ch. XI, App. to Sec. 1, Th. XI, 19, Ex. 33.

2.246 *Answer*: In the case $n = 3$ it is easy from the standard solution formula: $u(x,t) = (4\pi t)^{-1} \int_{|y-x|=t} u_1(y)\ dS(y) = (4\pi t)^{-1} \int_{|y-x|<t} \nabla_y \cdot (t^{-1}(y - x)u_1(y))\ dy = (4\pi t)^{-1} \int_{|y-x|<t} (|\nabla u_1| + 3t^{-1}|u_1|)\ dy$ and from $\int_{|y-x|<t} |u_1|\ dy \leq ct(\int |u_1|^{3/2}\ dy)^{2/3} \leq ct \int |\nabla u_1|\ dy$. [St] Ch. 1 (13). For a proof in arbitrary dimension, see B. Marshall, W. Strauss, S. Wainger, *J. Math. pures appl.* 59, 1980, 417–440.

2.247 *Hint*: Write explicitly the solution of the (generalized) wave equation with the initial data $u(0)$, $u_t(0)$ using the spectral theorem. [Go] Ch. 2, Th. 7. 12, Ex. 7. 14. 7. See also: R. Strichartz, *J. Functional Analysis* 40, 1981, 341–357, for a more precise result on spatiotemporal localization of the gradient of the solution to the wave equation: $\lim_{t\to\pm\infty} | xu_t/|x| \pm \nabla u |_2 = 0$.

2.248 [Go] Ch. 2, Th. 7. 13, Ex. 7. 14. 6.

2.249 P. Biler, *Math. Methods in the Appl. Sci.* 12, 1990, 95–103.

2.250 [Ty] Ch. X.

2.251 [Ty] Ch. IV, Sec. 4, Ex. 3.

2.252 [Ty] Ch. VIII, (2.3), Corollary 2. 3.

2.253 *Answer:* $u(x,t) = h((x + t)/(k + 1))(1 + k)^2/2(1 + k^2) + (1 - k^2)/2(k^2 + 1)^{1/2} \int_0^{(x+t)/(1+k)} g + h((x - t)/(1 - k))(1 - k)^2/2(1 + k^2) - (1 - k^2)/2(k^2 + 1)^{1/2} \int_0^{(x-t)/(1-k)} g$. If $k = 1$, then solutions either do not exist or they are not unique.

2.254 *Answer:* The general solution is $u(x,y) = f(x) - f(y/a)$ with a C^2 function f vanishing for $x \leq 0$. [PPJ] 10. 10.

2.255 *Answer:* These conditions are: $\varphi_1(0) = \varphi_3(0)$, $\varphi_2(0) = \varphi_3'(0)$, $\varphi_2'(0) = a\varphi_3''(0)$, and then the solution is defined by $u(x,y) = \varphi_1(x) - \varphi_1(0) + \varphi_3(y) - a \int_0^x \varphi_2(t) \, dt$ if $x \geq 0$, $y > 0$, and $\varphi_1(x) + a \int_x^{y/a} \varphi_2(t) \, dt$ if $x \geq 0$, $ax \leq y \leq 0$. [PPJ] 10. 11.

2.256 *Answer:* $u(x,y) = \int_0^x f(t)J_0(2i(y(x - t))^{1/2}) \, dt + \int_0^y g(t)J_0(2i(x(y - t))^{1/2}) \, dt + (f(0) + g(0))J_0(2i(xy)^{1/2})$, where J_0 is the Bessel function and f, g are arbitrary C^1 functions. [V12] 2. 5.

2.257 [Hi] Ex. 6. 5. 1.

2.258 *Answer:* $u(x,t) = \sum_{k=0}^{\infty}(t^{(2+\alpha)k}v_k(x) + t^{(2+\alpha)k+1}w_k(x))$, where $v_k(x) = -\Delta v_{k-1}/((2 + \alpha)k((2 + \alpha)k - 1))$, $w_k = -\Delta w_{k-1}/((2 + \alpha)k((2 + \alpha)k + 1))$. [Ya] Ch. 4, Sec. 2.

2.259 *Hint:* After the change of variables $\xi = x^{1/2}$ the equation reads $\partial(\xi \, \partial u/\partial\xi)/\partial\xi = 4\xi \, \partial^2 u/\partial t^2$. Separation of variables leads to the following form of the solution $u(x,t) = \sum_{k=1}^{\infty} (A_k \cos(\mu_k t/2) + B_k \sin(\mu_k x^{1/2}))J_o(\mu_k x^{1/2})$, where μ_k are the zeros of the Bessel function J_0. [KGS] Ch. XIV, Sec. 1; [V12] 1. 17.

2.260 [Hö] Th. 7. 3. 4; [J2] Ch. V; [Fa] Ch. IX, 4°.

2.261 *Hint:* $Z_0 = \delta_0$. [T] Ex. 8. 7.

2.262 *Hint:* Use separation of variables. D. Bourgin, *Duke Math. J.* 7. 1940, 97–120.

2.263 *Hint:* (i) Use the Fourier transform and the stationary phase method. (iii) Integrate over $|x| \leq t$, where the estimate (ii) can be used, and over $|x| \geq t$, where u is bounded by an arbitrary power of t^{-1}. [R-S] Ch. XI, Sec. 1, Appendix, Th. XI, 17, Corollary.

2.264 *Hint:* Change the variable $w(x,t) = \exp(-tb/a) u(x,t)$ in order to obtain the equation $u_{tt} = a^{-1/2}u_{xx} + (b^2 - ac)a^{-1/2}u$.
Remark: If $b^2 = ac$, then, according to O. Heaviside, the line is called distortion free. The interpretation is quite obvious. [KGS] Ch. VII, Sec. 2–5; C. A. Coulson, A. Jeffrey, *A Mathematical Approach to the Common Types of Wave Motion*, 2nd ed., Longman, London, 1977, Ch. 1. 9, 1. 10.

2.265 *Answer*: The system reads $u_t = v$, $v_t = \Delta u - 2\alpha v$ in Ω, $u = 0$ on $\partial\Omega$. The operator

$$\begin{bmatrix} 0 & I \\ \Delta & -2\alpha I \end{bmatrix}$$

generates a strongly continuous group in $H_0^1(\Omega) \times L^2(\Omega)$ or in $(H^2(\Omega) \cap H_0^1(\Omega)) \times H_0^1(\Omega)$. [Ha2] Th. 4. 6. 7, 4. 6. 8.

2.266 *Hint*: Consider separately the cases $\delta^2 < 4\alpha$, $\delta^2 = 4\alpha$, $\delta^2 > 4\alpha$ (which correspond, roughly, to parabolic and hyperbolic behavior of solutions). Take $v = 2u_t$ and $v = u$ in the identity satisfied by all weak solutions $d(u_t,v)/dt + (Au,v) + \delta(u_t,v) = 0$ and consider a linear combination of the energy equations obtained.
Answer: If $\delta^2 < 4\alpha$ then $\|u(t)\|^2 = \mathcal{O}(\exp(-\delta t))$ (in fact, one shows $\|u_t\|^2 + (\alpha - \delta^2/4)\|u\|^2 \le \|u_t\|^2 + (Au, u) + \delta(u_t, u) \le C \exp(-\delta t)$). If $\delta^2 = 4\alpha$ then $\|u(t)\|^2 = \mathcal{O}(t^2 \exp(-\delta t))$ (which follows from $\|u_t + \delta u/2\|^2 \le C \exp(-\delta t)$). If $\delta^2 > 4\alpha$ then $\|u(t)\|^2 = \mathcal{O}(\exp((-\delta + (\delta^2 - 4\alpha)^{1/2})t))$ (here first one proves $\|u_t + (\delta/2 + (\delta^2/4 - \alpha)^{1/2})u\|^2 \le C \exp((-\delta + (\delta^2 - 4\alpha)^{1/2})t))$. A generalization for nonlinear equations is in P. Biler, *Nonlinear Analysis* 10, 1986, 839–842.

2.267 *Hint*: Consider the evolution in time of the quantity $\|u_t + Bu - \rho u\|^2$ for certain $\rho > 0$. Write a suitable energy equation. Remember that $A = f(C)$, $B = g(C)$ for a selfadjoint positive operator C.
Answer: $\|u(t)\|^2 = \mathcal{O}(\exp(-\sigma t))$, where $\sigma = \min_{\xi \ge \zeta} g(\xi)(1 - \text{Re}(1 - 4f(\xi)/g^2(\xi))^{1/2})$ and $f(\zeta) = \alpha$, $g(\zeta) = \beta$, ζ is the least eigenvalue of C. Compare: P. Biler, *Nonlinear Analysis* 11, 1987, 841–849, and *Applicable Analysis* 32, 1989, 277–285, for similar nonlinear equations.

2.268 *Hint*: If $\xi = xy$, $\eta = y/x$, then the equation takes the form $2\xi \, \partial^2 u/\partial\xi \, \partial\eta = \partial u/\partial\eta$.
Answer: $u(x,y) = \Phi(xy) + x\Psi(y/x)$ with arbitrary functions Φ, Ψ. [V] Ch. XV, Sec. 288, 289.

2.269 *Hint*: Use the method of separation of variables or construct the Riemann functions.
Remark: The substitution $w = x^{\lambda/2}u$ transforms this equation into the selfadjoint equation $w_{xx} - w_{tt} + \frac{1}{4}\lambda(2 - \lambda)x^{-2} w = 0$. [G] Ch. 4, Sec. 4, Ex. 5, 6.

2.270 *Answer*: $u(x,t) = t^{1-n}\sigma_n^{-1} \int_{|y| = t} v(x - y) \, d\sigma(y)$ where the integration is over the $(n - 1)$-dimensional sphere. [C–S] Ch. 1, Th. 2. 1; [J2] Ch. V.

2.271 [Go] Ch. 2, Ex. 15. 26. 11.

2.272 [Sm2] Ch. I, Sec. 1, Example 2.

2.273 [Sm1] 31.

2.274 [Sm1] 32.

2.275 *Answer:* In the first case $u(x,y) = \partial^{\alpha+\beta-2}/\partial x^{\alpha-1} \partial y^{\beta-1})((F_1(x) - F_2(y))/(x - y))$; in the second case $u(x,y) = (y - x)^{1-\alpha-\beta} \int_0^1 F_1(x + (y - x)t) t^{-\alpha}(1 - t)^{-\beta} dt + \int_0^1 F_2(x + (y - x)t) t^{\beta-1}(1 - t)^{\alpha-1} dt$, where F_1 and F_2 are arbitrary functions. [Sm1] 33, 34; [V12] 2. 10.

2.276 [R-Y] Ch. I, Sec. 13.

2.277 *Hint:* The general solution of this system is $u_1(x,y) = (x + y)\varphi(x - y) + (x - y)\varphi_1(x + y) + \psi(x - y) + \psi_1(x + y)$, $u_2(x,y) = (x + y)\varphi(x - y) - (x - y)\varphi_1(x + y) + \psi(x - y) - \psi_1(x + y)$, where φ, φ_1, ψ, and ψ_1 are arbitrary twice-differentiable functions. [Bi] Ch. III, Sec. 2. 1.

2.278 [G] Ch. 3, Sec. 4, Ex. 6.

2.279 [G] Ch. 4, Sec. 2, Ex. 5.

2.280 *Answer:* This example is due to S. G. Piatkov; $u(x,t) = 24/(1-8(t-x/2))$ is a solution with smooth initial data $u(x,0) = 24/(1 + 4x)$, $u_t(x,0) = 192/(1 + 4x)^2$ and a discontinuity on the line $1 - 8t + 4x = 0$. [LNY] Ch. 4, Sec. 1. 8.

2.281 [V12] 14. 61; [Fs] Part V, Sec. 229.

2.282 *Hint:* Write the equation as a first-order system for the vector $\langle u_x, u_t\rangle$ and find solutions with $u_x = \vartheta$, $u_t = F(\vartheta)$.
Answer: $u = \pm\frac{1}{2}\vartheta^2 + h(x \pm (1 + \vartheta)t) + ct$ where c is a constant and ϑ is the solution of $\vartheta = h'(x \pm (1 + \vartheta)t)$. [J1] Ch. 2, Sec. 6, Problem.

2.283 **Remark:** The first invariant is, of course, the energy; the second one is called the conformal charge. There is a similar invariant for the equation $u_{tt} - \Delta u + |u|^{4/(n-1)} u = 0$ in $\mathbb{R}^n \times \mathbb{R}$ (which is also conformally invariant!).
Hint: One may use invariance properties of the equation under some groups of (the Lorentz) transformations and the Noether theorem. For instance, [R-S] Ch. XI, Sec. 13, Th. XI, 101, App. to Sec. 13. For a discussion of the conformal invariants and their applications to the asymptotic behavior of solutions see J. Ginibre, G. Velo, *Ann. Institut Henri Poincaré, Physique th.* 47, 1987, 221–261.

2.284 *Hint:* If, more generally, one considers $u_{tt} - \Delta u + f(u) = 0$ and $F' = f$, $F(0) = 0$, then the result follows from the dilation identity

$$\frac{d}{dt} \int \left(t\left(\frac{1}{2}u_t^2 + \frac{1}{2}| \nabla u |^2 + F(u)\right) + ru_r u_t + (n - 1)\frac{uu_t}{2} \right) dx =$$
$$-\frac{1}{2} \int ((n - 1)uf(u) - 2(n + 1)F(u)) dx$$

See the references to the preceding problem and [St] Ch. 2.

2.285 *Hint:* Show the identity

$$\frac{d^2\left(\int \frac{1}{2}u^2\,dx\right)}{dt^2} = \left(2 + \frac{\varepsilon}{2}\right)\int u_t^2\,dx + \frac{\varepsilon}{2}\int |\nabla u|^2\,dx +$$

$$\int\left(\left(2 + \frac{\varepsilon}{2}\right)F(u) - uf(u)\right)dx - \left(2 + \frac{\varepsilon}{2}\right)E(u)$$

Then deduce the inequality $II'' \geq (1 + \varepsilon/4)(I')^2$ where $I = \int \frac{1}{2}u^2\,dx$. This implies $J(t) \leq J(0) + tJ'(0)$ with $J = I^{-3/4}$ so $J(T) = 0$ for some $T > 0$ (as $J'(0) < 0$). This shows $\int u^2\,dx \to \infty$ as $t \to T$. [St] Ch. 4, Th. 2; H. Levine, *Trans. AMS* 192, 1974, 1–21.

2.286 *Hint:* Consider $F(t) = \int_{\mathbb{R}} |u(x,t)|^2\,dx$ and show that $F^{-\alpha}(t)$ is concave for some $\alpha > 0$ (for instance, $\alpha = 1/2$). Hence $F(t)$ would be negative, which is absurd.
Answer: If $\int u(x, 0)u_t(x, 0)\,dx < 0$ and $E(0) < \infty$, then any solution cannot be global in time. [R-S] Ch. X, Sec. 13; M. Balabane, M. Frisch, *Portugaliae Matematica,* to appear.

2.287 *Hint:* Show that φ must be positive on $[0, T)$ and then $\varphi'(2H(\varphi) + c)^{-1/2} \geq 1$ for $c = (\varphi'(0))^2 - 2H(\varphi(0))$. Then integrate. R. T. Glassey, *Math. Zeit.* 177, 1981, 323–340 and the last reference to the preceding problem.

2.288 *Hint:* Consider $F(t) = \int_{\mathbb{R}} u(x,t)^2\,dx$. Choose u_0, u_1 so that $(F(t)^{-\alpha})'' \leq 0$ for all $t \geq 0$ and $(F(t)^{-\alpha})' < 0$ for $t = 0$, $\alpha = (k - 1)/4$. $F(t)$ blows up in finite time if $E(0) = \frac{1}{2}\int(u_x^2 + u_t^2)\,dx - (k + 1)^{-1}\int u^{k+1}\,dx < 0$. [R-S] Ch. X. Sec. 13, Example; Ph. Korman, *Nonlinear Systems of Partial Differential Equations,* Part 2, Lectures in Applied Math. 23, ed. B. Nicolaenko, D. D. Holm, J. M. Hyman, AMS, Providence, 1986.

2.289 [K-C] Ch. 3, Sec. 1, Problem 1.

2.290 For instance, [Te] Ch. VII, Sec. 4. 3.

2.291 *Answer:* The bifurcation diagram is essentially the same as this for the parabolic equation $u_t - u_{xx} = \lambda f(u)$ ([Ha2] Ch. 4, Sec. 3. 3). For $n^2 < \lambda < (n + 1)^2$ there are exactly $2n + 1$ equilibrium points and orbits connecting some of them. For large α the flow on the global attractor is equivalent to the flow for the parabolic equation. [Ha2] Ch. 4, Sec. 7. 2.

2.292 *Answer:* For example, $\exp(te^{i\varphi}\Delta)$ with $0 < |\varphi| < \pi/2$ considered as the semigroup on $L^2(\mathbb{R}^n)$ has the norm $1/\cos\varphi$ for $t > 0$. This corresponds to the complex time $te^{i\varphi}$ in the heat equation on $\mathbb{R}^n$. Another example is given by the matrices

$$\begin{bmatrix} \cos t & 2\sin t \\ -(1/2)\sin t & \cos t \end{bmatrix}$$

on $\mathbb{R}^2$. Here, however, the norms of the operators in the semigroup tend to 1 as t tends to 0. [R-S] Ch. X, Sec. 8, Example 5.

2.293 *Answer:* The classical example of Lumer and Phillips is $Au = u'$, where $X = C[0, 1] \cap \{u(0) = u(1) = 0\}$ and $D(A) = X \cap \{u : Au \in X\}$. It is easy to see that $(I - A)D(A) \neq X$. [F] Ex. 3. 6. 5; G. Lumer, R. S. Phillips, *Pacific J. Math.* 11, 1961, 679–698.

2.294 [R-S] Ch. X, Ex. 79.

2.295 *Hint:* It is tempting to put $(Af)(x) = -f'(x) - V'(x)f(x)$. However, the function V need not be continuously differentiable, and if it is not there exist functions $f \in C_0^\infty$ which do not lie in $D(A)$. [D] Ex. 1. 11, Problem 1. 12.

2.296 *Answer:* Consider the semigroup $(T(t)u)(x) = u(t + x)$, $t, x \in \mathbb{R}$, in the space of bounded continuous functions on the real line $BUC(\mathbb{R})$. The Favard class of $T(t)$ is equal to $Lip(\mathbb{R}) \cap BUC(\mathbb{R})$, while the infinitesimal generator is defined for u such that $u'(x)$ exists for all $x \in \mathbb{R}$ and $u' \in BUC(\mathbb{R})$. [CHADP] Ch. 3, Sec. 4.

2.297 [Go] Ch. 2, Ex. 15. 26. 8.

2.298 *Answer:* No. However, $(t, f) \mapsto T(t)f$ is jointly continuous from $(0, \infty)$ $C[0, 1]$ to $C[0, 1]$.
Hint: Show that $T(t)f(0) = 0$ for all f and $t > 0$ and $\lim_{t\to 0} \|T(t)\| = \infty$. [D] Ex. 1. 26.

2.299 *Hint:* For some $t_n \to \infty$ $P_{t_n}f$ converges weakly to f_0, that is, a fixed point of P. Write $f \in L^1$ in the form $f = f - f_0 + f_0$ and show (using the Hahn-Banach theorem) that for every $\varepsilon > 0$ the function $f - f_0$ may be written as $f - f_0 = Pg - g + r$, where $g \in L^1$ and $\|r\| \leq \varepsilon$. A. Lasota, M. C. Mackey, *Probabilistic Properties of Deterministic Systems,* Cambridge University Press, Cambridge, 1985.

2.300 *Hint:* Assume that $|f| \leq cf_0$ for some $c > 0$. Show that $e^{t(P-I)}f$, $t \geq 0$, is weakly precompact and use the result of the preceding problem.

2.301 *Hint:* The generator A of $T(t)$ is given formally by $(Af)(x) = (ix^2 - \log(1 + x^2))f(x)$. Although $T(t)$ is norm continuous for $0 < t < \infty$, for a general $f \in C_0(\mathbb{R})$, $T(t)f$ lies in $D(A^n)$ only for $n < t < \infty$.
Remark: As is well known (for instance [D], Th. 1. 28) $T(t)f \in D(A)$ for $0 < t < \infty$ implies C^∞ smoothness of $T(t)$ for $0 < t < \infty$. [D] Ex. 1. 29.

2.302 [CHADP] Ex. 5. 22.

2.303 *Hint*: $S(z) = \sum_{n=0}^{\infty} (z - t)^n S^{(n)}(t)/n! = \sum_{n=0}^{\infty} (z - t)^n t^{-n} n^n (n!)^{-1}$ $(tn^{-1}S'(tn^{-1}))^n$ and this series converges for $|z - t|/t < 1 + \delta$ for some $\delta > 0$. Hence $\lim_{t\to 0} \|S(t) - I\| = 0$ and A must be bounded. For an example consider $S(t) a = (e^{-nt}a_n) \in \ell^2$. [Pa] Ch. 2, Th. 5. 3, Example 5. 4.

2.304 *Hint*: Analyticity is equivalent to the following condition: there exist a complex number ζ, $|\zeta| = 1$, and positive constants δ, ε such that $\|(\zeta I - S(t))f\| \geq \varepsilon \|f\|$ for every f and $0 < t < \delta$. [Pa] Ch. 2, Corollary 5. 7, Th. 5. 6c).

2.305 *Answer*: Define the one-parameter semigroup $T(t)$ on the space of continuous functions on $[0, \infty)$ vanishing at 0 and ∞ by $(T(t)f)(x) = f(x - t)$ if $0 \leq t \leq x$ and $= 0$ if $t > x$. The function $(T(t)f)(a) = (T(t)f, \delta_a)$, where $f(a) \neq 0$, cannot be analytically extended to any neighborhood of $\mathbb{R}^+$. [D] Ex. 1. 50. See also [R-S] Ch. X, Sec. 6, for criteria of existence of analytic and entire vectors.

2.306 [CHADP] Ex. 5. 19, 5. 20.

2.307 *Answer*: For instance, $S(t)f(x) = f(x + t)$ for f in the space $\{f: [0, \infty) \to \mathbb{R} : \|f\|_p + \int_0^\infty/e^t |f(t)| \, dt < \infty\}$, $1 < p < \infty$. [Pa] Ch. 4, Example 4. 2.

2.308 [F] Example 3. 5. 6.

Remark: It can be proved that, under the assumption $|A_k(x)| \leq \rho(|x|)$ for an increasing function ρ, $\int_1^\infty \rho(r)^{-1} \, dr = \infty$, the closure of the operator $L = \sum_{k=1}^n A_k(x) \, \partial/\partial x_k + B(x)$ restricted to $C_0^\infty(\mathbb{R}^n)$ is equal to the adjoint of its formal adjoint $(L')^*$. The dissipativity assumption is $\mathrm{Re}(B(x) - \frac{1}{2}\sum_{k=1}^n (\partial A_k/\partial x_k)(x)) \leq 0$, but it is irrelevant here. [F] Th. 3. 5. 2.

2.309 [D] Th. 2. 3. 1; [Go] Ch. 1, Sec. 9. 2; [He] Ch. 1, Sec. 4; [Pa] Ch. 2, Sec. 6; K. Yosida, *Functional Analysis*, Springer, Berlin, 1965, Ch. IX, Sec. 11.

2.310 T. G. Kurtz, M. Pierre, *J. Differential Equations* 52, 1984, 407–414.

2.311 *Answer*: $C_0(t) = \frac{1}{2}(T(t) + T(-t))$, that is, the well-known d'Alembert formula; $C_a(t)x = C_0(t)x + at \int_0^t (t^2 - s^2)^{-1/2} I_1(a(t^2 - s^2)^{1/2} C_0(s)x \, ds$, where is the modified Bessel function of order 1. [Go] Ch. 2, Sec. 8. 11, Remark.

2.312 [Go] Ch. 2, Th. 8. 7.

2.313 *Answer*: The Kaniel example is: B is the diagonal operator multiplying the kth coordinate by $ik + \delta$. The function $u(t) = \sum_{k=1}^\infty e^{ikt} \varphi_k$, where φ_k is the eigenvector of B, solves the equation $u_t + (B - \delta)u = 0$ and it is nondifferentiable. It suffices to construct a function f which is compact and C^1 such that $f = \delta I$ on the set $\{\sum_{k=1}^\infty b_k \varphi_k, |b_k| \leq 1/k, k = 1, 2, \ldots\}$. To do this consider a smooth function ζ on $\mathbb{R}^+$, $\zeta(r) = r$ if $0 \leq r \leq 1$, ζ' decreasing to zero on $[1,2]$, $\zeta(r) = \zeta(2)$ for $r \geq 2$, and let $f(\sum_{k=1}^\infty a_k \varphi_k) = \delta \sum_{k=1}^\infty k^{-1} \zeta(ka_k)\varphi_k$. [Ha2] Proposition 3. 10. 3.

2.314 *Hint*: If $u(t) \neq 0$, then $(\log \|u(t)\|^2)'' = -4\|u(t)\|^{-4}(Au(t), u(t))^2 + 4\|u(t)\|^{-2}\|Au(t)\|^2$. [F] Ex. 6. 6. 1.

Remark: More general results on logarithmic convexity for second-order equation like $Pu_t + Mu_t + Nu = 0$ can be found in H. Levine, *Math. Zeit.* 126, 1972, 345–360.

2.315 *Hint*: Consider a differential inequality satisfied by $\Lambda(t) = (A\varphi, \varphi)/\|\varphi\|^2$. See for this abstract result and earlier examples: J.-M. Ghidaglia, *J. Differential Equations* 61, 1986, 268–294. For another applications see P. Biler, *Nonlinear Analysis* 11, 1987, 841–849 and *Applicable Analysis* 32, 1989, 277–285.

2.316 *Hint*: Observe that for $\Lambda(t) = \|A^{1/2}w(t)\|^2 / \|w(t)\|^2$ one has the inequality $\Lambda' \leq 2k^2\Lambda$. Argue by contradiction, assuming that $w(t_0) \neq 0$ for some t_0, and consider $\log\|w(t)\|$. [Te] Ch. III, Sec. 6, Lemma 6. 2; C. Bardos, L. Tartar, *Arch. Rational Mech. Anal.* 50, 1973, 10–25; J.-M. Ghidaglia, *Nonlinear Analysis* 10, 1986, 777–790.

Remark: Such results are applicable to a large class of (nonlinear) parabolic problems and they are useful in a study of asymptotic behavior of solutions as $t \to \infty$ via dynamical systems methods.

2.317 *Hint*: Change the independent variable putting $s = \int_{t_0}^t dt/c(t)$, $t > 0$. Consider separately the cases: (i) $\int_{t_0}^{0^+} dt/c(t)$ finite. (ii) this integral diverges but there is no purely imaginary element in the spectrum of $A - f_x(0)$. [He] Ch. 5, Sec. 2, Ex. 1; [C-S].

2.318 *Hint*: Consider the characteristics $bx - ay = $ const. [K-C] Ch. 4, Sec. 1. 1.

2.319 *Hint*: The equation in polar coordinates takes the form $\varepsilon(u_{rr} + u_r/r + u_{\vartheta\vartheta}/r^2) = ru_r$. Then introduce the transformation $v(r,\vartheta) = \exp(-r^2/4\varepsilon)u(r,\vartheta)$ to eliminate the first derivative term. Seek for a variational formulation for the new variable v. [K-C] Ch. 4, Sec. 1. 1.

2.320 [K-C] Ch. 4, Sec. 1, Problem 1.

2.321 *Answer*: $u(x,t;\varepsilon) = F(x - at/b) + \varepsilon(b^{-1}(1 - a^2/b^2)tF''(x - at/b) + ab^{-2}F'(x - at/b) + b^{-1}G(x - at/b)) + \mathbb{O}(\varepsilon^2)$ as $\varepsilon \to 0$. [K-C] Ch. 4, Sec. 1. 1.

2.322 [G] Ch. 5, Sec. 1, Ex. 14. For a more general equation $\varepsilon^2 u_{tt} + u_t = u_{xx} + cu + f$ see, for instance, W. Fulks, R. B. Guenther, *Czechoslovak Math. J.* 21, 1971, 683–695.

2.323 *Hint*: The nonlinear term can be eliminated by setting $u = -2\varepsilon v_x/v$. [K-C] Ch. 4, Sec. 1. 3.

2.324 [V12] 14. 56, 14. 59.

2.325 *Hint*: Represent $u_\varepsilon(t)$ as $S(t/\varepsilon)u(0) + \varepsilon^{-1}\int_0^t S(s/\varepsilon)f(t - s)\,ds = S(t/\varepsilon)u(0) + \varepsilon^{-1}\int_0^t S(s/\varepsilon)(f(t - s) - f(t))\,ds + S(t/\varepsilon)A^{-1}f(t) - A^{-1}f(t)$.

Remark: It can be shown that convergence is uniform on compact subsets of $(t(\varepsilon), \infty)$ if $t(\varepsilon)/\varepsilon \to \infty$ as $\varepsilon \to 0$. [F] Example 5. 9. 2; [Kn] Ch. 4.

2.326 *Hint:* Use the Fourier transformation. [F] Ex. 5. 9. 4.

2.327 [Go] Ch. 2, Th. 11. 2.

2.328 [Go] Ch. 2, Th. 11. 3.

2.329 [Go] Ch. 2, Example 11. 8. For some related results for nonlinear equations and systems see: S. H. Schochet, M. I. Weinstein, *Comm. Math. Phys.* 106, 1986, 569–580; P. Biler, *SIAM J. Math Anal.* 21, 1990, 1190–1212.

2.330 *Answer:* The integral surfaces are $a^2/x + b^2/y + c^2/z = k$. [Sn] Ch. 1, Sec. 6, Example 8.

2.331 *Answer:* The solution is determined by the relation $xyz = c(x+y+z)$. [Sn] Ch. 1, Sec. 6, Example 10.

2.332 *Answer:* The coordinates of N are $(x + pz, y + qz, 0)$, $p = z_x$, $q = z_y$, hence $\tan NOP = ((y + qz)/(x + pz) - y/x)/(1 + y(y + qz)/x(x + pz)) = \tan \varphi$ $=$ const. In other words $(\cos \varphi) z(qx - py) = (\sin \varphi)(x^2 + y^2 + z(px + qy))$. The characteristics are described by the equation $dx/(-x \sin \varphi - y \cos \varphi) = dy/(x \cos \varphi - y \sin \varphi) = z\, dz/(x^2 + y^2) \sin \varphi$ and the general solution is given by $\arctan(y/x) = -(\cotan \varphi) \log(x^2 + y^2)^{1/2} + f(x^2 + y^2 + z^2)$ with an arbitrary function f. [Ju] 11.

2.333 *Hint:* Observe that $(x - a)(q - y) = (y - b)(p - x)$. Hence in the new variables $X = x - a$, $Y = y - b$, $Z = z -(x^2 + y^2)/2$, $P = \partial Z/\partial X = p - x$, $Q = \partial Z/\partial Y = q - y$ one has $XQ - YP = 0$, that is, the equation of surfaces of revolution $Z = f(R)$, $X = R \cos \varphi$, $Y = R \sin \varphi$.
Answer: After determination of f one obtains $(z - ax - by + c)^2 = h(2z - x^2 - y^2)$ with a constant h. [Ju] 29.

2.334 *Answer:* $zp/x = $ const, $zq/y = $ const are the first integrals of the associated system $dx/(p/x^2) = dy/(q/y^2) = dz/(p^2/x^2 + q^2/y^2) = -dp/(-p^2/x^3 + 2p/z^3) = -dq/(-q^2/y^3 + 2q/z^3)$. Let $p/x = 2^{1/2} \cos \varphi/z$, $q/y = 2^{1/2} \sin \varphi/z$, so $2^{-1/2}z\, dz = (\cos \varphi)x\, dx + (\sin \varphi)y\, dy$. The general solution is obtained as the envelope of the surfaces $2^{-1/2}z^2 = x^2 \cos \varphi + y^2 \sin \varphi + h(\varphi)$ with an arbitrary function h. [Ju] 46.

2.335 *Hint:* The trajectories orthogonal to the surface $\vartheta(x,y,z) = C$ are determined by the differential equations $dx/\vartheta_x = dy/\vartheta_y = dz/\vartheta_z$. $(\vartheta_x, \vartheta_y, \vartheta_z)$ is constant as $d\vartheta_x = \vartheta_{xx}\, dx + \vartheta_{xy}\, dy + \vartheta_{xz}\, dz = (\vartheta_{xx} \vartheta_x + \vartheta_{xy} \vartheta_y + \vartheta_{xz} \vartheta_z)\, d\vartheta \equiv 0$. [Ju] 53.

2.336 *Answer:* An integral curve is the envelope of characteristics: $V(x,y,z,a) = b$. The characteristics are represented by this equation with a relation between a and b: $V_a(x,y,z,a) = c$. So the integral curve is the envelope of the characteristics $V(x,y,z,a) = \varphi(a)$, $V_a(x,y,z,a) = \psi(a)$ and its coordinates (x,y,z) are determined from the relations $V_a(x,y,z,a) = \varphi'(a)$, $V_{aa}(x,y,z,a) = \psi'(a)$. Note that $\psi(a) = \varphi'(a)$ is the compatibility condition. If one replaces x, y, z in $V(x,y,z,a_0) = b_0 = \varphi(a_0)$ by their parametrizations by a, then one gets a function $H(a)$ on the right-hand side. Second-order tangency mean $H'(a_0) = H''(a_0) = 0$. Since $H'(a_0) = V_x(x,y,z,a_0)x_a + V_y(x,y,z,a_0)y_a + V_z(x,y,z,a_0)z_a$, and (differentiating $V = \varphi(a)$) $V_x x_a + V_y y_a + V_z z_a = 0$, so $H'(a_0) = 0$. Now $H''(a) = V_x x_{aa} + V_y y_{aa} + V_z z_{aa} + V_{xx}(x_a)^2 + 2V_{xy}x_a y_a + \cdots + V_{zz}(z_a)^2$ and $V_x x_{aa} + V_y y_{aa} + V_{xx}(x_a)^2 + 2V_{xy}x_a y_a + \cdots + V_{zz}(z_a)^2 + V_{xa}x_a + V_{ya}y_a + V_{za}z_a = 0$. The last three terms vanish (to see this differentiate $V_a = \varphi'(a)$ and use $V_{aa} = \varphi''(a)$), which proves the theorem. [Ju] 55.

2.337 *Answer*: These are helical surfaces $x_1 = t_1$, $x_2 = t_2$, $x_3 = \text{sgn } t_2 \cdot \arccos(t_1/(t_1^2 + t_2^2)^{1/2}) + \text{const.}$

2.338 [Fa] Ch. VII, Ex. 2. 2.

2.339 [K] Ch. VII, III, Sec. 9, Ex. 10.

2.340 [G] Ch. 2, Sec. 1, Ex. 2.

2.341 *Answer*: $u = \arctan(x/y) + \int (u((a/c)^2 - 1) - a^2)^{1/2}(2u)^{-1} \, du + b$. [V] Ch. XV, 277.

2.342 *Answer*: $u(x,y) = 2x^{3/2}(1 - 27y)^{-1/2}$, but it is not a regular solution. [J1] Ch. 1, Sec. 9, Ex. 2.

2.343 *Answer*: $u(x,y) = 1 - (1 - 2x - 2y)^{1/2}$. [Fo] Ch. 1, Sec. B, Ex. 4.

2.344 *Hint*: The characteristics (lines with the tangent coefficient equal to $1/u(x)$) meet at $x = 0$, $t = 1$. [Ty] Ch. IV, Sec. 5, (5.34); see also Sec. 7, Ex. 6–20.

2.345 [S] Ch. 15, Sec. B.

2.346 [Sv] Sec. 7. 5; [B-W] 9. 199.

2.347 *Hint*: Suitable weak solutions of the first-order equation $u_t + f(u)_x = 0$ can be approximated by (more regular) solutions of the generalized Burgers equation $u_t + f(u)_x = \varepsilon u_{xx}$, $\varepsilon > 0$. For the latter equation observe that $\int u_{xx} \text{ sgn } u = \lim_{t \to 0} t^{-1}(\exp(t \, \partial^2/\partial x^2) u - u) \text{ sgn } u \le 0$ since $\int(\exp(t \, \partial^2/\partial x^2)u - u) \text{ sgn } u \le \int |\exp(t \, \partial^2/\partial x^2)u| - \int |u| \le 0$ from the L^1 contraction property of the (linear) semigroup associated with the heat equation. Finally let $\varepsilon \to 0$. This formal argument can be made precise as in: B. Keyfitz-Quinn, *Comm. Pure Appl. Math.* 24, 1971, 125–132.

2.348 **Remark:** This is not true in the case of nonconvex nonlinearity f. P. D. Lax, *Comm. Pure Appl. Math.* 10, 1957, 537–566, Th. 2. 1, Th. 2. 2; L. Tartar, *Nonlinear Analysis and Mechanics, Heriot-Watt Symp. vol. IV*, R. J. Knops ed., Research Notes in Math. 39, Pitman, 1979, 136–212, Th. 26.

2.349 *Hint*: Localize u_k, v_k by multiplying them by a smooth compactly supported function. Pass to their Fourier transforms. Use the following identity: $\xi_j(W,T) = \xi_j \sum_{i=1}^n W_i \overline{T_i} = \overline{T_j} \sum_{i=1}^n \xi_i W_i + \sum_{i=1}^n \xi_i (T_{i-} - \xi_i \overline{T_j}) W_i$ for every $W, T \in \mathbb{C}^n$, $\xi \in \mathbb{R}^n$.

Remark: This theorem (the so-called div-curl lemma) and its generalizations and ramifications have numerous applications to the analysis of hyperbolic type problems. F. Murat, *Ann. Scuola Norm. Sup. Pisa* 5, 1978, 489–507.

2.350 *Hint*: The question is: Does $f(u^\varepsilon) \rightharpoonup f(u)$ in L^∞ weak*?
Answer: In general no, unless f is an affine function. However, if some additional properties are imposed for a solution (the so-called entropy conditions), then the answer is positive. L. Tartar, the reference to 2.348, Th. 26.

2.351 *Hint*: $u(x,t) = \exp(\lambda t)v(x \exp(-t))$, $0 \le x < \infty$. A. Lasota, *Rend. Sem. Mat. Univ. Padova* 61, 1979, 39–48. For further development of invariant measure and ergodicity concepts for first-order partial differential equations see: A. Lasota, *Nonlinear Analysis* 5, 1981, 1181–1193; P. Brunovský, *ibidem* 7, 1983, 167–176; K. Łoskot, *J. Differential Equations* 58, 1985, 1–14; R. Rudnicki, *Ergodic Th. Dynam. Sys.* 5, 1985, 437–443; *J. Math. Anal. Appl.* 133, 1988, 14–26.

2.352 [G] Ch. 4, Sec. 3, Ex. 8.

2.353 *Hint*: Verify that $u(x,t) = z((t - x + 1)/2)$, $v(x,t) = z((t + x + 1)/2)$, $0 \le x \le 1$, $t \ge 0$. Yu. L. Maĭstrenko, E. Yu. Romanenko, A. N. Sharkovskiĭ, *Mathematical Mechanisms of Turbulence*, Inst. Mat. AN USSR, Kiev, 1986, 74–91.

2.354 [G] Ch. 4, Sec. 3, Ex. 7.

2.355 T. G. Kurtz, *Trans. AMS* 186, 1973, 259–272. For the Carleman system see also: [Mt] Ch. IX, Sec. 6, Ex. 20.

2.356 *Hint*: Show that the operator $Au = \sum_{j=1}^n a_j(x) \, \partial u/\partial x_j + b(x)u$ is the infinitesimal generator of a group of continuous operators in $(L^2(\mathbb{R}^n))^N$. The main steps in proving this are based on the following estimates: $\|(A - \lambda)u\| \ge (|\lambda| - \beta)\|u\|$, $\|(A' - \lambda)u\| \ge (|\lambda| - \beta')\|u\|$ for positive numbers β, β' and sufficiently large $|\lambda|$, where $A'u = -\sum_{j=1}^n (\partial/\partial x_j)(a_j(x)u) + \overline{b(x)}u$ is the formal adjoint of the operator A. For instance: [Ta] Ch. 3, Sec. 5, Application 1.

2.357 [A-M] 5. 5. B.

2.358 *Hint*: Show that $\mathrm{Re}(Lu,u) = (Ku, u)$ for every $u \in H^1(\mathbb{T}^n)$, and even if u, $Lu \in L^2(\mathbb{T}^n)$. Deduce the inequalities $|u| \le C |Lu|$, $|u| \le C |L*u|$. Using the Gårding inequality prove that, under the condition (P_s), $\forall u \in H^{s+1}$ $\mathrm{Re}(\langle D\rangle^s Lu, \langle D\rangle^s u) \ge \frac{1}{2}c_s \|u\|^2 - C |u|^2$, where $\langle D\rangle^s$ is the operator of the symbol $(1 + |\xi|^2)^{s/2}$, and then show $\|u\|_s \le C\|Lu\|_s$. Consider for $\varepsilon > 0$ the regularized operator $L_\varepsilon = L_s - \varepsilon \Delta I$ and solve $L_\varepsilon u_\varepsilon = f$. Observe that $u_\varepsilon \in H^{s+2}$ and (u_ε) is bounded in H^s for $\varepsilon > 0$. Note that if (P_s) is satisfied for all s, then A_j must be constant matrices. [A-G] Ex. II, C, 7, and III. C, 3. 5. For the Gårding inequality see for instance: [Hö] Ch. 18, Sec. 6, or S. Agmon, *Lectures on Elliptic Boundary Value Problems*, Van Nostrand, Princeton, 1965.

2.359 [Tr1] Ch. V, Ex. 12.

2.360 *Hint:* The change of variables $u = e^v$, $u = v^{1/(1-\gamma)}$ if $\gamma = 1$, $\gamma \neq 1$ respectively, leads to the linear equation $xyv_{xy} - y^2 v_{yy} - 2xv_x + 2yv_y - 2v = 0$. Then take $\xi = x$, $\eta = xy$, $v = \xi^{-1}\eta^2 V(\xi,\eta)$ to get $V_{\xi\eta} = 0$. [Bi] Ch. V, Sec. 2. 1; J. L. Reid, P. B. Burt, *J. Math. Anal. Appl.* 47, 1974, 520–530.

2.361 *Answer:* $E = (c\omega\varepsilon)^{-1} \int_G (-k^2\varphi j + \mathrm{grad}\ \varphi\ \mathrm{div}\ j)\ dm - (4\pi)^{-1}$ $\int_\Sigma (ik_0\mu(v \times H)\varphi + ((v \times E) \times \mathrm{grad}\ \varphi) + (vE)\ \mathrm{grad}\ \varphi)\ d\sigma$, $H = c^{-1}$ $\int_G (j \times \mathrm{grad}\ \varphi)\ dm + (4\pi)^{-1} \int_\Sigma (ik_0\varepsilon(v \times E) - ((v \times H) \times \mathrm{grad}\ \varphi) - (vH)$ $\mathrm{grad}\ \varphi)\ d\sigma$, where $\varphi = e^{ikr}/r$, $k = \omega c^{-1} (\varepsilon\mu)^{1/2}$, $k_0 = \omega c^{-1}$. [BST] VII. 75.

2.362 *Answer:* The Hadamard example is: The integral $I = \int F(x,y,u,p,q)\ dx\ dy$ could vanish for some $u = f(x,y)$ without (the existence of) the second derivatives. Let $F = p^2 - q^2$, $u = f(x,y)$ for $f \in C^1\backslash C^2$. One has $\delta I = 2\iint f'(x + y)\ (\eta_x - \eta_y)\ dx\ dy$ and taking $x + y = u$, $x - y = v$, $\eta(x,y) = \varphi(u,v)$, $\delta I = -2\iint f'(u)\varphi_v(u,v)\ du\ dv$. The double integral vanishes but it is not necessary that f' has the derivative (see the Green formula). [Gr] Ch. XXXIV, Ex. 8.

2.363 *Answer:* One should minimize $T = \int_A^B dt$ under the condition $(x' - u)^2 + (y' - v)^2 = 1$; that is, the Euler functional is $1 + \frac{1}{2}\lambda((x' - u)^2 + (y' - v)^2 - 1)$ with some $\lambda = \lambda(t)$. This leads to the system

$$\frac{d(\lambda \cos \varphi)}{dt} + \lambda \left(\cos \varphi\ \frac{\partial u}{\partial x} + \sin \varphi\ \frac{\partial v}{\partial x} \right) = 0$$

$$\frac{d(\lambda \sin \varphi)}{dt} + \lambda \left(\cos \varphi\ \frac{\partial u}{\partial y} + \sin \varphi\ \frac{\partial v}{\partial y} \right) = 0$$

$$\frac{dx}{dt} = \cos \varphi + u, \qquad \frac{dy}{dt} = \sin \varphi + v$$

where φ is the angle of the relative velocity and the x-axis. Eliminating λ, one gets $\partial\varphi/\partial t = -(\partial u/\partial y) \cos^2 \varphi + (\partial u/\partial x - \partial v/\partial y) \sin \varphi \cos \varphi + (\partial v/\partial x) \sin^2 \varphi$. [Fa] Ch. XIII, Ex. 2.

2.364 *Answer:* The potential V and temperature T in a body with electrical and thermal conductivities $\sigma(T)$, $k(T)$ satisfy the equations $\nabla \cdot (\sigma \nabla V) = 0$, $\nabla \cdot (k \nabla T) + \sigma \mid \nabla V \mid^2 = 0$ and the boundary condition on the electrodes $V =$

V_0 on S_0, $V = V_1$ on S_1, $T = T_0$ on $S_0 \cup S_1$, while on the insulated surface, $\partial V/\partial v = \partial T/\partial v = 0$. One may verify by direct substitution that the pair of functions V, $T(V)$ satisfy the equations and boundary conditions provided that $T(V)$ is given by $\int_{T_0}^{T(V)} k(T)(\sigma(T))^{-1} \, dT = \frac{1}{2}(V - V_0)(V_1 - V)$. The maximum temperature, obtained by differentiation with respect to V, is given by $\int_{T_0}^{T_{max}} k(T)(\sigma(T))^{-1} \, dT = \frac{1}{8}(V_1 - V_0)^2$, which is the desired result. J. A. Lewis, *SIAM Review* 13, 1971, 123–125, Problem 69–11.

2.365 [A-K] M. Berger, Ch. VI, Sec. 11, Example 1; Ch. VII, Sec. 13.

2.366 *Hint*: Consider the associated potential energy functional $\|w\|^2 + \|\frac{1}{2}C(w,w) + L_1 w\|^2 - \lambda(Lw,w) - \lambda(Z',w)$ on H_0^2. Here C and L are determined from the transformed (equivalent) system $F = -\frac{1}{2}C(w,w) - Lw$, $w = C(F,w) + \lambda C(F_0,w) + Lw + \lambda Z'$; $(C(\omega,g), \varphi) = \int((g_{xy}\omega_y - g_{yy}\omega_x)\varphi_x + (g_{xy}\omega_x - g_{xx}\omega_y)\varphi_y)$, $Lu = C(u,F_0)$. [Be] (2.5.7), (4.3.21)–(4.3.27), Ch. 6, Sec. 6. 2B.

2.367 [Sn] Ch. 3, Sec. 6, Problem 3.

2.368 [R-Y] Ch. 1, Sec. 13.

2.369 [G] Ch. 14, Sec. 1, Ex. 8.

2.370 [G] Ch. 14, Sec. 1, Ex. 9.

2.371 *Hint*: Pass to the Fourier coefficients of u, v, w.
Answer: There exists a decreasing sequence (μ_n) of the values of μ corresponding to bifurcation points. The asymptotic behavior of solutions for $\mu_n > \mu > \mu_{n+1}$ is determined by an invariant n-dimensional torus. [A-K] Ch. VI, Sec. 3, Example 6; [He] Ch. 6, Sec. 4, Ex. 2; E. Hopf, *Comm. Pure Appl. Math.* 1, 1948, 303–322.

2.372 *Hint*: Show that $\frac{1}{2}d|u|^2/dt + |u|^2 = u_1 \le \frac{1}{2}(|u|^2 + 1)$. This means that the unit ball in $\mathbb{R}^3$ contains the global attractor of the dynamical system associated with the Minea system.
Answer: $u_0 = (1, 0, 0)$ and the circle $u_1 = 1/\delta$, $u_2^2 + u_3^2 = (\delta - 1)/\delta^2$ for $\delta > 1$ are the stationary solutions. u_0 is stable for $\delta < 1$; for $\delta > 1$ u_0 has a two-dimensional unstable manifold. There are two negative eigenvalues for $1 < \delta < 9/8$ and two conjugate with a negative real part for $\delta > 9/8$ as well as one eigenvalue 0 at the other stationary points. [Te] Ch. I, Sec. 2. 2; Gh. Minea, *Rev. Roumaine Math. Pures Appl.* 21, 1976, 1071–1075.

2.373 *Hint*: Show that $d\|u\|_{L^2}^2/dt + v\|u\|_{H^1}^2 \le (v\lambda_1)^{-1}\|f\|_{L^2}^2$ (where λ_1 is the first eigenvalue of the Stokes operator on Ω) and use the Gronwall inequality. Observe that $(v\lambda_1)^{-1}\|f\|_{L^2}$ can be taken as C. [Te] Ch. III, Sec, 2. 2.
Remark: A similar estimate can be proved in the stronger norm of $H^1(\Omega)$. This shows that the two-dimensional Navier-Stokes system possesses the global attractor compact in $L^2(\Omega)$ norm. [Te] Ch. III, Th. 2. 2.

2.374 [HKW] Ch. 2, Sec. 2, Example 3.

Remark: The full system may be studied with the Neumann or Dirichlet boundary conditions on a bounded open set.

2.375 [He] Ch. 9, Sec. 3, Ex. 1; N. Alikakos, *J. Differential Equations* 33, 1979, 201–225.

2.376 [M-MC] Remark (8A. 34).

2.377 *Answer:* The function $v(x,t) = (t + 1)^{-1/2} \exp(-x^2/4(t + 1))$ satisfies the heat equation $v_t = v_{xx}$. Take K such that $\exp(-T_0^2/4K^2) > (T_0 + 1)^{-1/2}$ and let $\Pi_{T_0} = \mathbb{R} \times [0, T_0)$, $\Pi^+ = \Pi_{T_0} \cap \{t > -Kx + T_0\}$, $\Pi^- = \Pi_{T_0} \cap \{t > Kx + T_0\}$, $\Delta = \Pi_{T_0} \backslash (\Pi^+ \cup \Pi^-)$. There exists $u \in C^\infty(\Pi_{T_0})$ such that u, u_t, $u_{xx} \in L^\infty(\Pi_T)$ for all $T \in (0, T_0)$ and $u(x,t) = 1/x$ in Π^+, $u(x,t) = -1/x$ in Π^-, and $u(x,t) \geq K/(T_0 - t)$ in Δ. For $v \in (v_0, 1]$, let $v_0 = (T_0 + 1)^{-1/2}$, $E_v = \{(x,t) \in \Delta : v(x,t) = v\}$. Observe that $\sup_{E_v} u < \infty$, $\sup_{E_v}(u_t - u_{xx}) < \infty$ for all $v \in (v_0, 1)$ ($E_v \subset \Pi_T$). Furthermore, $u_t - u_{xx} < 0$ in $\Pi_{T_0} \backslash \Delta$ and for all $(x,t) \in \Delta$ $(u(x,t), v(x,t))$ is in the set where lim sup and lim inf with respect to the v-axis are equal to lim at $v = v_0$; $\lim_{v \to v_0 + 0} \inf_{(x,t) \in E_v} u(x,t) = \lim_{v \to v_0 + 0} \sup_{(x,t) \in E_v} u(x,t) = \infty$. Hence there is $f(u,v) \in C^\infty(\mathbb{R}^2)$ strictly positive and bounded on lines parallel to the u-axis such that $u_t - u_{xx} \leq f(u,v)$ for all $(x, t) \in \Pi_{T_0}$. V. V. Churbanov, *Doklady AN SSSR* 310, 6, 1990, 1308–1309.

2.378 *Hint:* Observe that $d(\int_\Omega n^2 \, dx)/dt \leq 0$ and $nf(T)$ is bounded in L^2. Study the system linearized at $(0, 1)$. [He] Ch. 5, Sec. 1, Example 1; Ch. 10, Sec. 2; I. M. Gelfand, *Uspekhi Mat. Nauk* 14:2, 1959, 87–158.

2.379 [Te] Ch. III, Sec. 1. 1. 3, Example 1. 4.

2.380 *Hint:* The nonlinear terms are sublinear, so the existence of solutions is standard. Then use the maximum principle. [He] Ch. 4, Sec. 1, Example 2.

2.381 *Answer:* $3C \cosh^{-2}(C^{1/2}/(2(1 + C)^{1/2})(x - (C + 1)t))$ where $C > 0$ is a parameter specifying the amplitude and velocity of the wave. For instance: J. L. Bona, *Applications of Nonlinear Analysis in the Physical Sciences,* H. Amann, N. Bazley, K. Kirchgässner eds., Pitman, London, 1981, 183–205.

2.382 E. Conway, D. Hoff, J. Smoller, *SIAM J. Appl. Math.* 35, 1978, 1–16. For similar relationships of solutions of systems of partial differential equations to solutions of ordinary differential systems, see, for instance, P. Constantin, C. Foias, B. Nicolaenko, R. Temam, *Integral Manifolds and Inertial Manifolds for Dissipative Partial Differential Equations,* Springer, New York, 1989.

2.383 *Answer:* $\lim_{\xi-\text{fixed},t \to -\infty}(u_2(x,t) + 8 \cosh^{-2}(2x - 32t - 2\xi_1)) = 0$, where $\exp(-4\xi_1) = 3$ and $\lim_{\xi-\text{fixed},t \to \infty}(u_2(x,t) + 8 \cosh^{-2}(2x - 32t - 2\xi_1'))$ $= 0$, where $\exp(4\xi_1') = 3$, $\lim_{\zeta-\text{fixed},t \to -\infty}(u_2(x,t) + 2 \cosh^{-2}(x - 4t - \zeta_1)) =$ 0, where $\exp(2\zeta_1) = 3$ and $\lim_{\zeta-\text{fixed},t \to \infty}(u_2(x,t) + 2 \cosh^{-2}(x - 4t - \zeta_1')) =$ 0, where $\exp(-2\zeta_1') = 3$. This justifies the name "two-soliton solution." For instance: P. L. Bhatnagar, *Nonlinear Waves in One-Dimensional Dispersive*

Systems, Clarendon Press, Oxford, 1979, Ch. 3, Sec. 6. See also M. J. Ablo-
witz, H. Segur, *Solitons and the Inverse Scattering Transform,* SIAM, Phila-
delphia, 1981, Ch. 1, Sec. 7.

2.384 *Hint*: Observe that $\|u(.,t)\|_2$ can be unbounded as $t \to \infty$. Hence
consider the equation for the difference $u(x,t) - \int_0^1 u(x,t)\, dx$: $u_t + vu_{xxxx} + u_{xx}$
$+ \frac{1}{2}(u_x)^2 = \frac{1}{2}\int_0^1 (u_x)^2\, dx$. The linear part of the above equation generates a
compact analytic semigroup and the nonlinearity is an analytic function. A
suitable space is $H^2(0, 1) \cap \{\varphi \in L^2(0, 1) : \int_0^1 \varphi(x)\, dx = \int_0^1 \varphi(x) \sin nx\, dx =$
0 for $n = 1, 2, \ldots \}$. The solution operators $T(t)$, $u(t) = T(t)u(0)$, are compact
for $t > 0$. The crucial estimate in the proof of the existence of the global attractor
is $\lim_{t \to \infty} \sup\|u_x(t)\|_2 \leq$ const. [Ha2] 4. 9. 5; [Te] Ch. III, Sec. 4. 1; B.
Nicolaenko, B. Scheurer, R. Temam, *Physica D* 16, 1985, 155–183.

References

[A-M] R. Abraham, J. E. Marsden, *Foundations of Mechanics*, Benjamin/Cummings, Reading, Massachusetts, 1978.

[A-R] R. Abraham, J. Robbin, *Transversal Mappings and Flows*, Benjamin, New York, 1967.

[A-G] S. Alinhac, P. Gérard, *Opérateurs pseudodifférentiels et théorème de Nash-Moser*, Service de publications, Université de Paris-Sud, Orsay, 1989.

[ALGM1] A. A. Andronov, E. A. Leontovich, I. I. Gordon, A. G. Mayer, *Qualitative Theory of Second-Order Dynamic Systems*, Nauka, Moskva, 1966; Wiley, New York, 1973.

[ALGM2] A. A. Andronov, E. A. Leontovich, I. I. Gordon, A. G. Mayer, *Theory of Bifurcations of Dynamic Systems on a Plane*, Nauka, Moskva, 1967; IPST, Jerusalem, 1971.

[AVK] A. A. Andronov, A. A. Vitt, S. E. Khaĭkin, *Theory of Oscillators*, Nauka, Moskva, 1981; Pergamon Press, Oxford, 1966.

[A-K] S. Antman, J. B. Keller, ed., *Bifurcation Theory and Nonlinear Eigenvalue Problems*, Benjamin, New York, 1969.

[Ap] P. E. Appell, *Traité de mécanique rationnelle, t.1*, Gauthier-Villars, Paris, 1953.

[A1] V. I. Arnold, *Ordinary Differential Equations*, 3rd ed., Nauka, Moskva, 1984.

[A2] V. I. Arnold, *Supplementary Chapters of the Theory of Ordinary Differential Equations*, Nauka, Moskva, 1978. Translation: *Geometrical Methods in the Theory of Ordinary Differential Equations*, Springer, New York, 1988.

[A3] V. I. Arnold, *Mathematical Methods of Classical Mechanics*, Nauka, Moskva, 1979. Translation: Springer, New York, 1978.

[A-P] D. K. Arrowsmith, C. M. Place, *Ordinary Differential Equations, Approach with Applications*, Chapman and Hall, London, 1982.

[At] R. H. Atkin, *Classical Dynamics*, Heinemann, London, 1959.

[BSW] P. B. Bailey, L. F. Shampine, P. E. Waltman, *Nonlinear Two Point Boundary Value Problems*, Academic Press, New York, 1968.

[Ba] C. Bandle, *Isoperimetric Inequalities and Applications*, Pitman, Boston, 1980.

[Bn1] E. A. Barbashin, *Introduction to the Theory of Stability*, Nauka, Moskva, 1967; Wolters-Nordhoff, Groningen, 1970.

[Bn2] E. A. Barbashin, *Method of Sections in the Theory of Dynamical Systems*, Nauka i Tekhnika, Minsk, 1971.

[B-L] N. N. Bautin, E. A. Leontovich, *Methods of Qualitative Study of Dynamical Systems in the Plane*, Nauka, Moskva, 1976.

[Bl] V. V. Beleckiĭ, *Essays on Motions of Celestial Bodies*, Nauka, Moskva, 1972.

[B1] R. Bellman, *Stability Theory of Differential Equations*, Dover, New York, 1959.

[B2] R. Bellman, *Perturbation Techniques in Mathematics, Physics and Engineering*, Holt, Rinehart and Winston, New York, 1964.

[B3] R. Bellman, *Methods of Nonlinear Analysis*, Academic Press, New York, 1973.

[B-V] V. V. Belov, E. M. Vorob'ev, *Problem Book on Supplementary Chapters of Mathematical Physics*, Vysshaya shkola, Moskva, 1978.

[B-S] F. A. Berezin, M. A. Shubin, *The Schrödinger Equation*, Izd. MGU, Moskva, 1983.

[Be] M. Berger, *Nonlinearity and Functional Analysis*, Academic Press, New York, 1977.

[BJS] L. Bers, F. John, M. Schechter, *Partial Differential Equations*, Interscience, New York, 1964.

[B-W] P. Biler, A. Witkowski, *Problems in Mathematical Analysis*, Dekker, New York, 1990.

[Bi] A. V. Bitsadze, *Some Classes of Partial Differential Equations*, Nauka, Moskva, 1981.

[B-K] A. V. Bitsadze, D. F. Kalinichenko, *Problem Book on the Equations of Mathematical Physics*, Nauka, Moskva, 1985.

[B-P] V. N. Bogayevsky, A. Ya. Povzner, *Algebraic Methods in Nonlinear Perturbation Theory*, Nauka, Moskva, 1987.

[Bo] O. I. Bogoyavlenskiĭ, *Methods in the Qualitative Theory of Dynamical Systems in Astrophysics and Gas Dynamics*, Nauka, Moskva, 1980; Springer, New York, 1985.

[Br] M. Braun, *Differential Equations and Their Applications. An Introduction to Applied Mathematics*, 3rd ed., Springer, New York, 1983.

[Bu] J. C. Burkill, *The Theory of Ordinary Differential Equations*, Oliver and Boyd, Edinburgh, 1956.

[BST] B. M. Budak, A. A. Samarskiĭ, A. N. Tikhonov, *Problem Book in Mathematical Physics*, Nauka, Moskva, 1972.

[BNF] N. V. Butenin, Yu. I. Neĭmark, N. A. Fufaev, *Introduction to the Theory of Nonlinear Oscillations*, Nauka, Moskva, 1987.

[BVGN] B. F. Bylov, R. E. Vinograd, D. M. Grobman, V. V. Nemytskiĭ, *Theory of Liapunov Exponents*, Nauka, Moskva, 1966.

[C] J. Cannon, *The One-Dimensional Heat Equation*, Addison-Wesley, Reading, Massachusetts, 1984.

[C-S] R. W. Carroll, R. E. Showalter, *Singular and Degenerate Cauchy Problems*, Academic Press, New York, 1976.

[Ca] H. Cartan, *Calcul différentiel*, Hermann, Paris, 1967. Translation: *Differential Calculus*, Hermann, Paris, 1971.

[Ce] L. Cesari, *Asymptotic Behavior and Stability Problems in Ordinary Differential Equations*, Springer, Berlin, 1959.

[Ch-H] K. W. Chang, F. A. Howes, *Nonlinear Singular Perturbation Phenomena: Theory and Applications*, Springer, New York, 1984.

[CHADP] Ph. Clément, H. J. A. M. Heijmans, S. Angenent, C. J. van Duijn, B. de Pagter, *One-Parameter Semigroups*, North Holland, Amsterdam, 1987.

[C-L] E. A. Coddington, N. Levinson, *Theory of Ordinary Differential Equations*, McGraw-Hill, New York, 1955.

[C-H] R. Courant, D. Hilbert, *Methods of Mathematical Physics*, vol. II, *Partial Differential Equations*, Interscience, New York, 1962.

[D-K] Yu. L. Daleckiĭ, M. G. Kreĭn, *Stability of Solutions of Differential Equations in Banach Space*, Nauka, Moskva, 1970; AMS, Providence, Rhode Island, 1974.

[D-L] R. Dautray, J.-L. Lions, éd., *Analyse mathématique et calcul numérique pour les sciences et les techniques*, Masson, Paris, 1984–85. Translation: *Mathematical Analysis and Numerical Methods for Science and Technology*, vol. 1, Springer, Berlin, 1990.

[D] E. B. Davies, *One-Parameter Semigroups*, Academic Press, New York, 1980.

[Da] H. T. Davis, *Introduction to Nonlinear Differential and Integral Equations*, Dover, New York, 1962.

[De] B. P. Demidovich, *Lectures on Mathematical Theory of Stability*, Nauka, Moskva, 1967.

[Di] J. Dieudonné, *Foundations of Modern Analysis*, Academic Press, New York, 1960.

[Du] G. N. Duboshin, *Celestial Mechanics*, 2nd ed., Nauka, Moskva, 1978.

[D-G] J. Dugundji, A. Granas, *Fixed Point Theory*, vol. 1, PWN, Warszawa, 1982.

[Eg] Yu. V. Egorov, *Linear Differential Equations of Principal Type*, Nauka, Moskva, 1984; Consultants Bureau, New York, 1986.

[E] L. E. El'sgol'ts, *Introduction to the Theory of Differential Equations with Deviating Arguments*, Nauka, Moskva, 1964; Holden Day, San Francisco, 1966.

[Ep] B. Epstein, *Partial Differential Equations*, McGraw-Hill, New York, 1962.

[Er] N. P. Erugin, *The Riemann Problem*, Nauka i Tekhnika, Minsk, 1982.

[F] H. Fattorini, *The Cauchy Problem*, Addison-Wesley, Reading, Massachusetts, 1983.

[Fa] J. Favard, *Cours d'analyse de l'École Polytechnique*, t. 3, Gauthier-Villars, Paris, 1962.

[Fi] A. F. Filippov, *Recueil de problèmes d'équations différentielles*, Nauka, Moscou, 1976. Translation of: *Problem Book in Differential Equations*, Nauka, Moskva, 1973.

[Fo] G. Folland, *Introduction to Partial Differential Equations*, Princeton University Press, Princeton, New Jersey, 1976.

[Fs] A. R. Forsyth, *Theory of Differential Equations*, I–VI, Cambridge University Press, Cambridge, 1890–1906.

[Fr1] A. Friedman, *Generalized Functions and Partial Differential Equations*, Prentice-Hall, Englewood Cliffs, New Jersey, 1963.

[Fr2] A. Friedman, *Partial Differential Equations of Parabolic Type*, Prentice-Hall, Englewood Cliffs, New Jersey, 1964.

[Gv] F. D. Gakhov, *Boundary Value Problems*, Nauka, Moskva, 1977; Pergamon Press, Oxford, 1966.

[Gl] A. S. Galiullin, *Methods of Solution of Inverse Problems of Dynamics*, Nauka, Moskva, 1986.

[Ga] G. Gallavotti, *The Elements of Mechanics*, Springer, New York, 1983.

[G] P. Garabedian, *Partial Differential Equations*, Wiley, New York, 1964.

[G-T] D. Gilbarg, N. S. Trudinger, *Elliptic Partial Differential Equations of Second Order*, 2nd ed., Springer, Berlin, 1983.

[GDO] E. B. Gledzer, P. V. Dolzhanskiĭ, A. M. Obukhov, *Systems of Hydrodynamic Type and Their Applications*, Nauka, Moskva, 1981.

[Gd] H. Goldstein, *Classical Mechanics*, Heinemann, London, 1959.

[Go] J. Goldstein, *Semigroups of Linear Operators and Applications*, Oxford University Press, Clarendon Press, Oxford, 1985.

[Gr] É. Goursat, *Cours d'analyse mathématique*, cinquième èd., Gauthier-Villars, Paris, 1956.

[Gb] E. A. Grebenikov, *Averaging Method in Applied Problems*, Nauka, Moskva, 1986.

[G-H] J. Guckenheimer, P. Holmes, *Nonlinear Oscillations, Dynamical Systems and Bifurcations of Vector Fields*, Springer, New York, 1983.

[Ha1] J. K. Hale, *Theory of Functional Differential Equations*, Springer, New York, 1977.

[Ha2] J. K. Hale, *Asymptotic Behavior of Dissipative Systems*, Math. Surveys and Monographs 25, AMS, Providence, Rhode Island, 1988.

[Hr] A. Haraux, *Nonlinear Evolution Equations—Global Behavior of Solutions*, Lecture Notes in Math. 841, Springer, Berlin, 1981.

[H] P. Hartman, *Ordinary Differential Equations*, Wiley, New York, 1964.

[HKW] B. D. Hassard, N. D. Kazarinoff, Y.-H. Wan, *Theory and Application of Hopf Bifurcation*, Cambridge University Press, Cambridge, 1981.

[He] D. Henry, *Geometric Theory of Semilinear Parabolic Equations*, Lecture Notes in Math. 840, Springer, Berlin, 1981.

[Hi] E. Hille, *Ordinary Differential Equations in the Complex Domain*, Wiley, New York, 1976.

[H-S] M. W. Hirsch, S. Smale, *Differential Equations, Dynamical Systems and Linear Algebra*, Academic Press, New York, 1974.

[Ho] H. Hochstadt, *Integral Equations*, Wiley, New York, 1973.

[Hö] L. Hörmander, *Analysis of Partial Differential Operators*, 1–4, Springer, Berlin, 1983–85.

[I] E. L. Ince, *Ordinary Differential Equations*, Dover, New York, 1956.

[I-J] G. Iooss, D. D. Joseph, *Elementary Stability and Bifurcation Theory*, Springer, New York, 1980.

[Ir] M. C. Irwin, *Smooth Dynamical Systems*, Academic Press, New York, 1980.

[J1] F. John, *Partial Differential Equations*, 4th ed., Springer, New York, 1982.

[J2] F. John, *Plane Waves and Spherical Means Applied to Partial Differential Equations*, Interscience, New York, 1955.

[Ju] G. Julia, *Exercices d'analyse* III, IV, Gauthier-Villars, Paris, 1948.

[K] O. D. Kellogg, *Foundations of Potential Theory*, Frederick Ungar, New York, 1929.

[K-C] J. Kevorkian, J. D. Cole, *Perturbation Methods in Applied Mathematics*, Springer, New York, 1981.

[KNK] E. Khorozov, N. Nikiforov, G. Karadzhov, *Problem Book on Ordinary Differential Equations*, Sofijski Universitet "Kliment Ohridski", Sofia, 1984.

[KGS] N. S. Koshlakov, E. B. Gliner, M. M. Smirnov, *Partial Differential Equations of Mathematical Physics*, Vysshaya shkola, Moskva, 1970.

[K-S] A. G. Kostyuchenko, I. S. Sargsyan, *Distributions of Eigenvalues*, Nauka, Moskva, 1979.

[KPPZ] M. A. Krasnoselskiĭ, A. I. Perov, A. I. Povolockiĭ, P. P. Zabreĭko, *Vector Fields in the Plane*, GIFML, Moskva, 1963.

[Kr] M. L. Krasnov, *Ordinary Differential Equations*, Mir Publishers, Moscow, 1987.

[KKM1] M. L. Krasnov, A. I. Kiselev, G. I. Makarenko, *Problems and Exercises: Integral Equations*, Nauka, Moskva, 1968.

[KKM2] M. L. Krasnov, A. I. Kiselev, G. I. Makarenko, *Problem Book in Ordinary Differential Equations*, Vysshaya shkola, Moskva, 1973.

[Kn] S. G. Kreĭn, *Linear Differential Equations in Banach Spaces*, Nauka, Moskva, 1967; AMS, Providence, Rhode Island, 1971.

[LNY] N. A. Lar'kin, V. A. Novikov, N. N. Yanenko, *Nonlinear Equations of Variable Type*, Nauka, Novosibirsk, 1983.

[Le] S. Lefschetz, *Geometric Theory of Differential Equations*, Interscience, New York, 1957.

[Li] R. B. Lindsay, *Physical Mechanics,* Chapman and Hall, London, 1933.

[Lo] S. A. Lomov, *An Introduction to the General Theory of Singular Perturbations*, Nauka, Moskva, 1981.

[L] W. V. Lovitt, *Linear Integral Equations*, McGraw-Hill, New York, 1924.

[Ma] R. Mañé, *Ergodic Theory and Differentiable Dynamics*, Springer, Berlin, 1987.

[M-MC] J. E. Marsden, M. Mc Cracken, *The Hopf Bifurcation and Its Application,* Springer, New York, 1976.

[Mt] R. H. Martin, *Nonlinear Operators and Differential Equations in Banach Spaces*, Wiley, New York, 1976.

[Ms] V. P. Maslov, *Asymptotic Methods and Perturbation Theory*, Nauka, Moskva, 1988.

[M-S] J. L. Massera, J. J. Schäffer, *Linear Differential Equations and Function Spaces*, Academic Press, New York, 1966.

[M] V. P. Mikhaĭlov, *Partial Differential Equations*, Nauka, Moskva, 1983.

[Mn] S. G. Mikhlin, *Linear Partial Differential Equations*, Vysshaya shkola, Moskva, 1977.

[M1] R. K. Miller, *Nonlinear Volterra Equations*, Benjamin, New York, 1971.

[Mi] C. Miranda, *Equazioni alle derivate parziali di tipo ellittico*, Springer, Berlin, 1955.

[N] E. Nelson, *Topics in Dynamics I: Flows*, Math. Notes, Princeton University Press, Princeton, New Jersey, 1969.

[N-S] V. V. Nemytskiĭ, V. V. Stepanov, *Qualitative Theory of Differential Equations*, Princeton University Press, Princeton, New Jersey, 1960.

[Ni] Z. Nitecki, *Differentiable Dynamics*, MIT Press, Cambridge, Massachusetts, 1971.

[P-M] J. Palis, W. de Melo, *Geometric Theory of Dynamical Systems. An Introduction*, Springer, New York, 1982.

[Pa] Yu. G. Pavlenko, *Hamiltonian Methods in Electrodynamics and Quantum Mechanics*, Izd. MGU, Moskva, 1985.

[Pz] A. Pazy, *Semigroups of Linear Operators and Applications to Partial Differential Equations*, Springer, New York, 1983.

[P1] I. G. Petrovskiĭ, *Lectures on the Theory of Ordinary Differential Equations*, GITTL, Moskva, 1949; Prentice-Hall, Englewood Cliffs, New Jersey, 1966.

[P2] I. G. Petrovskiĭ, *Lectures on the Theory of Partial Differential Equations*, Fizmatgiz, Moskva, 1961; Interscience, New York, 1954.

[PTKY] E. S. Piatnickiĭ, N. M. Trukhan, Yu. I. Khanukaev, G. N. Yakovenko, *Problem Book on Analytical Mechanics*, Nauka, Moskva, 1980.

[PSV] L. C. Piccinini, G. Stampacchia, G. Vidossich, *Ordinary Differential Equations in $\mathbb{R}^n$, Problems and Methods*, Springer, New York, 1984.

[P1] V. A. Pliss, *Integral Sets of Periodic Systems of Differential Equations*, Nauka, Moskva, 1977.

[Po] H. Pollard, *Mathematical Introduction to Celestial Mechanics*, Prentice-Hall, Englewood Cliffs, New Jersey, 1966.

[P-S] G. Pólya, G. Szegö, *Problems and Theorems in Analysis*, I, II, Springer, Berlin, 1972, 1976.

[PPJ] P. Popivanov, N. Popivanov, J. Jordanov, *Problem Book on Partial Differential Equations*, 2nd ed., Izkustvo, Sofia, 1985.

[P-T] J. Pöschel, E. Trubowitz, *Inverse Spectral Theory*, Academic Press, New York, 1987.

[P-W] M. H. Protter, H. F. Weinberger, *Maximum Principles in Differential Equations*, Springer, New York, 1984.

[Ra] R. Rabczuk, *Elements of Differential Inequalities* (in Polish), PWN, Warszawa, 1976.

[R-S] M. Reed, B. Simon, *Methods of Modern Mathematical Physics*, 1–4, Academic Press, New York, 1972–1979.

[RSC] R. Reissig, G. Sansone, R. Conti, *Nichtlineare Differentialgleichungen höherer Ordnung*, Edizioni Cremonese, Roma, 1969.

[Re] L. E. Reĭzin', *Liapunov Functions and Identification Problems*, Zinatne, Riga, 1986.

[R] M. Roseau, *Équations différentielles*, Masson, Paris, 1976.

[RHL] N. Rouche, P. Habets, M. Laloy, *Stability Theory by Liapunov's Direct Method*, Springer, New York, 1971.

[R-M1] N. Rouche, J. L. Mawhin, *Équations différentielles ordinaires*, t. 1, Masson, Paris, 1973.

[R-M2] N. Rouche, J. L. Mawhin, *Ordinary Differential Equations*, Pitman, Boston, 1980.

[R-Y] B. L. Rozhdestvenskiĭ, N. N. Yanenko, *Systems of Quasilinear Equations and their Applications to Gas Dynamics*, Nauka, Moskva, 1978; AMS, Providence, Rhode Island, 1983.

[SGKM] A. A. Samarskiĭ, V. A. Galaktionov, S. P. Kurdyumov, A. P. Mikhaĭlov, *Strained Regimes in Problems for Quasilinear Parabolic Equations*, Nauka, Moskva, 1987.

[S-V] J. A. Sanders, F. Verhulst, *Averaging Methods in Nonlinear Dynamical Systems*, Springer, New York, 1985.

[Sa] G. Sansone, *Equazioni differenziali nel campo reale*, N. Zanichelli, Bologna, 1948–49.

[S-C] G. Sansone, R. Conti, *Non-linear Differential Equations*, Pergamon Press, Oxford, 1964.

[S1] G. R. Sell, *Topological Dynamics and Ordinary Differential Equations*, Van Nostrand Reinhold, New York, 1971.

[Sv] G. E. Shilov, *Mathematical Analysis. The Second Special Course*, Izd. MGU, Moskva, 1984.

[Si] K. S. Sibirskiĭ, *Introduction to Topological Dynamics*, Shtinca, Kishinev, 1970; Noordhoff International Publishing, Leyden, 1975.

[S-M] C. L. Siegel, J. K. Moser, *Lectures on Celestial Mechanics*, Springer, Berlin, 1971.

[Sm1] M. M. Smirnov, *Problems on the Equations of Mathematical Physics*, Nauka, Moskva, 1975; Noordhoff, Groningen, 1966.

[Sm2] M. M. Smirnov, *Degenerate Hyperbolic Equations*, Vysheĭshaya shkola, Minsk, 1977.

[S] J. Smoller, *Shock Waves and Reaction-Diffusion Equations*, Springer, New York, 1983.

[Sn] I. Sneddon, *Elements of Partial Differential Equations*, McGraw-Hill, New York, 1957.

[So] J. Sotomayor, *Lições de equações diferenciais ordinárias*, IMPA, Rio de Janeiro, 1979.

[Sp] C. Sparrow, *The Lorenz Equations: Bifurcations, Chaos and Strange Attractors*, Springer, New York, 1982.

[Sg] M. R. Spiegel, *Applied Differential Equations*, 3rd ed., Prentice-Hall, Englewood Cliffs, New Jersey, 1981.

[Se] E. M. Stein, *Singular Integrals and Differentiability Properties of Functions*, Princeton University Press, Princeton, New Jersey, 1970.

[St] W. A. Strauss, *Nonlinear Wave Equations*, CBMS Regional Conference Series 73, AMS, Providence, Rhode Island, 1989.

[Sb] R. A. Struble, *Nonlinear Differential Equations*, McGraw-Hill, New York, 1962.

[Sz] J. Szarski, *Differential Inequalities*, PWN, Warszawa, 1967.

[Sk] W. Szlenk, *An Introduction to the Theory of Dynamical Systems*, PWN, Warszawa, 1982.

[Tm] I. Tamura, *Topology of Foliations* (in Japanese), Iwanami Shoten, Tokyo, 1976. Russian translation: Mir, Moskva, 1979.

[Ta] H. Tanabe, *Equations of Evolution*, Pitman, London, 1979.

[Ty] M. E. Taylor, *Pseudodifferential Operators*, Princeton University Press, Princeton, New Jersey, 1981.

[Te] R. Temam, *Infinite-Dimensional Dynamical Systems in Mechanics and Physics*, Springer, New York, 1988.

[T-P] M. Tenenbaum, H. Pollard, *Ordinary Differential Equations*, Harper and Row, New York, 1964.

[T] F. Treves, *Basic Linear Partial Differential Equations*, Academic Press, New York, 1975.

[Tr1] F. G. Tricomi, *Lezioni sulle equazioni a derivate parziali*, Editrice Gheroni, Torino, 1954.

[Tr2] F. G. Tricomi, *Integral Equations*, Wiley, New York, 1957.

[V] G. Valiron, *Équations fonctionnelles et applications*, Masson, Paris, 1945.

[V-K] N. I. Vasil'ev, Yu. A. Klokov, *Foundations of the Theory of Boundary Value Problems for Ordinary Differential Equations*, Zinatne, Riga, 1978.

[Ve] I. N. Vekua, *New Methods for Solving Elliptic Equations*, OGIZ, Moskva, 1948; North Holland, Amsterdam, 1967.

[Vl1] V. S. Vladimirov, *Equations of Mathematical Physics*, Nauka, Moskva, 1981; Dekker, New York, 1971.

[Vl2] V. S. Vladimirov et al., *Problem Book on the Equations of Mathematical Physics*, Nauka, Moskva, 1974.

[W] J. Wermer, *Potential Theory*, Lecture Notes in Math. 408, Springer, Berlin, 1974.

[Wi] D. Widder, *The Heat Equation*, Academic Press, New York, 1975.

[Ya] A. I. Yanushauskas, *Analytic Theory of Elliptic Equations*, Nauka, Novosibirsk, 1979.

[Y] K. Yosida, *Lectures on Differential and Integral Equations*, Interscience, New York, 1960.

[Z] C. Zuily, *Problèmes de distributions avec solutions détaillées*, Hermann, Paris, 1986. Translation: *Problems in Distributions and Partial Differential Equations*, North Holland, Amsterdam, 1988.

Index

Following each entry is the problem/answer number in which the entry may be found. This index does not contain key words from the titles of (sub)sections.